T0391770

Soft Computing Techniques in Connected Healthcare Systems

This book provides an examination of applications of soft computing techniques related to healthcare systems and can be used as a reference guide for assessing the roles of various techniques. *Soft Computing Techniques in Connected Healthcare Systems* presents soft computing techniques and applications used in healthcare systems, along with the latest advancements. The authors examine how connected healthcare is the essence of combining a practical operative procedure of interconnectedness of electronic health records, mHealth, clinical informatics, electronic data exchange, practice management solutions, and pharmacy management. The book focuses on different soft computing techniques, such as fuzzy logic, ANN, and GA, which will enhance services in connected health systems, such as remote diagnosis and monitoring, medication monitoring devices, identifying and treating the underlying causes of disorders and diseases, improved access to specialists, and lower healthcare costs. The chapters also examine descriptive, predictive, and social network techniques and discuss analytical tools and the important role they play in enhancing the services to connected healthcare systems. Finally, the authors address real-time challenges with real-world case studies to enhance the comprehension of topics. This book is intended for under graduate and graduate students, researchers, and practicing professionals in the field of connected healthcare. It provides an overview for beginners while also addressing professionals in the industry on the importance of soft computing approaches in connected healthcare systems.

Biomedical and Robotics Healthcare

Series Editors: Utku Kose, Jude Hemanth, and Omer Deperlioglu

Artificial Intelligence for the Internet of Health Things
Deepak Gupta, Eswaran Perumal and K. Shankar

Biomedical Signal and Image Examination with Entropy-Based Techniques
V. Rajinikanth, K. Kamalanand, C. Emmanuel and B. Thayumanavan

Mechano-Electric Correlations in the Human Physiological System
A. Bakiya, K. Kamalanand and R. L. J. De Britto

Machine Learning and Deep Learning in Medical Data Analytics and Healthcare Applications
Om Prakash Jena, Bharat Bhushan, and Utku Kose

Applied Artificial Intelligence: A Biomedical Perspective
Swati V. Shinde, Varsha Bendre, D. Jude Hemanth, and MA Balafar

Combating Women's Health Issues with Machine Learning: Challenges and Solutions
Jude Hemanth and Meenu Gupta

Soft Computing Techniques in Connected Healthcare Systems
Moolchand Sharma, Suman Deswal, Umesh Gupta, Mujahid Tabassum, and Isah A. Lawal

For more information about this series, please visit: https://www.routledge.com/Biomedical-and-Robotics-Healthcare/book-series/BRHC

Soft Computing Techniques in Connected Healthcare Systems

Edited by
Moolchand Sharma, Suman Deswal,
Umesh Gupta, Mujahid Tabassum,
and Isah A. Lawal

CRC Press is an imprint of the
Taylor & Francis Group, an **informa** business

Designed cover: © Shutterstock

First edition published 2024
by CRC Press
2385 NW Executive Center Drive, Suite 320, Boca Raton FL 33431

and by CRC Press
4 Park Square, Milton Park, Abingdon, Oxon, OX14 4RN

CRC Press is an imprint of Taylor & Francis Group, LLC

ISBN: 978-1-032-51347-8 (hbk)
ISBN: 978-1-032-52129-9 (pbk)
ISBN: 978-1-003-40536-8 (ebk)

DOI: 10.1201/9781003405368

Typeset in Times
by codeMantra

Dedication

Mr. Moolchand Sharma *would like to dedicate this book to his father, Shri. Naresh Kumar Sharma, and his mother, Smt. Rambati Sharma, for their constant support and motivation, and his family members, including his wife, Ms. Pratibha Sharma, and son, Dhairya Sharma. He also thanks the publisher and his other co-editors for believing in his abilities.*

Dr. Suman Deswal *would like to dedicate this book to her father, Late Shri. V. D. Deswal, and mother, Smt. Ratni Deswal, who taught her to never give up, and her husband, Mr. Vinod Gulia, and daughters, Laisha and Kyna, for always loving and supporting her in every endeavor of life. She also thanks the publisher and co-editors who believed in her capabilities.*

Dr. Umesh Gupta *would like to dedicate this book to his mother, Smt. Prabha Gupta and his father, Shri. Mahesh Chandra Gupta, for their constant support and motivation, and his family members, including his wife, Ms. Umang Agarwal, and son, Avaya Gupta. He also thanks the publisher and his other co-editors for believing in his abilities. Before beginning and after finishing this endeavor, he appreciates the Almighty God, who provides him with the means to succeed.*

Mr. Mujahid Tabassum *would like to dedicate this book to his father, Mr. Tabassum Naeem Ahmed, and mother, Mrs. Razia Tabassum (late), for their constant support, prayers, and motivation. He would also like to give his special thanks to the publisher and his co-editors for their collaboration and mutual support.*

Dr. Isah A. Lawal *would like to dedicate this book to his late parents, and his family members, including his wife, Ms. Marakisiya Ado, and children, Muhsinah, Muhsin, and Aisha, for their constant support and motivation. He also thanks the publisher and his other co-editors for believing in his abilities.*

Contents

Editors' Profile

Moolchand Sharma is currently an Assistant Professor in the Department of Computer Science and Engineering at Maharaja Agrasen Institute of Technology, GGSIPU, Delhi. He has published scientific research publications in reputed International Journals and Conferences, including SCI indexed and Scopus indexed Journals such as *Cognitive Systems Research* (Elsevier), *Physical Communication* (Elsevier), *Intelligent Decision Technologies: An International Journal, Cyber-Physical Systems* (Taylor & Francis Group), *International Journal of Image & Graphics* (World Scientific), *International Journal of Innovative Computing and Applications* (Inderscience) & *Innovative Computing and Communication Journal* (Scientific Peer-reviewed Journal). He has authored/co-authored chapters with international publishers like Elsevier, Springer, Wiley, and De Gruyter. He has authored/edited four books with a national/international level publisher (CRC Press, Bhavya Publications). His research area includes Artificial Intelligence, Nature-Inspired Computing, Security in Cloud Computing, Machine Learning, and Search Engine Optimization. He is associated with various professional bodies, such as IEEE, ISTE, IAENG, ICSES, UACEE, Internet Society, Universal Inovators research lab life membership, etc. He possesses teaching experience of more than 10 years. He is the co-convener of ICICC, DOSCI, ICDAM & ICCCN Springer Scopus Indexed conference series and ICCRDA-2020 Scopus Indexed IOP Material Science & Engineering conference series. He is also the organizer and Co-Convener of the International Conference on Innovations and Ideas toward Patents (ICIIP) series. He is also the Advisory and TPC committee member of the ICCIDS-2022 Elsevier SSRN Conference. He is also the reviewer of many reputed journals, such as Springer, Elsevier, IEEE, Wiley, Taylor & Francis Group, IJEECS, and World Scientific Journal, and many springer conferences. He has also served as a session chair in many international Springer conferences. He is currently a doctoral researcher at DCR University of Science and Technology, Haryana. He completed his Post Graduation in 2012 from SRM UNIVERSITY, NCR CAMPUS, GHAZIABAD, and Graduated in 2010 from KNGD MODI ENGG. COLLEGE, GBTU.

Prof. Suman Deswal holds a Ph.D. from DCR University of Science and Technology, Murthal, India. She completed her M.Tech. (CSE) from Kurukshetra University, Kurukshetra, India, and B.Tech. (Computer Science and Engg.) from CR State College of Engg., Murthal, India, in 2009 and 1998, respectively. She possesses 21 years of teaching experience and is presently working as a Professor in the Department of Computer Science and Engineering at DCR University of Science and Technology, Murthal, India. Her research area includes Wireless Networks, Heterogeneous Networks, Distributed Systems, Machine Learning, Deep Learning Approaches, and Bioinformatics. She has many research papers to her credit in reputed journals, including SCI indexed, and Scopus Indexed Journals and conferences. Her publications have more than 143 citations. She is also the reviewer of many reputed journals, such as Springer, Elsevier, IEEE, Wiley, and International IEEE and Springer Conferences. She is a member of IAENG and Computer Society of India (CSI).

Dr. Umesh Gupta is currently an Associate Professor at Bennett University, India. He received a Doctor of Philosophy (Ph.D.) (Machine Learning) from the National Institute of Technology, Arunachal Pradesh, India. He was awarded a gold medal for his Master of Engineering (M.E.) from the National Institute of Technical Teachers Training and Research (NITTTR), Chandigarh, India, and Bachelor of Technology (B.Tech.) from Dr. APJ, Abdul Kalam Technical University, Lucknow, India. His research interests include SVM, ELM, RVFL, Machine Learning, and Deep Learning Approaches. He has published over 35 referred journal and conference papers of international repute. His scientific research has been published in reputable international journals and conferences, including SCI-indexed and Scopus-indexed journals like *Applied Soft Computing* (Elsevier) and *Applied Intelligence* (Springer), each of which is a peer-reviewed journal. His publications have more than 158 citations with an h-index of 8 and an i10-index of 8 on Google Scholar as of March 1, 2023. He is a senior Member of IEEE (SMIEEE) and an active member of ACM, CSTA, and other scientific societies. He has also reviewed papers for many scientific journals and conferences in the United States and other foreign countries. He led the sessions at the 6th International Conference (ICICC-2023), 3rd International Conference on Data Analytics and Management (ICDAM 2023), the 3rd International Conference on Computing and Communication Networks (ICCCN 2022), and other international conferences like Springer ETTIS 2022 and 2023. He is currently supervising two Ph.D. students. He is the co-principal investigator (co-PI) of two major research projects. He published three patents in the years 2021–2023. He also published four book chapters with Springer, CRC.

Mujahid Tabassum is a lecturer at Noroff University College (Noroff Accelerate), Kristiansand, Norway. He has completed a Master of Science (Specialization in Computer System Engineering) degree from Halmstad University, Sweden, and a bachelor's degree from the University of Wollongong, Australia. He has worked in various international universities in Malaysia and the Middle East, making his profile well reputed. He has managed and led several student and research projects and published several research articles in well-known SCI journals and Scopus conferences. He is a qualified "Chartered Engineer—CEng" registered with the Engineering Council, UK. He has 13 years of teaching experience. He is a Cisco, Microsoft, Linux, Security, and IoT-certified instructor. His research interests include Computer Networks, AI, Wireless Sensor networks, IoT, Security, and Applications. He has published several Scopus papers, journals, and book chapters. He is a Member of IEEE, a Member of the Institution of Engineering and Technology, a Member of IAENG, a Member of the Australia Computing Society (ACS), and a Member of MBOT Malaysia. He is an active member of the Society of IT Engineers and Researchers, UK.

Dr. Isah A. Lawal was an Erasmus Mundus Joint Doctorate Fellow with over 10 years of professional work experience, including teaching and research. He has participated in several collaborative multidisciplinary research projects at different universities, including in Europe (Italy and United Kingdom) and the Middle East (Saudi Arabia). He has authored several articles in peer-reviewed journals and conferences ranging from data-driven predictive modeling to machine learning for intelligent systems. In addition to actively engaging in research, he has taught Data Mining, Innovative Systems, and Artificial Intelligence courses at both undergraduate and postgraduate levels. Dr. Isah's research interests include the multidisciplinary application of Machine Learning Techniques, Data Mining, and Smart Systems. He has supervised and examined several undergraduate projects and master's thesis in Statistical Data Analysis, Data Visualisation, and Natural Language Processing. Dr. Isah is currently participating in the EEA granted data-driven public administration project as a consultant for the Department for Strategic Development and Coordination of Public Administration, Ministry of the Interior of the Czech Republic. The project involves using big data analytics to analyze public mobility from mobile positioning data to efficiently plan and distribute public services and public administration in the Czech Republic.

Contributors

Aparna Agarwal
Middle East College, Oman

Tanveer Ahmed
Bennett University, India

Anshika Arora
Bennett University, India

H. D. Arora
Amity University, India

Tarkeshwar Barua
REGex Software Services, India

Manish Bhardwaj
KIET Group of Institutions, India

M. P. S. Bhatia
Netaji Subhas University of Technology, India

Amit Kumar Bhuyan
Michigan State University, MI, USA

Pinaki Chakraborty
Netaji Subhas University of Technology, India

Hrishikesh Dutta
Michigan State University, MI, USA

Naomi Tia Chantelle Freeman
Noroff University College, Norway

Manoj Kumar Gupta
Shri Mata Vaishno Devi University, India

Sakshi Gupta
Amity University Noida, India

Umesh Gupta
Bennett University, India

Khoula Al Harthy
Middle East College, Oman

A. Hepzibah Christinal
Karunya Institute of Technology and Sciences, India

Yu-Chen Hu
Providence University, Taiwan

Sairamya Nanjappan Jothiraj
Karunya Institute of Technology and Sciences, India

Ashish Kumar
I.T.S Engineering College, India

Vijay Kumar
Manav Rachna International Institute of Research and Studies, India

Vivek Kumar
ABES Engineering College, India

Divya Kumari
Bennett University, India

Shambhavi Mishra
Bennett University, India

Manorama Mohapatro
Amity Institute of Information Technology, Amity University Ranchi, India

Arambam Neelima
National Institute of Technology, Nagaland, India

Samad Noeiaghdam
Irkutsk National Research Technical University, Russia

Olutayo Oyeyemi Oyerinde
University of the Witwatersrand, South Africa

Kiran Pal
Delhi Institute of Tool Engineering, India

J. Prasanna
Karunya Institute of Technology and Sciences, India

Asadi Pushpa
Sri Venkateswara University, India

Siva Ram Rajeyyagari
Shaqra University, Saudi Arabia

CV Ravikumar
Vellore Institute of Technology University, India

Ansu Elsa Regi
Karunya Institute of Technology and Sciences, India

C. Ruth Vinutha
Karunya Institute of Technology and Sciences, India

Mriganka Sarmah
National Institute of Technology, Nagaland, India

Abuzar Sayeed
Bennett University, India

Ashish Seth
INHA University, Uzbekistan

Jyoti Sharma
KIET Group of Institutions, India

Sumit Kumar Sharma
Ajay Kumar Garg Engineering College, India

Heisnam Rohen Singh
National Institute of Technology, Nagaland, India

Kshatrapal Singh
Krishna Engineering College, India

Goddindla Sreenivasulu
Sri Venkateswara University, India

Asadi Srinivasulu
BlueCrest University, Liberia

M. S. P. Subathra
Karunya Institute of Technology and Sciences, India

Madhusudana Subramanyam
Koneru Lakshmaiah Education Foundation, India

S. Thomas George
Karunya Institute of Technology and Sciences, India

Robin Zarine
Middle East College, Oman

Preface

We are delighted to launch our book entitled *Soft Computing Techniques to Enhance Services in Connected Healthcare Systems* under the book series *Biomedical and Robotics Healthcare*, CRC Press, Taylor & Francis Group. The healthcare industry is constantly evolving, with new technologies and techniques being developed to improve patient care and outcomes. Connected healthcare systems, which allow for the collection and sharing of patient data between healthcare providers, are becoming increasingly popular. However, the large amount of data collected can be overwhelming, and traditional methods of data analysis may not be sufficient. Soft computing techniques, which include artificial neural networks (ANNs), fuzzy logic, genetic algorithms (GAs), machine learning (ML), and natural language processing (NLP), can provide solutions to the challenges faced in connected healthcare systems. These techniques can be used to analyze large amounts of patient data, identify patterns, and make accurate predictions. It is a valuable source of knowledge for researchers, engineers, practitioners, and graduate and doctoral students working in the same field. It will also be helpful for faculty members of graduate schools and universities. Around 60 full-length chapters have been received. Amongst these manuscripts, 16 chapters have been included in this volume. All the chapters submitted were peer-reviewed by at least two independent reviewers and provided with a detailed review proforma. The comments from the reviewers were communicated to the authors, who incorporated the suggestions in their revised manuscripts. The recommendations from two reviewers were considered while selecting chapters for inclusion in the volume. The exhaustiveness of the review process is evident, given that a large number of articles were received addressing a wide range of research areas. The stringent review process ensured that each published chapter met the rigorous academic and scientific standards.

We would also like to thank the authors of the published chapters for adhering to the schedule and incorporating the review comments. We extend our heartfelt gratitude to the authors, peer reviewers, committee members, and production staff whose diligent work shaped this volume. We especially want to thank our dedicated peer reviewers who volunteered for the arduous and tedious step of quality checking and critiquing the submitted chapters.

Moolchand Sharma

Suman Deswal

Umesh Gupta

Mujahid Tabassum

Isah A. Lawal

About the Book

Connected healthcare systems are becoming increasingly popular because of their ability to collect and share patient data between healthcare providers, allowing for more comprehensive and coordinated care. However, the large amount of data collected can be overwhelming, and traditional methods of data analysis may not be sufficient to fully utilize these data. Soft computing techniques, which include artificial neural networks (ANNs), fuzzy logic, genetic algorithms (GAs), machine learning (ML), and natural language processing (NLP), can provide solutions to the challenges faced in connected healthcare systems.

Soft computing techniques are a subset of artificial intelligence (AI) that deals with uncertainty, imprecision, and incomplete information. They are designed to mimic human thinking and reasoning, and can handle complex and nonlinear problems. These techniques have been applied in various fields, including finance, engineering, and manufacturing, and have shown promising results in healthcare. This book explores the application of soft computing techniques in connected healthcare systems, with a focus on enhancing services and improving patient outcomes. The book is divided into chapters, with each chapter covering a different soft computing technique and its applications in healthcare. The chapters provide practical examples and case studies to demonstrate the effectiveness of each technique, as well as discussing the benefits and limitations.

The book concludes with a discussion on the future of soft computing techniques in connected healthcare systems. We believe that soft computing techniques have the potential to revolutionize healthcare by providing more personalized and accurate services to patients, improving the accuracy of diagnoses, and optimizing treatment plans. We hope that this book will inspire further research and innovation in the field of connected healthcare systems, and provide a valuable resource for healthcare professionals, researchers, and students.

1 Automation in Healthcare Forecast and Outcome

A Case Study

Kshatrapal Singh
KIET Group of Institutions

Ashish Kumar
I.T.S. Engineering College

Manoj Kumar Gupta
Shri Mata Vaishno Devi University

Ashish Seth
INHA University

1.1 INTRODUCTION

As indicated by some reports of the United States Census Bureau in 2016, another segment pattern has arisen. People over the age of 65 years will outnumber children under the age of 5 years by 2020. What's more, by 2050, the population of individuals aged 65 years and over will be more than twice the population of little youngsters. According to the United Nations, accelerated aging will become a serious problem globally (in percentage terms between 2010 and 2050, see Figure 1.1).

This supposed "segment seismic tremor" is relied upon to begin producing results on the worldwide work markets. It means, among others, the number of people who require medical help is also on the rise. The healthcare industry has to find the resources to adapt to this new demographic trend [1–4]. Process automation in healthcare should thus be taken full advantage of, in the attempt to provide sufficient medical care to a huge number of people. The constructive outcomes of automation in the medical services industry stretch out past this perspective. Moreover, it can make a critical commitment to reducing expenses and to giving better client care [5,6]. Note that these are unequivocally the medical care needs referenced by 75% of emergency clinic CEOs, as per a Health Leaders report from 2016.

The advantages of utilizing measure automation for smoothing out cycles are nothing new, while that among smoothed out cycles, productivity, and decreased expenses are commonsensical. Considering the normal development of 6.5% for medical services costs, as indicated by UiPath, and the fact that a KPMG investigation

DOI: 10.1201/9781003405368-1

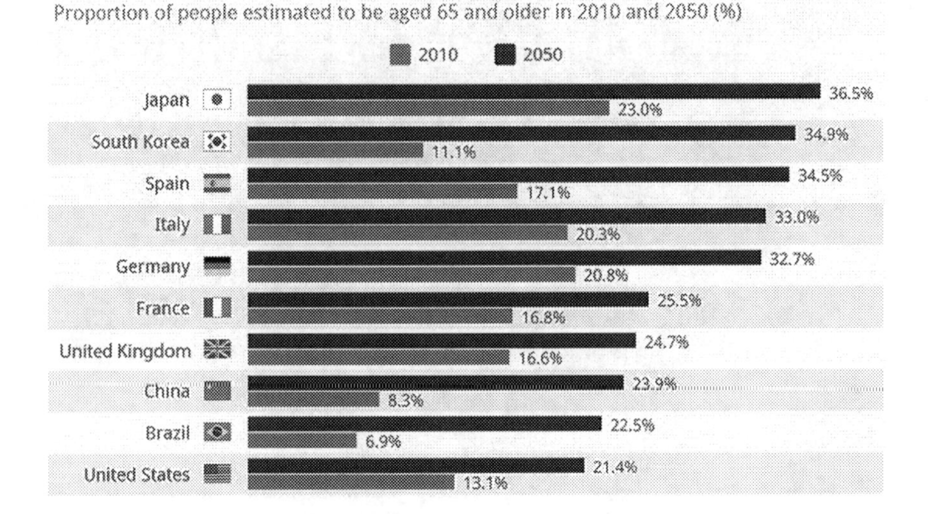

FIGURE 1.1 Rapid aging problem as per UN data.

demonstrates that mechanical cycle computerization can increase savings by up to 50%, you should consider the capacity of automation to enhance cost reduction [7–10]. One alternate way that Prior Authorization (PA) may add to reserve funds is by more precisely distinguishing the medical services guarantees that don't meet the prerequisites (around 30%–40%, as per the previously mentioned study), and along these lines recuperating a great deal of cash that may have in any case gone unpaid.

Also, if back-office, routine assignments are given to programming robots, it would permit specialists to utilize their human abilities to more readily address their patients' necessities as opposed to filling in reliably names, dates, and addresses. PA may encourage client care in medical services. As per the measurements accessible in *McKinsey Quarterly*, 36% of the medical services assignments—generally administrative and back-office—are agreeable to mechanization [11–15]. So let us presently spread out these development and productivity openings that mechanical cycle mechanization could bring to medical services.

The advantages of PA in medical care mean expanded operational proficiency, efficiency, and cost-sharpness. Indeed, there is more than this quantitative way to deal with the same issue; we ought not to fail to remember the subjective component of PA in medical services. For example, Peter Nichol likewise points out that "the discussion has extended past cost decrease to quality, commitment, and advancement" [16–18]. In any case, let us first explain the fundamentals, with the focus on information from the HfS Blueprint Report. Figure 1.2 shows data about the countries that prioritize healthcare.

The KPMG report that we previously cited recommends a substantial amount of PA for the entirety of an emergency clinic's revenue cycle due to the practicality of automation for organized information, with specified boundary sets [19–23]. The income cycle incorporates authoritative and clinical capacities fundamental for the administration of patients' records, from pre-registration to charge installment.

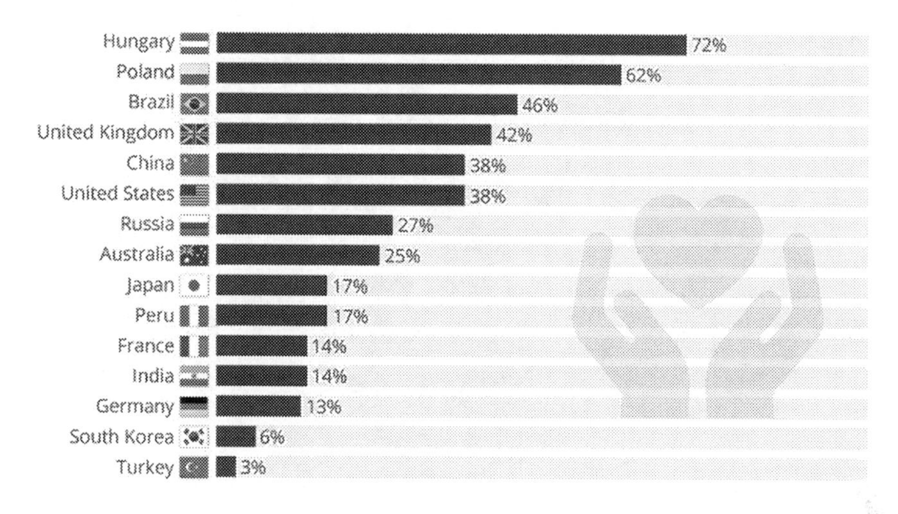

FIGURE 1.2 List of countries that prioritize healthcare.

Elements of the income cycle, for example, new patient arrangement demands, persistent pre-appearance and appearance, or guarantee refusals are especially appropriate for computerization. Simply consider how exorbitant blunders like broken information sections might be! So then why not go for a blunder-free accomplice, similar to what PA vows to be? Moreso, PA can encourage consistency with the enormous number of medical care guidelines, which are crucial when managing claims dissents. Errands consolidating different arrangements, for example, patients booking, claims assessment, organization and the board, or clinical document can be without mistakes as well as quick whenever performed by bots [24–29]. As a result, the rehandling expenses may be gotten rid of, which would diminish general expenses of medical services.

Figure 1.3 clearly illustrates the effects of automation at work, advantages as well as business risks associated with it. UiPath presents another example of overcoming adversity to support the case of the usefulness of cycle automation in medical care. The primary explanation behind embracing PA was to oversee its tasks, which was thusly expected to contribute altogether as per the general inclination of their higher number of patients. The usage of PA procedure was a success—winning circumstance for both supplier as well as patients [30–34]. To a greater extent, smoothed out inventory measures (like cases checking or charging) in view of 80% automation prompted a cost decrease for every case. Upward of 75% of the patients were glad to utilize computerized administration, such as planning interviews, admittance to clinical history, or charging data, as it restricted availing clinical help to obtain such data before PA enablement. Last yet positively not least, we want to mention that PA ought not to be viewed as an extreme contender to be dreaded in employment rivalry [35–37]. This section is, in reality, an alternate perspective on the preceding point.

Maybe in medical care it is clearer than in different enterprises that computerization isn't implied as "substitute for headcount," rather a facilitator of a legitimate categorization of work. Finally, if bots assume control of mundane authoritative

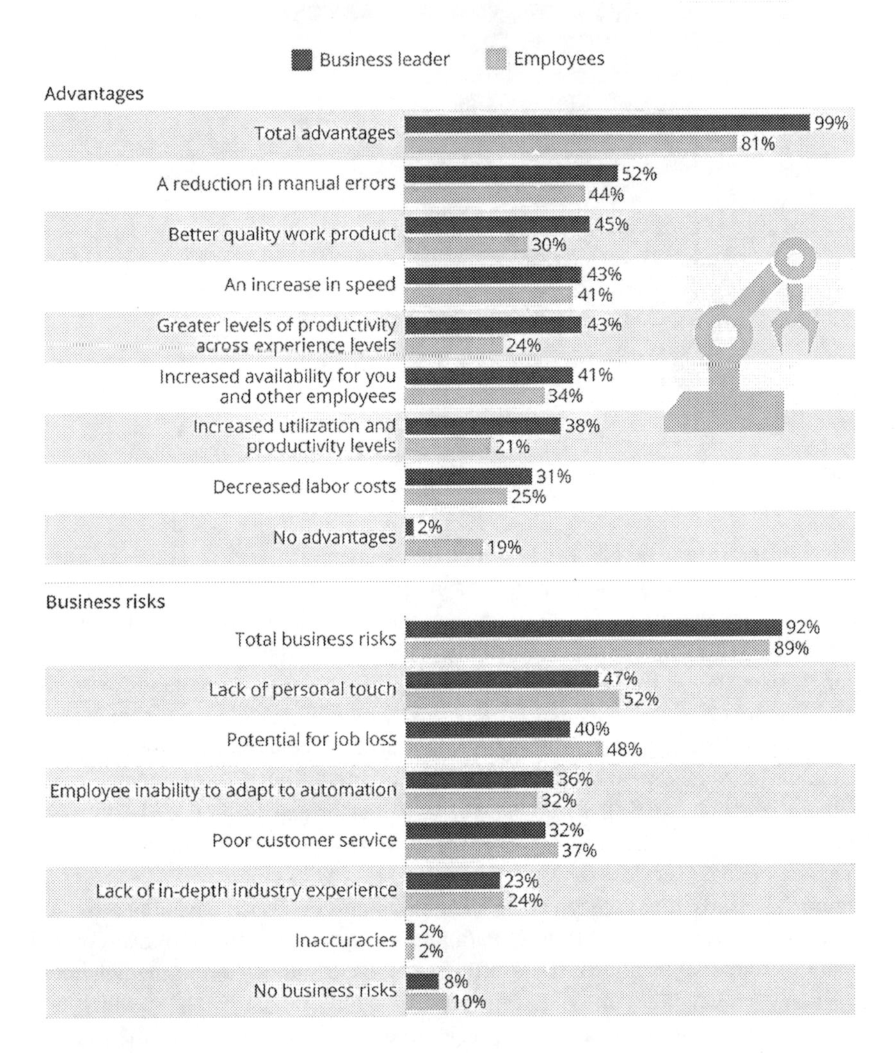

FIGURE 1.3 Automation effects at work.

positions, the human staff can maximize their extraordinarily human limitations, such as unambiguous articulation and differential symptomatic reasoning.

1.2 METHODS AND MATERIALS

This draws from the outcomes from a different contextual analysis, with the end goal of examining emergency clinic inward coordination frameworks and deciding how automation could be helpful in enhancing them. In responding to the "how to" questions, a contextual analysis arrangement was chosen, despite the fact that it never gave bases for a measurable speculation yet depended on perception and investigation as the reason for ends.

To meet the objectives, the examination consists of three cases with distinct purposes, points, and approaches, each of which is illustrated below. The principle rationale of directing the various contextual investigations was to fabricate an extensive arrangement, where the discoveries from the distinctive sub-concentrates together form an absolute information mass of basic perspectives. Subsequently, the three sub-contemplates were chosen to supplement each other. Case 1 is the most significant exhaustive case and was the investigation's center of attention. Following Case 1, Case 2 was included to cover a more extensive scope of clinics with various requirements and level of enforced automation.

1.2.1 Case 1

The stand of the various examinations was a contextual analysis at a Noida medical clinic containing 350 beds. The investigation inquiries in Case 1 were: (1) How was the internal coordination's framework organized and how were calculated exercises completed and (2) what improvement regions are there and what are the possibilities for automation to better these? This was responded to by breaking down the particular method of work, identifying zones requiring improvement, and examining how the calculated framework could be better utilizing computerization. Case emergency clinic 1 was selected because its physical and authoritative conditions testify to its size, age, and innovation imperatives from the perspective of the executives. Case 1 was an engaging report with a time period of half an year. Three primary techniques for information assortment were utilized to locate data: perceptions, interviews (open and semi-organized), and authentic record/archive.

The perceptions were based on ten events at nursing wards as well as the transportation division. In between the perceptions, the researchers were directed through the activities as well as the contemplated day-by-day tasks of attendants and the transportation staff. With direct perceptions, the investigation could continue events and their settings continuously and also mention objective facts that the staff engaged in the exercises on a daily basis do not notice. Exercises that were tedious, ergonomically imperfect, or requiring improvement with regard to other aspects were recognized, and notes were made separately on the perceptions and afterward thought after each event. During the examination, the perception notes were filtered for simple subjects referenced and furthermore were contrasted with the outcomes of other related meetings.

Twenty-six semi-organized and open meetings were conducted with the medical attendants, the heads of the three nursing wards, the overseer of the ward distribution, the medical clinic organizer, the chief of the transport division as well as workers at the transport division. Most meetings were conducted with one interviewee, yet there were meetings of a conversational nature with a few witnesses. The meetings went from 15 to 30 minutes, and at any rate two specialists were gathering notes between particular meetings. The discussions encompassed all aspects of general practice, including scheduling, time management, improvement areas, and aspirations for future outcomes. Data materials, reports, and other interior documents were used to locate or additionally explore perspectives that arose during perceptions and meetings. Also, a period analysis was done at the ten nursing wards that were examined. The medical personnel scheduled the time study exercises like a clock, day and night, for 10 days on a structure with 25 predefined exercises. The material was collected by the scientists, depicting which exercises were most tedious, and during the investigation it was contrasted with the outcomes from different sections of the examination.

1.2.1.1 Current State of Logistic Activities and Transportation Department

Case hospital 1 has a transport office, functioning as an external entity from which the hospital avails administration services. This division is responsible for all transportation activities that have a place with the inward help of the hospital, for example, transport of waste material, clothing, food, drugs, and patients. In addition, they are accountable for correspondence management and the primary issue room in each ward, which contains supply requests and appropriations. The division just works during day-shift (09.00–17.00) and utilizes 25 laborers as well as two supervisors with an employable job. In the present document, a concise overview of the key components of the transportation sector is provided, drawing upon meetings and observations.

1.2.1.1.1 Patient Transport

The wards book quiet vehicles from the transport division by giving the date, time, and spot for pickup, and if important, data with respect to the patients' status and so on. The transportation section facilitates approximately 8,000 patient transportation trips annually, accounting for 47% of all patient vehicular journeys. The remaining patient transport arrangements are done by the nursing staff either because of a surprising bit of news or no accessible transport staff. Booking of patient vehicles should be done by telephone before 09.00 AM. Every morning the transport staff plan and split the day's vehicles between the staff. Be that as it may, only 40% of the appointments are made ahead of time: 40% are made by telephone between the day and the leftover 20% back transport patients to the wards.

One trouble communicated was that the staff did not have any correspondence gear, which works in the ducts. Typically they use cell phones. As these are not permitted or working in every section of the medical clinic, it is hard to oversee as well as arrange transportation or staff effectively on tasks. This deters the arranging, brings down the adaptability, and expands the lead time, as much time is missed when staff should get back to the transport office to get new work orders. The well-known methods for transport involve the transportation staff physically driving the bed/wheelchair. There are anyway helps, for instance, a carriage which can be associated with the modern truck as well as furthermore a "bed pusher," which lifts the bed and makes it simpler to move. However, these devices don't fit in the lifts and can therefore be utilized only on the ground floor and in the ducts associated with the clinic structures. The carriage is the preferred mode of transportation, particularly on the courses, where long distances and precipitous inclines make these vehicles particularly demanding. Nonetheless, currently, one carriage is being used to its most extreme limit.

1.2.1.1.2 Equipment

The transportation office possesses all the requisite gear, for example, modern truck and carriage utilized for interior vehicles. The transport office has a contract with the hospital for a long-term period. As a result, the director conveyed concern regarding responsibility for speculations, fostering a protracted perspective.

1.2.1.1.3 Waste Materials and Laundry

Two kinds of waste are dealt with: ordinary waste, for example, pieces of food and center waste, for example, organic waste, needles, and other sharp or possibly tainted material. For the nonclinical waste, there are chutes in a few areas in the clinic where the nursing staff drop packs of waste from the ward. These chutes go down to the cellar where the waste winds up in dumpsters. The dumpsters should be physically supplanted when full; however, there is no sign of this happening. Rather the transportation staff consistently controls them to keep away from clogging in the lines or flood of waste. The system is the equivalent for clothing sacks: chutes at the wards boost the clothing packs to roller confines that should be supplanted when full. Around one ton of clothing is taken care of every day at the emergency clinic.

1.2.1.1.4 Food Management

The transport division gets food carriages from the emergency clinic kitchen and transports them to the wards thrice a day (breakfast, lunch, and supper). The carriages are shipped to the lift passage closest to each ward. The nursing staff brings the food carriages into the ward. After every meal, the nursing staff takes the carriages back to the lift foyer where the transport office gathers them to take to the kitchen.

1.2.1.1.5 Collection of Samples

Rounds for getting tests at the wards and carrying them to the research facility are made a few times daily. Test transport is frequently done along with mail transportation, but not with any fixed timetable. The medical caretakers likewise convey numerous examples, particularly critical examples, to the research center. There is a pneumatic cylinder framework between the research facility and wards that every day hold countless examples (for instance, for the trauma center). The framework is traditional and not changed for the present design. A couple of associations exist and the limit is low prompting utilized just for dire examples.

1.2.1.1.6 Goods

All bundles more than 2 kg shipped off/from the emergency clinic are taken care of as products. The clinic gets around 18,000 bundles per year. Upon bundle appearance, two duplicates consisting of the conveyance notes or address mark are generated. One is taped to the crate and the other is placed in a folio. At the point when the bundle is conveyed, the beneficiary signs the conveyance note on the case, which is carried back and put with the other duplicate in the fastener as confirmation that the bundle has been conveyed. This strategy was among the perceptions experienced as perplexing and not completely trustworthy, a view that was upheld between interviews. The undertaking is tedious, confounded, and offers low detectability because conveyance notes regularly disappear during the process cycle.

1.2.1.2 Time-Consuming Logistic Activities and the Nurses' Perspective

During interviews, the nursing staff provided some information about issues impacting their workload. One reoccurring concern was the huge number of patient vehicles driven by the nursing staff, despite it being the duty of the transport office. The transport office was seen as incapable of dealing with all vehicles and not generally showing up as expected for booked vehicles, leaving the attendants to drive out the vehicles. Transporting is tedious and actually stressing because of significant distances. Other tedious but not straightforwardly persistent related exercises were shipping tests to the research center, gathering/leaving food carriages in the lift corridor, and additionally data-calculated exercises, for example, arrangements before rounds, releases as well as paper work. The time with the patients was seen as a small part of the complete time.

1.2.1.3 Quantitative Time Analysis of Nursing Activity

The exercises in the time study were classified into four categories. Direct consideration refers to exercises completed vis-à-vis with the patient, for instance, medicines and assessments. Indirect consideration shows restraint-related exercises not acted in direct contact with the patient, for instance, calculated exercises, and getting medicine and food ready. Administration alludes to authoritative errands that are not straightforwardly understanding related, for example, financial arrangements or instruction. Personal time incorporates all delays other than lunch, namely, for utilizing the bathroom, short breathers and so on inside the controlled working hours.

Figure 1.4 depicts the normal time appropriation for medical caretakers during the day shifts. The roundabout consideration class is the biggest classification possessing 48% of the medical attendants' absolute time. The immediate consideration possesses 34% of the time traced by authoritative time (10%) as well as individual time (8%).

1.2.1.4 Betterment Areas and Future State Vision

Betterment needs with respect to foundation reoccurred during the meetings with the medical caretakers. The staff encountered a great deal of "going around": procuring supplies for patients, going among work areas as well as patient rooms and so forth. The inward coordinations were communicated as jumbled and scattered. A particular zone requiring betterment was clothing and waste dealing. A pack averages 10–12 kg, but can weigh up to 25 kg, necessitating heavy lifts when hurling them down chutes. All wards are independently answerable for its funds. Purchase and acquirement are anyway overseen for the whole medical clinic. This implies that the lone route for a ward to control its costs is fundamentally through the degree of staff.

An overall assessment among medical caretakers was that the outstanding burden had expanded both with respect to the number of assignments and additionally

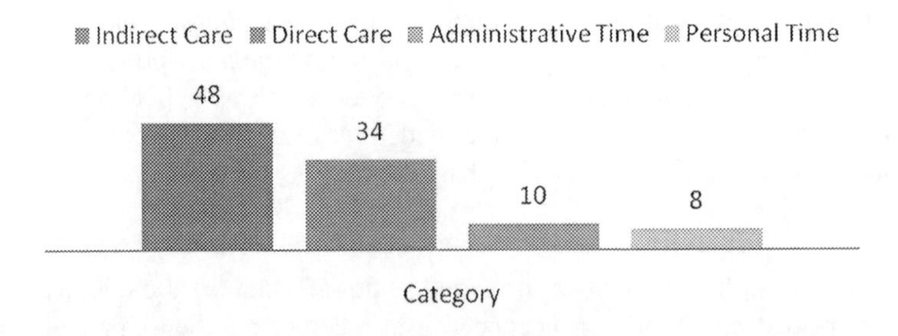

FIGURE 1.4 Normal time appropriation for medical caretakers.

genuinely over the most recent decade because the patients are more established and less versatile. The tension in the medical services was communicated as expanded while the number of staff was diminished. During a meeting with the ward superintendent and the HODs of two nursing wards, a 5-year plan for medical care and clinic coordination was formulated. The main expansive vision was that the nursing staff ought to be freed from all nonnursing undertakings, for example, supply renewal, unloading clothing as well as placing it in cupboards, dealing with food, doing dishes, doing tasks for patients, and so on. Different perspectives were aired, for example, extraordinary design with singular patient rooms and good methods of correspondence among patients as well as medical caretakers to abstain from reducing the "going around" for the attendants.

1.2.2 Case 2

The ensuing case was an in-depth report on the subject of upgrades within the calculated framework at a large medical clinic with two locations. Case hospital 2 is one of Delhi's biggest medical clinics, with 775 beds at more than two destinations, and was chosen because it effectively works with improving the calculated framework, utilizing, for example, automation. The medical clinic is additionally in the process of arranging and planning another site. The examination question was: How can and should robotized enhancements of emergency clinic inside coordination's frameworks be organized?

In Case 2, being a spellbinding report, information for investigation was collected basically by perceptions as well as meetings. The two emergency clinic locations were observed once each for three hours, during which the specialists observed and guided the inventory administrator through the working method. The examined zones pertain to the material handling and interior transportation of both patients and goods. Semi-organized meetings were led by the chief of provisions and the people responsible for the extraordinary computerized frameworks noticed. The fundamental viewpoints finished were the method of daily working, with enhancements of the present calculated framework. The venture director for the calculated framework in the new medical clinic was additionally met to discuss how the emergency clinic works with planning another calculated framework. Other than perceptions and interviews, data materials, reports, and other internal documents were assembled furthermore, concentrated to check or foster examining zones talked about.

1.2.2.1 Present State Description of Internal Logistic Solution

Case hospital 2 was prioritized in addition to Case 1 in terms of hospital internal logistics development in a large hospital reference with two sites.

1.2.2.1.1 Patient Transportations at Site 1

Site 1 comprises six structures circulated over a huge zone with 1 km of associated courses. The emergency clinic utilizes an Information Technology (IT)-framework for requesting, arranging, and dealing with the vehicles. The nursing staff state get spot, objective, and most recent reason of appearance, as well as need potential comments with respect to the patient current state. The framework designs the vehicles by organizing them after appearance time, need, and objective and attempts to coordinate transportations to limit pointless travel. Manual changes are anyway fundamental and a transport organizer arranges the vehicles relying upon accessible staff. All strategic staff convey a situating hand Primary Care (PC) permitting the organizer to appropriate the requests. The organizer likewise sends new transport requests to the staff that acknowledges or dismisses it relying upon their remaining task at hand.

1.2.2.1.2 Pneumatic Send Off at Site 1

In late 2010, another pneumatic sendoff framework was placed for use at Site 1. There are 50 sendoff stations from where more than 1,000 shipments are sent on work days and roughly 500 are sent on weekends. Only authorized personnel can view the shipments, and the framework is used for tests and blood as well as medication, administrative work, and currency. The research center is by a wide margin the biggest beneficiary of dispatches. Subsequently, modern robots can handle the shipments by opening the cylinders and exhausting the substance on various transports (subject is expressed when transporting). The robot cell likewise handles the back shipments of the cylinders. This cell was extraordinarily refreshing as it handled a profoundly redundant assignment. This framework was depicted as modern and has got a lot of recognition and furthermore been the motivation for numerous such establishments globally. The greatest focal points are the diminished lead time, the expanded adaptability, and the unwavering quality of the framework.

1.2.2.1.3 Automatic Guided Vehicles at Site 2

Site 2 comprises just two structures. This has created a viable condition for the utilization of 20 Computerized Guided Vehicles (AGVs) that transport a significant portion of food, cloth, clothing, waste, and merchandise. Certain exceptional vehicles are still operated manually. Automated Guided Vehicles (AGVs) employ designated lifts and allocated pickup/drop-off places, both within lift tunnels and at the transportation office located on the ground floor.

1.2.2.1.4 Small Amount of Goods at Site 2

At Site 2, the lone pneumatic send off line is between the trauma center and the research center. A significant portion of the transportation of little products, for example, test specimens, blood, and drugs, is made by the nursing staff. Some booked vehicles consist of modern trucks manned by the staff from the coordinations division. As per a study started by the emergency clinic, all little product transportations involve 35 full-time workers on an ordinary work day. Because of the high numbers of little product transports, Site 2 is planning a pneumatic send off framework like Site 1. The projected duration for the venture's restitution is five years, contingent solely on the attendees' savings for the event.

1.2.2.2 Future Hospital Logistics

Case hospital 2 is currently arranging and building another site supplanting Site 1 that was opened in 2017. A vital investigation of the calculated requirements was conducted as a stage in the arranging. The outcome was presented in a report, affirmed by the clinic executives as rules for future interior strategic framework. Programmed transports will be an establishing guideline for future coordinations, because of expanding requests for effectiveness, availability, more limited lead times, and non-stop assistance. AGVs, pneumatic dispatch frameworks, and air compelled squander/clothing frameworks, are necessities later on in the location. A serious level of computerization is strived for, for example, by IT frameworks for requesting materials, following hardware, and staff. Hardware for taking and passing medication for patients is likewise to be coordinated with the pneumatic dispatch framework. There is additionally an overall wish to "professionalize the calculated cycles." The strategic streams are essential yet too costly to even think about customizing to each ward. A normalized method of working with the center around a productive stream is wanted. A few lean standards are additionally referenced as client center (part of TQM) Just-In-Time conveyances as well as constant enhancements.

1.3 DISCUSSION

In this part, the observations found from each case are analyzed with respect to posed research questions. The section is characterized by a prevalent discourse on technology management.

1.3.1 Reflections from Case 1

The inquiries in Case 1 were: (1) how is the inside strategic framework organized and conveyed out and (2) what improvement zones are there and how is it possible that automation would be utilized to improve these? In the past segment, the main inquiry was expounded. Case emergency clinic 1 has a lot to gain from Case emergency clinic 2, and the accompanying improvement zones could be seen in a nutshell.

1.3.1.1 Patient Transports

A superior arranging framework (ideally computerization based) for the quiet vehicle like the one in Case 2 would in all probability build the portion of transportations performed by the transport division, and not by the nursing staff. Additional and more specialized hardware for transporting the beds, such as equipment that can be accommodated in elevators, would also enhance the efficiency of the tasks.

1.3.1.2 Internal Communication and Planning

A superior framework for interior correspondence among the transports staff, for example, by utilization of hand PCs, would lessen pointless travel to get new requests, make it conceivable to more readily accessible design undertakings, and consequently render more effective utilization of the staff.

1.3.1.3 Indicate for Full Dumpsters and Roller Cages

Presenting a sensor framework showing when the gathering compartments for clothing as well as waste are full would free a part of time. The sensor framework ought to be incorporated with a correspondence framework imparting signs straightforwardly to the transportation staff. A further developed methodology is to follow a few carriages that naturally drive the full compartment away and place an unfilled one under the line.

1.3.1.4 Small Goods and Mail Handling

A broader pneumatic send off framework would be efficient for the two medical attendants and transportation staff in the transportation of little products, for example, mail, test specimens, and blood. However, it is significantly more expensive and a more complex process to install a pneumatic dispatch system in an existing hospital building, particularly one with a partially low ceiling height and fragile internal walls, compared to integrating it from the beginning.

1.3.1.5 Handling of Packages

A product framework coordinated with standardized tags and scanner tag per users would both lessen the extent of manual care of patients and increment the possibility of following products with less exertion.

1.3.1.6 Lifting Aids for Laundry and Waste Bags

The tangible burden associated with maintaining garments is correspondingly diminished. The transportation of squander sacks can be facilitated by the utilization of automated lifting equipment or manual lifting methods for the purpose of depositing them into chutes.

1.3.2 Reflections from Case 2

The inquiries between Case 2 were: How can robotic enhancements of emergency hospital interior co-ordination frameworks be organized? The way toward arranging and building a new coming emergency hospital includes a huge extent of work and boundaries to be considered. It is of solid significance that all capacities, including internal coordinations, are associated with the beginning stages when planning the new hospital. The hospital has played out an assessment of the present method of attempting to distinguish improvements needs and made a vision for the future medical hospital within a strategic framework. It is generally satisfying and critical to help the hospital in gathering information on the pertinent necessities of patients and the public. A dream and clear objectives for the strategic capacity have been lacking in the case of other emergency hospitals. As discovered, Site 2 in fact was fabricated in an advanced medical hospital; however, the fixed arrangements made it difficult to redesign, giving a fairly outdated arrangement. Adaptability and versatility should be the cornerstones when addressing future necessities and requests.

1.3.3 General Discussion

For more case-specific reflection on the logistic activities, five problem areas with respect to technology managing as well as the uses of automation in the internal logistic system were identified.

1.3.3.1 Ownership

Between the meetings, the absence of possession as well as clear obligation regarding new, particularly long haul, automation speculation was uncovered. As the internal coordinations involve frequently re-appropriated work, nobody assumes the requisite responsibility for building up the strategic exercises. This is like different medical hospitals where the coordination work is more divided. It is anyway evident that medical hospitals need to zero in on the administration of the internal production network to boost administration levels.

1.3.3.2 Overall Strategic View

The medical care process lacks a fundamental requirement, specifically the effective management of internal logistics, including technology. The wards are separately capable of managing their expenditure, yet left with fundamentally just a single method to restrict costs: the level of staff. The vision perspectives on the coming health sector verbalized in the Case 1 meetings were very unobtrusive with just little subtleties varying from the current circumstance. It is critical to consider the thoughts and recommendations from the staff that work in the cycles. Studies have indicated that particularly little emergency clinics are only occasionally engaged with quality improvement yet there are very much recorded strategies for advancement work, frequently with source from the assembling business, which have demonstrated to be helpful additionally in the medical services area.

1.3.3.3 Division of Work

The time concentrates in Case 1 indicated that 34% of the medical caretakers' time was consumed on direct care exercises. The outcome is likewise in accordance with the outcome from past examinations by Hendrich et al. and furthermore appraisal from other Swedish medical clinics. Most of the attendants' time is spent on aberrant consideration exercises, frequently associated with coordination. A significant number of these exercises are not the center ability of the medical attendants. As expressed by Poulin, numerous exercises that could be completed by the help work force are regularly on the rundown of obligations performed by the medical services staff. In this manner, these undertakings should be moved or allotted to, for example, the transportation division or another in-ward administration situation that can do nonpatient-related errands. This with the utilization of computerized arrangements would free nurses' time for understanding related exercises, which is a highly proficient method of utilizing the assets.

1.3.3.4 What Are the Physical Limitations?

Contrasting Case hospital 1 and Case emergency clinic 2, it is apparent that there is accessible automated innovation and methods to work that Case medical hospital 1 would profit from. Aside from enhancements recommended over, an AGV framework would give a much highly adaptable transports plan not restricted to the accessible man power as well as furthermore permit more incessant conveyances. Notwithstanding, the actual structure of the hospital is a constraint when attempting to adjust to new innovations, for example, AGVs, in old structures with, for example, steep grades/decreases in courses and restricted lifts. This is one motivation behind why, despite developing innovation, it is still an exorbitant venture.

1.3.3.5 Transfer of Facts

There are various mechanized arrangements and help accessible for medical clinic exercises today. Different lines of business can likewise be a motivation because the inward coordinations store network is essentially equivalent to, for instance, the exchange area or industry. The delivery of medical services, in terms of hierarchy and specialized adaptation, bears resemblance to the structure of the state in the healthcare business a decade ago. The principal issue isn't the absence of accessible innovation and good examples, but instead how the information is moved and how to embrace the information and innovation to best suit the requests and needs.

1.4 CONCLUSION

The always expanding tension on the medical care framework the hospitals like some other businesses compelled to zero in on utilizing their assets in the most effective manner. This investigation validates the assertion that the nursing staff dedicates a significant portion of their time to indirect care activities, many of which are connected to administrative tasks. A few of these undertakings ought to rather be completed by the coordination division or non-nursing staff in the wards, with legitimate innovation backing and automation, to provide medical attendants the ideal opportunity for direct consideration exercises. There are various methods to enhance the internal logistical system by streamlining the tasks performed, hence increasing efficiency. Automation of the actual stream and the progression of data have ended up being one fruitful approach to accomplish this. This study indicates that there is existing innovation that could yield improvement in clinic interior coordination and significant information and encounters are found inside and outside the medical care area. The trouble lies in finding the correct type and level of computerization to suit physical and hierarchical requirements.

The analysis revealed that the internal logistical system typically lacks a crucial responsibility, resulting in inadequate management and development of the system. In the realm of entrepreneurship, it is very uncommon for aspirations and goals pertaining to calculated capability to fall short, resulting in ambiguous ownership and hindrances in the establishment of new companies. The utilization of automation as an essential method of taking care of, seeing, and improving the strategic framework is an approach to ensure proficient medical services.

REFERENCES

[1] New Scientist, AI interns: Software already taking jobs from humans. *New Scientist*, 2015.

[2] HfS Research, *Robotic Automation Emerges as a Threat to Traditional Low Cost Outsourcing*. HfS Research, 2012.

[3] London School of Economics, *Nine Likely Scenarios Arising from the Growing Use of Software Robots*. London School of Economics, 2015.

[4] Aguirre S, Rodriguez A. Automation of a Business Process Using Robotic Process Automation (RPA): A Case Study. In: Figueroa-García, J., López-Santana, E., Villa-Ramírez, J., Ferro-Escobar, R. (eds), *Applied Computer Sciences in Engineering*. WEA 2017. Communications in Computer and Information Science, vol. 742. Springer, Cham, 2017. https://doi.org/10.1007/978-3-319-66963-2_7.

[5] Robotic Process Automation for Healthcare, 2017.

[6] Ratia M, Myllärniemi J, Helander N. Robotic process automation—creating value by digitalizing work in the private healthcare? In: *Proceedings of the 22nd International Academic Mindtrek Conference*, 2018.

[7] Baird B, Charles A, Honeyman M, Maguire D, Das P. *Understanding Pressures in General Practice*. The King's Fund, 2016. https://www.kingsfund.org.uk/publications/pressures-in-general-practice [accessed 2019 January 24].
[8] Martin S, Davies E, Gershlick B. *Under Pressure: What the Commonwealth Fund's 2015 International Survey of General Practitioners Means for the UK*. The Health Foundation, 2016. https://www.health.org.uk/publications/under-pressure [accessed 2019 January 24].
[9] Autor DH. The "task approach" to labor markets: An overview. *J Labour Market Res* 2013;46(3):185–199. doi:10.1007/s12651-013-0128-z.
[10] Novek J. Hospital pharmacy automation: Collective mobility or collective control? *Soc Sci Med* 2000;51(4):491–503.
[11] Anand V, Carroll A, Downs S. Automated primary care screening in pediatric waiting rooms. *Pediatrics* 2012;129(5):e1275–e1281. doi:10.1542/peds.2011-2875.
[12] Moran W, Davis K, Moran T, Newman R, Mauldin P. Where are my patients? It is time to automate notification of hospital use to primary care practices. *South Med J* 2012;105(1):18–23. doi:10.1097/SMJ.0b013e31823d22a8.
[13] Greenes RA. Features of computer-based clinical decision support. In: *Clinical Decision Support: The Road Ahead*. Academic Press, Boston, MA, 2007.
[14] Pope C, Halford S, Turnbull J, Prichard J, Calestani M, May C. Using computer decision support systems in NHS emergency and urgent care: Ethnographic study using normalisation process theory. *BMC Health Serv Res* 2013;13:111. doi:10.1186/1472-6963-13-111.
[15] O'Sullivan D, Fraccaro P, Carson E, Weller P. Decision time for clinical decision support systems. *Clin Med (Lond)* 2014;14(4):338–341.
[16] Goddard K, Roudsari A, Wyatt J. Automation bias: A hidden issue for clinical decision support system use. *Stud Health Technol Inform* 2011;164:17–22
[17] Grewal D, Motyka S, Levy M. The evolution and future of retailing and retailing education. *J Market Edu* 2018;40(1):85–93. doi:10.1177/0273475318755838
[18] Occupational ESI. *Occupational Employment Statistics*. United States Department of Labor—Bureau of Labor Statistics https://www.bls.gov/oes/ [accessed 2019 January 24].
[19] Pooler M. Amazon robots bring a brave new world to the warehouse. *Financial Times*, 2017. https://www.ft.com/content/916b93fc-8716-11e7-8bb1-5ba57d47eff7 [accessed 2018 January 25]
[20] Frey C, Osborne M. The future of employment: How susceptible are jobs to computerisation? *Technol Forecast Soc Change* 2017;114:254–280. doi:10.1016/j.techfore.2016.08.019.
[21] Autor D, Dorn D. The growth of low-skill service jobs and the polarization of the US labor market. *Am Econ Rev* 2013;103(5):1553–1597. doi:10.1257/aer.103.5.1553
[22] Arntz MT, Gregory T, Zierahn U. The risk of automation for jobs in OECD countries: A Comparative analysis. In: *OECD Social, Employment and Migration Working Papers*. OECD, 2016, pp. 1–35. doi:10.1787/5jlz9h56dvq7-en.
[23] Jemma I, Borthakur A. *From Brawn to Brains: The Impact of Technology on Jobs in the UK*. Deloitte, 2015. https://www2.deloitte.com/content/dam/Deloitte/uk/Documents/Growth/deloitte-uk-insights-from-brawns-to-brain.pdf [accessed 2019-01-24] [WebCite Cache ID 75fZPQGfi]
[24] Bowles J. *The Computerisation of European Jobs*. Bruegel, 2014. https://bruegel.org/2014/07/the-computerisation-of-european-jobs/ [accessed 2019 January 24].
[25] Weiner J, Yeh S, Blumenthal D. The impact of health information technology and e-health on the future demand for physician services. *Health Aff (Millwood)* 2013;32(11):1998–2004. doi:10.1377/hlthaff.2013.0680.
[26] NHS Providers. *The Sate of the NHS Provider Sector*. NHS, 2016. https://nhsproviders.org/news-blogs/news/nhs-trust-leaders-warn-staff-shortages-now-outweigh-fears-over-funding [accessed 2019 January 24]
[27] Bakhshi H, Downing J, Osborne M, Schneider P. *The Future of Skills: Employment in 2030*. Pearson Future Skills, 2017. https://futureskills.pearson.com/research/assets/pdfs/technical-report.pdf [accessed 2019 January 24].

[28] Clifton L, Clifton D, Pimentel M, Watkinson P, Tarassenko L. Gaussian processes for personalized e-health monitoring with wearable sensors. *IEEE Trans Biomed Eng* 2013;60(1):193–197. doi:10.1109/TBME.2012.2208459.

[29] Alaa A,Yoon J, Hu S, van der Schaar M. Personalized risk scoring for critical care prognosis using mixtures of gaussian processes. *IEEE Trans Biomed Eng* 2018;65(1):207–218. doi:10.1109/TBME.2017.2698602.

[30] Chu W, Ghahramani Z. Gaussian processes for ordinal regression. *J Mach Learn Res* 2005;6:1019–1041.

[31] Liaw A, Wiener M. Classification and regression by random forest. *RNews*, 2002. https://www.r-project.org/doc/Rnews/Rnews_2002-3.pdf [accessed 2019 January 25].

[32] Manyika J, Chui M, Miremadi M, Bughin J, George K, Willmott P, et al. *A Future That Works: Automation, Employment, and Productivity.* McKinsey Global Institute, 2017. https://www.mckinsey.com/~/media/mckinsey/featured%20insights/Digital%20Disruption/Harnessing%20automation%20for%20a%20future%20that%20works/MGI-A-future-that-works-Executive-summary.ashx [accessed 2019 January 25]

[33] Grace K, Salvatier J, Dafoe A, Zhang B, Evans O. *When Will AI Exceed Human Performance? Evidence from AI Experts.* Cornell University, 2017. https://arxiv.org/abs/1705.08807 [accessed 2019 January 25]

[34] Dahl T, Boulos M. Robots in health and social care: A complementary technology to home care and telehealthcare? *Robotics* 2013;3(1):1–21. doi:10.3390/robotics3010001.

[35] Onwuegbuzie AJ, Johnson B, Collins K. Call for mixed analysis: A philosophical framework for combining qualitativeand quantitative approaches. *Int J Multip Res Approach* 2014;3(2):114–139. doi:10.5172/mra.3.2.114.

[36] Handel M. The O*NET content model: Strengths and limitations. *J Labour Market Res* 2016;49(2):157–176. doi:10.1007/s12651-016-0199-8.

[37] Kim H, Ghahramani Z. *Presented at: Proc Fifteenth Int Conf Artif Intell Stat.* Bayesian Classifier Combination, Singapore, 2012, p. 619. https://proceedings.mlr.press/v22/kim12.html

2 Optimizing Smartphone Addiction Questionnaires with Smartphone Application and Soft Computing

An Intelligent Smartphone Usage Behavior Assessment Model

Anshika Arora
Netaji Subhas University of Technology

Pinaki Chakraborty and M.P.S. Bhatia
Netaji Subhas University of Technology

2.1 INTRODUCTION

The ubiquitous incorporation of smartphones into the daily routine has rendered them indispensable, given their multifunctional capabilities that include communication, social networking, media consumption, and information retrieval. Smartphone has been owned and used universally in all facets of society at present. Studies confirm that more than one billion people worldwide own smartphones, which are being used for a variety of tasks, from phone calling and social interaction to surfing and information retrieval, and for entertainment (Do et al., 2011; Tossell et al., 2015). It has been found that web is accessed more from smartphones than any other digital device in the recent times (Tossell et al., 2015). Within the following 5 years, according to a 2018 prediction from the Consumer Technology Association, household TV ownership rates (96%) might be reached by smartphone ownership (Harris et al., 2020). Despite the numerous advantages attributed to the technology (Do et al., 2011; Mizuno et al., 2021), there has been growing concern about the potential for excessive smartphone use to become problematic in nature. Engaging in various activities on smartphones for prolonged periods of time can potentially lead to a dependency

DOI: 10.1201/9781003405368-2

on them for multiple purposes. The high dependency on smartphones may lead to addictive usage and development of associated psychological disorders (Arora et al., 2021a). In addition, excessive smartphone use raises the risk of stress and sleep disruption, which makes people feel less active, eventually resulting in sadness and decline in mental health causing depression (Vernon et al., 2018). Smartphone addiction, alternatively known as Problematic Smartphone Use (PSU), has been related to social emotional distress and is affected by personality (Volungis et al., 2020). New psychological variables, including the fear of missing out and nomophobia, that stem from smartphone addiction have also been investigated (Arora et al., 2021a; Arpaci, 2019; Arpaci et al, 2019; Wang et al., 2019). Millennials are increasingly very susceptible to smartphone addiction and the psychological illnesses that go along with it. The convenient accessibility of smartphones and internet connectivity, along with the flexible schedules of young individuals, are contributing factors to their excessive and addictive usage behaviors (Arora et al., 2021b; Mason et al., 2022).

With the prevalent concern about consequences of smartphone usage extensive research has been conducted to assess, categorize, and identify problematic smartphone use through the development of various scales, which consider smartphone usage behavior. A major concern is that all existing scales assessing smartphone depend on self-report, and therefore cannot reliably measure actual phone usage. According to studies, this is a research restriction that must be addressed (Harris et al., 2020). The other drawbacks of scales relying on questionnaires include unconscientious responses, missing responses, and incorrect question interpretations that produce false information. Moreover, people are often biased and influenced by social desirability leading them to report experiences that are deemed to be favorable or acceptable by society. Completion of questionnaires can be a burdensome task for participants because of the excessive number of questions. Self-report data must not be utilized alone because it is biased, according to experts in psychology research and diagnosis (Althubaiti, 2016), and a study is best done when self-report data are combined with other data, such as a person's behavior or physiological attributes. A multi-modal assessment provides an accurate picture of the subject. Specialized smartphone applications designed to track phone usage attributes can offer an intervention by providing accurate measurements of real-time usage patterns. The data retrieved through smartphone applications can be combined with the self-reported data to utilize a multi-modal approach.

This study aims to optimize the size of questionnaire-based smartphone addiction scales and replace the need to fill long questionnaires by developing purpose-built smartphone application with the capability to track real-time smartphone usage patterns to aid in the assessment of smartphone addiction. A mobile application, "UsageStats," has been created specifically for Android devices to monitor and record patterns of smartphone usage. This includes a range of features such as the duration of device usage, Internet data consumption, and the utilization patterns of various categories of applications installed on the device within the last 30 days. Forty-three distinct questions are identified via four commonly used smartphone addiction scales having pre-defined cut-off values, namely, Smartphone Addiction Scale-Short Version (SAS-SV) designed by Kown et al. (2013b), Problematic Smartphone Use Scale (PSUS-R) introduced by Valderrama (2014), Smartphone Overuse Screening Questionnaire (SOS-Q) developed by Lee et al. (2017), and Smartphone Addiction Inventory (SPAI-SF) developed by Lin et al. (2017). An optimal set of questions comprising the best questions among all scales is selected the using two-step filter-wrapper feature selection method. Filter techniques before using any classification algorithm choose the feature subset by eradicating the least significant attributes utilizing statistical properties of features. Generally, wrapper methods are implemented with predetermined learning algorithm and performance criteria, which choose the features in accordance with the precision of the training data, learn the classification model, and then apply it to the test data. The first step uses Information Gain (IG) as the filter method, which is used to rank the questions as per their importance and low ranked questions are discarded as they do not contribute to the assessment of smartphone addiction. The next step uses Particle Swarm Optimization Search (PSO Search) wrapper method to select the optimal questions set having the real-time smartphone features as captured by UsageStats. The questionnaire and android application are sent to undergraduate university students and the students are labeled as "addicted" and "nonaddicted" based on their responses. Real-time smartphone usage features along with the optimal questions set are sent as input to the Logistic Regression (LR) classifier and its performance has been evaluated for the prediction of smartphone addiction. It has also been evaluated how the optimal question set aids in improving the classification performance when added to real-time smartphone features for the assessment of smartphone addiction.

2.2 RELATED WORK

Problematic smartphone use involves the use of smartphones timelessly and choosing to use smartphones at the times of other important engagements such as attending classes, completing assignments, driving, meetings, spending time with family or friends, etc. It has been a prevalent area of study in recent decades. Gentina and Rowe (2020) relate materialism with problematic smartphone use and study the role of gender in it. Maier et al. (2020) study the risk of smartphone use while driving and study the personality traits in drivers using smartphones while driving. Hawi and Samaha (2016) have evidently investigated and concluded that smartphone addiction has a negative impact on academic performance where decline in academic performance has been linked to smartphone multitasking.

2.2.1 Conventional Smartphone Addiction Assessment

Primary and secondary studies applied to understanding smartphone addiction have leveraged survey-based and other self-report methods (Arora et al., 2021a; Harris et al., 2020). The traditional method for evaluating smartphone addiction involves utilizing smartphone addiction scales that rely on questionnaires, which are discussed in the subsequent subsections.

2.2.1.1 Scales and Questionnaires

With the prevalent concern about consequences of smartphone usage, extensive research has been conducted to assess, label, and categorize problematic smartphone use for the creation of several scales, which consider smartphone usage behavior. Early assessments of mobile phone addiction were based on mobile phone addiction scales, such as Mobile Phone Problem Use Scale devised by Bianchi and Phillips (2005) and Problematic Mobile Phone Use Questionnaire designed by Billieux et al. (2008). Smartphone addiction came into existence with the emergence of smartphones, and the first attempt to measure smartphone addiction was made by Kwon et al. (2013a), who created the Smartphone Addiction Scale, a 33-item scale designed for smartphone addiction assessment that includes six factors, including misuse, daily life disturbance, withdrawal, cyberspace-oriented interactions, tolerance, and positive anticipation. Later, they created the SAS-SV, a condensed version of SAS with ten items chosen based on content validity (Kwon et al., 2013b). Valderrama (2014) designed PSUS-R with 19 items and a defined cut-off value to identify smartphone addiction. Lin et al. (2014) designed a Smartphone Addiction Inventory with 26 items to assess smartphone addiction. Later, with the aim to develop its short version and to add a screening cut-off point, they designed Smartphone Addiction Inventory-Short Form (SPAI-SF) (Lin et al., 2017). Csibi et al. (2016) developed Smartphone Application–Based Addiction Scale with six items, which is an assessment of smartphones' application-based addictions. In a recent study, Lee et al. (2017) designed Smartphone Overuse Screening Questionnaire (SOS-Q) with 28 items and a cut-off value to mark smartphone overuse. In another recent study, Smartphone Use Questionnaires which differentiate general smartphone usage from absent-minded smartphone usage was proposed (Marty-Dugas et al., 2018). It uses the results of two ten-item scales called "general" and "absent-minded," where "general" emphasizes specific smartphone uses like how often people check social media apps and "absent-minded" deals with mindless usage like how often people check their phones without realizing why they're doing it. Various secondary studies have also been published summarizing the methods in existing literature for the assessment of smartphone addiction in people in different age groups (Al-Barashdi et al., 2015; Khan et al., 2021).

2.2.1.2 Emergence of Intelligent Data Analysis

Initial research incorporating intelligent data analysis methods for the assessment of smartphone addiction majorly used responses to smartphone addiction scales as the data source. Some studies have utilized soft computing algorithms on these self-reported data for technically advanced diagnosis of smartphone addiction. By using a clustering technique, Li et al. (2022) investigated how smartphone overdependence varies depending on use and presented a DT analysis technique to identify critical characteristics of smartphone usage in determining overdependence. Chaudhury and Tripathy (2018) analyzed questionnaire-based data using machine learning to study the effect of Internet addiction and smartphone addiction on academic performance of university students. Elhai et al. (2020) used machine learning algorithms in another study to identify the severity of PSU among Chinese undergraduate students. Mazumdar et al. (2020) employed machine learning algorithms to predict smartphone addiction on questionnaire-based data. Their model categorized the subjects in three clusters: highly addicted, moderately addicted, and nonaddicted. They also provided different facts about smartphone usage by male and female users. In a recent study, Lee et al. (2021) utilized responses to SAS by Korea Internet and Security Agency as data and implemented supervised machine learning algorithms including Decision Tree (DT), Random Forest (RF), and Extreme Gradient Boosting (XGBoost) to predict problematic smartphone use. In another recent study, Duan et al. (2021) analyzed questionnaire-based data to study risk factors associated with smartphone addiction in children and adolescents in China implementing a machine learning model. Giraldo-Jiménez et al. (2021) carried out a smartphone dependency test and associated risk factors assessment using supervised machine learning models, including DT, LR, RF, and Support Vector Machine (SVM). Koklu et al. (2021) implemented artificial neural networks for the prediction of smartphone addiction using data obtained from online surveys.

2.2.2 Automated Assessment with Real-Time Data and Soft Computing

Recent studies explored the real-time smartphone usage data and data from other digital devices, primarily smartphone apps to predict smartphone addiction. Recent research has indicated that tracking smartphone usage in real time can be a successful intervention for analyzing behavioral patterns and assessing problematic use (Arora et al., 2021a, b; Tossell et al., 2015). Efforts are being made to utilize soft computing techniques on real-time smartphone data in order to achieve a precise diagnosis of smartphone addiction. In order to create machine learning algorithms, Shin et al. (2013) employed a variety of mobile phone usage data to identify a number of features related to smartphone addiction to build machine learning models for the prediction of smartphone addiction. Another study by Lawanont et al. (2017) proposed smartphone addiction assessment based on supervised machine learning model models. Unsupervised machine learning models have also been explored by some researchers for this purpose. Ellis et al. (2019) employed an unsupervised machine learning algorithm to categorize users according to similar smartphone usage patterns. The researchers utilized a series of behavioral screen time metrics obtained from an iOS application for the usage behavior pattern analysis. In another study, Elhai et al. (2020) implemented classification methods and classification techniques to identify the severity of problematic smartphone use among Chinese undergraduate students. In a recent study (Arora et al., 2023), smartphone usage patterns extracted through an Android application have been used in combination with the responses of SAS-SV to efficiently predict smartphone addiction in university students using linear classification models, namely, SVM and LR. Researchers have utilized smartphone usage patterns and soft computing algorithms to evaluate additional psychological variables associated with smartphone addiction. Kim and Kang (2018) implemented Deep Belief Network, K-Nearest Neighbor, and SVM for the classification of smartphone addiction levels in individuals and conducted emotion analysis. For the examination of participant emotions, they also used EEG signals and determined that the high-risk group displayed greater emotional instability than the low-risk group, particularly when expressing the emotion "fear." Arora et al. (2021a) employed a Gaussian mixture clustering algorithm to identify nomophobic behavior among users with comparable smartphone usage patterns, which were extracted using a smartphone application. Arora et al. (2021b) conducted a separate study in which they obtained real-time smartphone data through an Android application to predict the severity of nomophobia, using supervised machine learning algorithms.

It is evident that real-time smartphone usage patterns can effectively track smartphone usage with precise measures of objective data and can provide an intervention for excessive use when utilized with machine learning algorithms. However, questions have been raised by researchers (Elhai et al., 2018; Harris et al., 2020) on whether objective smartphone use data collected through the use of smartphone applications is just a measure of smartphone use frequency or a diagnosis for smartphone addiction.

Therefore, this study combines the capabilities of smartphone application for the collection of real-time precise measures of smartphone usage with the capabilities of smartphone use scales for the assessment of smartphone addiction. Also, all of the studies employing soft computing techniques focus on machine learning algorithms for the prediction of addiction, and no study has employed other soft computing algorithms in an attempt to reduce the size of large questionnaires. Therefore, this study aims to implement soft computing algorithms in order to identify an optimal question set that can be combined with the application of extracted smartphone usage measures for the assessment of smartphone addiction with enhanced performance.

2.3 METHODOLOGY

2.3.1 Smartphone Application: UsageStats

A mobile application for android devices has been developed that automatically retrieves data from the smartphones of users, with the purpose of analyzing real-time statistics of smartphone usage among participants. The application has been developed using a UsageStatsManager for system total time duration. SQLiteOpenHelper is used for storing data in SQL inside the phone, and Firebase storage has been used for syncing data in an online database. UsageStats monitors both overall smartphone usage and the usage of various applications on the device, which includes both the pre-installed applications and any third-party applications that have been installed. Figures 2.1 and 2.2 present the developed smartphone application screen.

FIGURE 2.1 Smartphone usage during the day.

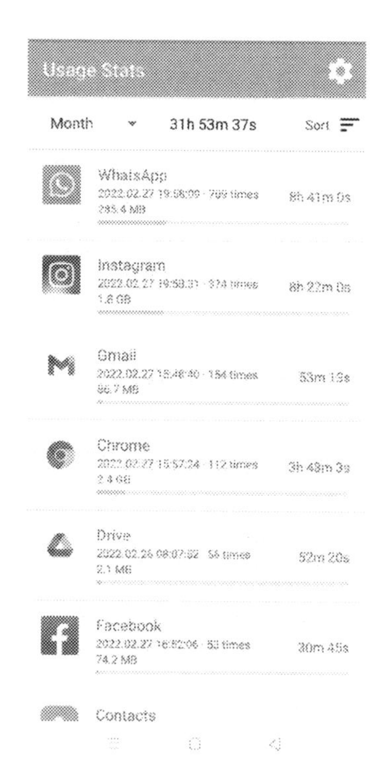

FIGURE 2.2 Smartphone usage during the month.

TABLE 2.1
Retrieved Smartphone Usage Attributes

S. No.	Attribute	Description
1.	Total-time	Total duration of smart phone use for the last 30 days
2.	Total-data-used	Sum of internet data used by all apps during the smart phone use in the last 30 days
		For Each Application
1.	Time-used	Total duration of the application usage for the last 30 days
2.	Times-open (sessions)	Total number of sessions of the application in the last 30 days
3.	Longest-session	Longest duration in the last 30 days for which the application is used in one launch
4.	Data-used	Sum of mobile and Wi-Fi Internet data that have been used during the application usage in the last 30 days
5.	Tag	Category to which the application belongs

The smartphone usage attributes as collected by UsgeStats is summarized in Table 2.1.

Following data acquisition, each application is categorized into one of the nine application tags (Arora et al., 2023). The default tags have been altered as follows:

- We separated the default tag, namely, Social & Communication to *Social Media and Communication.* We also moved phone from productivity as categorized in some smartphones to *Communication.*
- A new tag, namely *Entertainment*, was created.
- For this study, we combined *News & Surfing* applications into one category. We moved Google and Chrome from their default tags, i.e., Productivity and Social & Communication, respectively, to *News & Surfing.*
- *Work* tag was maintained for all the work and study-related applications. We moved Gmail and LinkedIn from their default tags, i.e., Social & Communication and Productivity, respectively, to the *Work* category.
- *Others* is the tag for applications belonging to none of the said categories.

Table 2.2 lists the tags with their corresponding applications.

Following the categorization and classification of applications, a dataset was generated that included attributes related to users' general smartphone usage, such as total_data_used (GB) and total_time (hours) as well as attributes for each tag, including number of times opened (sessions), longest session (minutes), data used (MB), and time_used (minutes).The extracted attribute set is normalized through the use of Min–Max normalization.

2.3.2 Smartphone Addiction Questionnaires

A number of mobile phone use and smartphone use scales have been reported in literature (Harris et al., 2020). Among the scales, 21 scales identify the problem of smartphone addiction or problematic smartphone use but still have their limitations,

TABLE 2.2
Tags with Their Respective Applications

S. No.	Tag	Applications
1.	Social media	TWITTER, TUMBLR, FACEBOOK, INSTAGRAM, REDDIT, etc.
2.	Communication	SNAPCHAT, WHATSAPP, TELEGRAM, TRUECALLER, messenger, DISCORD, contacts, phone, messaging, contacts and dialer, etc.
3.	Entertainment	Movies & videos and music & audio, SONYLIV, YOUTUBE, HOTSTAR, NETFLIX, PRIME VIDEO, FM RADIO, SPOTIFY, GAANA, etc.
4.	Productivity	PAYTM, GPAY, DOCS, SHEETS, etc.
5.	News and surfing	GOOGLE, CHROME, BROWSER, MI BROWSER, INTERNET INSHORTS, QUORA, GOOGLE NEWS, etc.
6.	Gaming	Battleground, etc.
7.	Work	LINKEDIN, GMAIL, MAIL, ZOOM, MEET, CLASSROOM, UNACADEMY, UDEMY, etc.
8.	Photos and camera	Albums, camera, photos, gallery, PICSART.
9.	Others	Files, drive, maps, shopping and food delivery applications, etc.

Note: System applications are given in lowercase while third-party applications are given in uppercase.

such as unreported item format, required partner (such as parents) to observe, no cut-off value defined, etc. Therefore, four scales with clear cut-off values have been selected for this study. These questionnaires are SAS-SV (Kown et al., 2013b), PSUS-R (Valderrama, 2014), SOS-Q (Lee et al., 2017), and SPI-SF (Lin et al., 2017). The questions belonging to different scales with similar semantics were merged and 43 distinct questions were identified, which were used to collect data from students about their dependence on smartphones. A common 4-point Likert scale was used initially for data collection.

After data collection, the responses on 4-point Likert scale were converted back to the original scale type of each questionnaire and the score for each questionnaire was calculated. The conversion of a value (X) on Likert scale 1 to a value (Y) on Likert scale 2 was done as represented by equation (2.1):

$$Y = (B - A) \times \frac{x - a}{b - a} + A \tag{2.1}$$

where a and b are minimum and maximum values, respectively, on the former Likert scale, and A and B are minimum and maximum values, respectively, on the latter Likert scale.

Based on the calculated score and the cut-off point of each scale, every user was labeled as addicted or nonaddicted for each scale.

2.3.3 Labeling

Each user instance of the dataset was labeled with either of the two classes, namely, "addicted" or "nonaddicted." The label was given based on the scores obtained on each of the four scales: SAS-SV, PSUS-R, SOS-Q, and SPI-SF. If a user was identified as Addicted on at least two scales, then the user was given a final label as Addicted.

2.3.4 Optimizing Smartphone Addiction Questions

Having real-time smartphone usage data, we attempted to reduce the number of questions to obtain an optimal set of questions for the assessment of smartphone addiction. For this, filter-wrapper feature selection model was developed, which selects the optimal question set from 43 distinct questions. Filter-wrapper models have been successfully used in studies to select optimal features for classification tasks (Kumar & Arora, 2019; Kumar et al., 2021). A filter method, IG, was used as a first step to rank the questions based on their significance followed by a wrapper method, namely, PSO Search, to select a subset of most relevant questions.

2.3.4.1 Filter Method: Information Gain

The IG with regard to the labeled class was measured using IG algorithm by calculating the contributions of each feature for reducing the total entropy. The IG score for every feature was evaluated as given in equation (2.2):

$$\text{InfoGain}(\text{Class, Attribute}) = H(\text{Class}) - H(\text{Class}|\text{Attribute}) \tag{2.2}$$

where $H(\text{Class})$ is the entropy of the class and is given by equation (2.3):

$$H(\text{Class}) = -\sum P_i * \log_2(P_i) \tag{2.3}$$

where P_i is the probability of class in the dataset.

The study used InformationGainAttributeEval attribute evaluator method of WEKA with the Ranker search method to rank the attributes.

2.3.4.2 Wrapper Method: Particle Swarm Optimization Search

PSO works by initializing a population of random particles representing the initial solutions. A particle represents a point in the search space, which has an associated memory to store its best position called the local best. Also, there exists a global best that represents the location of a particle with the best position among all particles. Figure 2.3 represents a flying particle.

Here, $x_i(t)$ is the location, $v_i(t)$ is the velocity, $P_i(t)$ is the best position of the particle *i*, and $g(t)$ is the global best experience of any particle, at time *t*.

This study uses WrapperSubsetEval method of WEKA with the PSO Search as the search method. PSO Search technique is a wrapper method for feature selection that works iteratively to select the best features that contribute to the class label. PSO Search is based on Geometric PSO (Fong et al., 2018), which varies from standard PSO in the sense that it does not have any velocity and contains mutation (Moraglio et al., 2007). A convex combination with parameters w_1, w_2, and w_3 forms an equation of position update. The parameters are nonnegative and add up to one. Figure 2.4 presents the Geometric PSO algorithm.

2.3.5 Classification: Logistic Regression

The application extracted real-time smartphone usage features in the normalized form with the optimal set of questions as selected by the filter-wrapper feature selection module are sent as input to the LR classifier for the prediction of smartphone addiction. LR is a classification algorithm for predicting a dichotomous dependent

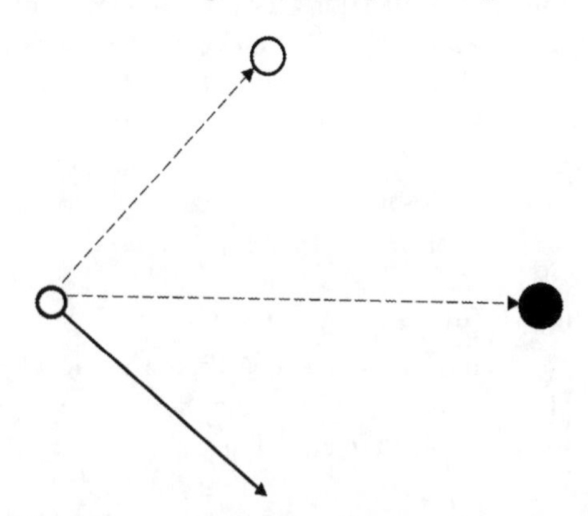

FIGURE 2.3 A flying particle.

For each particle *i* do

Initialize position x_i at random in the search space

End For

While stop criteria not met do

For each particle *i* do

Set personal best x_i as best position found so far by the particle

Set global best *g* as best position found so far by the whole swarm

End For

For each particle *i* do

Update position using a randomized convex combination

$x_i = CX(x_i, w_1), (g, w_2), (x_i, w_3)$

Mutate x_i

End For

End While

FIGURE 2.4 Geometric Particle Swarm Optimization (PSO) algorithm.

variable. LR works by modeling a sigmoid function in order to determine whether an outcome is present or absent based on values of a set of predictor variables. A model for p independent variables is presented in equation (2.4):

$$P(Y=1)=\frac{1}{1+e^{-(\beta_0+\beta_1x_1+\beta_2\beta_2+\cdots+\beta_p\beta_p)}} \tag{2.4}$$

where $P(Y=1)$ is the probability of the presence of smartphone addiction based on smartphone usage features and optimal set of questions and $\beta_0, \beta_1, \beta_2, \ldots, \beta_p$ are regression coefficients.

2.4 RESULTS

The model is validated on a dataset created using students' smartphone usage and questionnaire responses. Seventy-three undergraduate students participated in the study having an average age of 19.672 years (SD 1.246 years).

2.4.1 Filter Wrapper Feature Selection

At the first stage of the filter-wrapper feature selection, each attribute is ranked based on its IG score. A dataset having real-time smartphone usage features and 43 questions is given as input to the IG algorithm. As a result of this step, nine questions that have an IG score of 0 w.r.t. the class label have been eliminated, i.e., the questions

TABLE 2.3
PSO Search Hyperparameters

Classifier	Logistic Regression
Inertia weight	0.32
Individual weight	0.35
Social weight	0.32
Population size	20
Iterations	25

that have no role in the prediction of smartphone addiction. Hence, the size of the question set was reduced from 43 to 34 at the end of this step.

Following this, the PSO Search wrapper method was used to select the finest subset of questions. The PSO Search method was given an input, which contained real-time smartphone usage features with 34 questions selected at the first stage. The wrapper method used a preset learning algorithm, i.e., LR in this study to estimate the best feature set for that algorithm. The hyperparameters used for the PSO Search method are summarized in Table 2.3.

At the end of this stage, seven questions were selected, which formed the optimal question set for smartphone addiction assessment when having real-time smartphone usage features.

2.4.2 Classification

The performance of the LR classifier is observed on the dataset labeled based on the four smartphone addiction scales. The model is evaluated using five-fold cross validation and accuracy and F1 score performance metrics are used. The performance is observed with three feature sets, namely:

1. Feature set 1: Real-time smartphone features only,
2. Feature set 2: Real-time smartphone features with 43 distinct questions from four questionnaires,
3. Feature set 3: Real-time smartphone features and optimal question set as selected by the filter-wrapper feature selection model.

The comparative results are presented in Figures 2.5 and 2.6.

It can be observed that an accuracy of 88.88% is achieved when the labeled data are classified using real-time smartphone usage features obtained through the UsageStats application. The accuracy increases significantly on adding 43 smartphone addiction questions to the real-time smartphone features. The best performance is achieved when the optimal set of questions selected by the proposed methodology is added to real-time smartphone usage features. The proposed methodology achieves an excellent performance accuracy of 98.61% and an F1 score of 0.986.

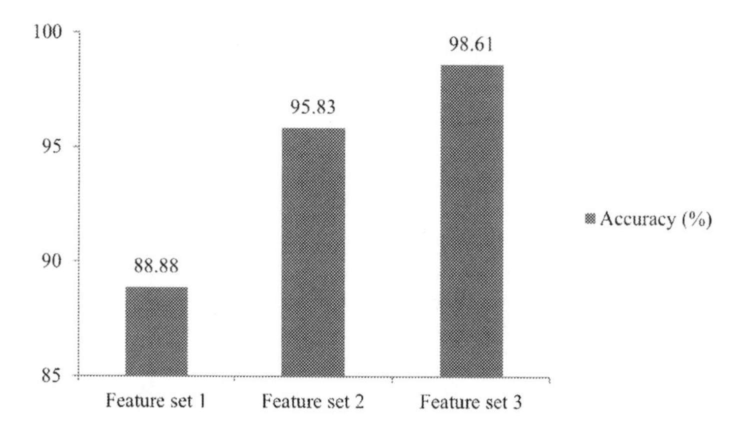

FIGURE 2.5 Accuracy of the proposed model.

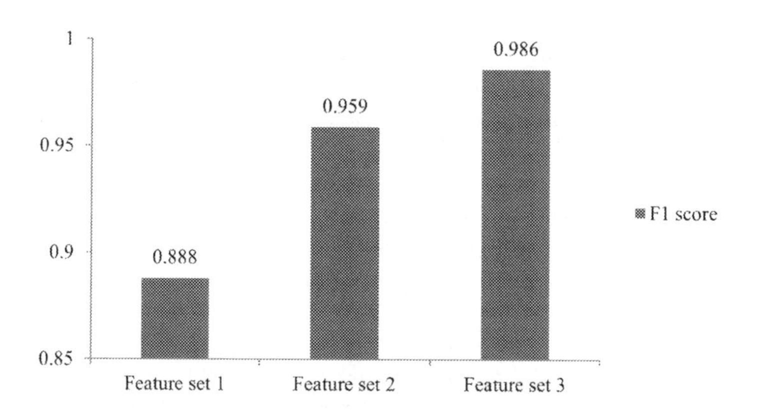

FIGURE 2.6 F1 score of the proposed model.

The findings of this study align with the beliefs of experts in behavioral and psychological research and diagnosis, which suggest that self-report data alone give inaccurate results (Althubaiti, 2016) while research is best done when combining self-report data with behavior or physiological data. In this study, self-reported data are obtained in the form of smartphone addiction questions and they are combined with behavioral data obtained through the android application with which the model achieves superlative performance.

2.4.3 The Optimized Question Set

This study concludes that, having real-time smartphone usage features, the size of the question set can be reduced by a significant amount. This study identifies an optimal set of seven questions that are necessary and sufficient for the assessment of smartphone addiction when having real-time smartphone usage features. The set of questions identified are presented in Table 2.4.

TABLE 2.4
The Optimal Question Set

Question No.	Question
Ql	When I use my smartphone, I lose track of how much time has passed.
Q2	Won't be able to tolerate not having a smartphone.
Q3	I find that I have been hooking on smartphone longer and longer.
Q4	I have spent increasing amounts of time on my smartphone to achieve satisfaction.
Q5	I end up using my smartphone longer than I had intended.
Q6	I have jeopardized or lost a significant relationship, job, or educational career opportunity because of my smartphone use.
Q7	Constantly and excessively checking my smartphone so as not to miss conversations on social media, e-mails, blogs, and other applications.

The optimal question set includes only seven questions. Hence, its size is the shortest among the existing smartphone addiction scales. The shortest and most commonly used among the existing questionnaires till now was SAS-SV, which has ten questions. Hence, this study proposes a smartphone addiction assessment model that avoids the cumbersome task of filling long questionnaires potentially leading to inaccurate results.

2.5 CONCLUSION

This study proposed a soft computing–based framework for the assessment of smartphone addiction using real-time smartphone usage data obtained through an android application and an optimal sized questionnaire. Four commonly used smartphone scales with clear cut-off values are identified, namely SAS-SV, PSUS, Smartphone Overuse Screening Questionnaire, and Smartphone Addiction Inventory. A mobile application for Android has been created to gather user data regarding smartphone usage. A two-stage filter-wrapper feature selection method was implemented to identify the optimal question set having objective smartphone usage data. The two-stage approach uses PSO as the wrapper method and IG as the filter method. This optimal question set with the application extracted smartphone usage features was used to train the LR model for the prediction of smartphone addiction, and the proposed methodology gave a performance accuracy of 98.61% and an F1 score of 0.986. This study identified an optimal question set that is the shortest among all existing smartphone addiction questionnaires comprising seven most important questions from all the scales. This framework accurately measures smartphone addiction with a performance accuracy of 98.61% in predicting smartphone addiction using the LR classifier and avoids the need to fill long questionnaires.

In future, behavioral addictions of participants can be monitored combining social media usage data (Arora et al., 2021c) and smartphone usage data. In addition, wearable devices such as smart watches and actigraphs enable extraction of physiological and psychological signals indicative of users' mental health (Kumar et al., 2022) that can be combined with usage data to assess overall behavioral health of participants.

REFERENCES

Al-Barashdi, H. S., Bouazza, A., & Jabur, N. H. (2015). Smartphone addiction among university undergraduates: A literature review. *Journal of Scientific Research and Reports*, *4*(3), 210–225.

Althubaiti, A. (2016). Information bias in health research: Definition, pitfalls, and adjustment methods. *Journal of Multidisciplinary Healthcare*, *9*, 211.

Arora, A., Chakraborty, P., & Bhatia, M. P. S. (2021a). Problematic use of digital technologies and its impact on mental health during COVID-19 pandemic: Assessment using machine learning. In: Arpaci, I., Al-Emran, M., A. Al-Sharafi, M., Marques, G., *Emerging Technologies During the Era of COVID-19 Pandemic* (pp. 197–221). Springer, Cham.

Arora, A., Chakraborty, P., & Bhatia, M. P. S. (2021b). Real time smartphone data for prediction of nomophobia severity using supervised machine learning. In: *AIJR Proceedings* (pp. 78–83). doi:10.21467/proceedings.114.11.

Arora, A., Chakraborty, P., Bhatia, M. P. S., & Mittal, P. (2021c). Role of emotion in excessive use of Twitter during COVID-19 imposed lockdown in India. *Journal of Technology in Behavioral Science*, *6*(2), 370–377.

Arora, A., Chakraborty, P., Bhatia, M. P. S., & Puri, A. (2023). Intelligent model for smartphone addiction assessment in university students using android application and smartphone addiction scale. *International Journal of Education and Management Engineering*, *13*(1), 29.

Arpaci, I. (2019). Culture and nomophobia: The role of vertical versus horizontal collectivism in predicting nomophobia. *Information Development*, *35*(1), 96–106.

Arpaci, I., Baloğlu, M., & Kesici, Ş. (2019). A multi-group analysis of the effects of individual differences in mindfulness on nomophobia. *Information Development*, *35*(2), 333–341.

Bianchi, A., & Phillips, J. G. (2005). Psychological predictors of problem mobile phone use. *Cyberpsychology and Behavior*, *8*(1), 39–51.

Billieux, J., van der Linden, M., & Rochat, L. (2008). The role of impulsivity in actual and problematic use of the mobile phone. *Applied Cognitive Psychology*, *22*(9), 1195–1210.

Chaudhury, P., & Tripathy, H. K. (2018). A study on impact of smartphone addiction on academic performance. *International Journal of Engineering and Technology*, *7*(2.6), 50–53.

Csibi, S., Demetrovics, Z., & Szabo, A. (2016). Hungarian adaptation and psychometric characteristics of brief addiction to smartphone scale (BASS). *Psychiatria Hungarica*, 31(1), 71–77.

Do, T. M. T., Blom, J., & Gatica-Perez, D. (2011). Smartphone usage in the wild: A large-scale analysis of applications and context. In: *Proceedings of the 13th International Conference on Multimodal Interfaces (ICMI '11)* (pp. 353–360). Association for Computing Machinery, New York. https://doi.org/10.1145/2070481.2070550

Duan, L., He, J., Li, M., Dai, J., Zhou, Y., Lai, F., & Zhu, G. (2021). Based on a decision tree model for exploring the risk factors of smartphone addiction among children and adolescents in China during the COVID-19 pandemic. *Frontiers in Psychiatry*, *12*, 897.

Elhai, J. D., Levine, J. C., Alghraibeh, A. M., Alafnan, A. A., Aldraiweesh, A. A., & Hall, B. J. (2018). Fear of missing out: Testing relationships with negative affectivity, online social engagement, and problematic smartphone use. *Computers in Human Behavior*, *89*, 289–298.

Elhai, J. D., Yang, H., Rozgonjuk, D., & Montag, C. (2020). Using machine learning to model problematic smartphone use severity: The significant role of fear of missing out. *Addictive Behaviors*, *103*, 106261.

Ellis, D. A., Davidson, B. I., Shaw, H., & Geyer, K. (2019). Do smartphone usage scales predict behavior?. *International Journal of Human-Computer Studies*, *130*, 86–92.

Fong, S., Biuk-Aghai, R. P., & Millham, R. C. (2018, February). Swarm search methods in weka for data mining. In: *Proceedings of the 2018 10th International Conference on Machine Learning and Computing* (pp. 122–127). Association for Computing Machinery, New York. https://doi.org/10.1145/3195106.3195167

Gentina, E., & Rowe, F. (2020). Effects of materialism on problematic smartphone dependency among adolescents: The role of gender and gratifications. *International Journal of Information Management*, *54*, 102134.

Giraldo-Jiménez, C. F., Gaviria-Chavarro, J., Urrutia-Valdés, A., Bedoya-Pérez, J. F., & Sarria-Paja, M. O. (2021). Machine-learning predictive models for dependency on smartphones based on risk factors. doi:10.21203/rs.3.rs-886633/v1.

Harris, B., Regan, T., Schueler, J., & Fields, S. A. (2020). Problematic mobile phone and smartphone use scales: A systematic review. *Frontiers in Psychology*, *11*, 672.

Hawi, N. S., & Samaha, M. (2016). To excel or not to excel: Strong evidence on the adverse effect of smartphone addiction on academic performance. *Computers and Education*, *98*, 81–89.

Khan, B., Janjua, U. I., & Madni, T. M. (2021). The identification of influential factors to evaluate the kids smartphone addiction: A literature review. In: *Proceedings of 4th International Conference on Computing and Information Sciences* (ICCIS) (pp. 1–6). Karachi. doi: 10.1109/ICCIS54243.2021.9676392

Kim, S. K., & Kang, H. B. (2018). An analysis of smartphone overuse recognition in terms of emotions using brainwaves and deep learning. *Neurocomputing*, *275*, 1393–1406.

Koklu, N., Taspinar G., & Sulak S. A. (2021). Estimating the level of smartphone addiction using artificial neural networks. In: *Proceedings of International Congress of Science Culture and Education* (INCES-2021) (pp. 28–33). Antalya.

Kumar, A., & Arora, A. (2019, February). A filter-wrapper based feature selection for optimized website quality prediction. In: *2019 Amity International Conference on Artificial Intelligence (AICAI)* (pp. 284–291). IEEE, New York, NY.

Kumar, A., Bhatia, M. P. S., & Sangwan, S. R. (2021). Rumour detection using deep learning and filter-wrapper feature selection in benchmark twitter dataset. *Multimedia Tools and Applications*, *81*(24), 34615–34632.

Kumar, A., Sangwan, S. R., Arora, A., & Menon, V. G. (2022). Depress-DCNF: A deep convolutional neuro-fuzzy model for detection of depression episodes using IoMT. *Applied Soft Computing*, *122*, 108863.

Kwon, M., Lee, J. Y., Won, W. Y., Park, J. W., Min, J. A., Hahn, C., Gu, X., Choi, J. H., & Kim, D. J. (2013a). Development and validation of a smartphone addiction scale (SAS). *PLoS One*, *8*(2), 56936.

Kwon, M., Kim, D. J., Cho, H., & Yang, S. (2013b). The smartphone addiction scale: Development and validation of a short version for adolescents. *PLoS One*, *8*(12), 83558.

Lawanont, W., & Inoue, M. (2017, June). A development of classification model for smartphone addiction recognition system based on smartphone usage data. In: Czarnowski, I., Howlett, R., & Jain, L. (eds.), *Intelligent Decision Technologies 2017. IDT 2017. Smart Innovation, Systems and Technologies* (vol. 73, pp. 3–12). Springer, Cham. https://doi.org/10.1007/978-3-319-59424-8_1

Lee, H. K., Kim, J. H., Fava, M., Mischoulon, D., Park, J. H., Shim, E. J., Lee, E. H., Lee, J. H., & Jeon, H. J. (2017). Development and validation study of the Smartphone Overuse Screening Questionnaire. *Psychiatry Research*, *257*, 352–357.

Lee, J., & Kim, W. (2021). Prediction of problematic smartphone use: A machine learning approach. *International Journal of Environmental Research and Public Health*, *18*(12), 6458.

Li, X., Kim, B. T., Yoon, D. W., & Hwang, H. (2022). Differences in smartphone overdependence by type of smartphone usage: Decision tree analysis. In: *Proceedings of Future of Information and Communication Conference* (pp. 12–21).

Lin, Y. H., Chang, L. R., Lee, Y. H., Tseng, H. W., Kuo, T. B., & Chen, S. H. (2014). Development and validation of the smartphone addiction inventory (SPAI). *PLoS One*, *9*(6), 98312.

Lin, Y. H., Pan, Y. C., Lin, S. H., & Chen, S. H. (2017). Development of short-form and screening cutoff point of the Smartphone Addiction Inventory (SPAI-SF). *International Journal of Methods in Psychiatric Research*, 26(2), 1525.

Maier, C., Mattke, J., Pflügner, K., & Weitzel, T. (2020). Smartphone use while driving: A fuzzy-set qualitative comparative analysis of personality profiles influencing frequent high-risk smartphone use while driving in Germany. *International Journal of Information Management*, *55*, 102207.

Marty-Dugas, J., Ralph, B. C., Oakman, J. M., & Smilek, D. (2018). The relation between smartphone use and everyday inattention. *Psychology of Consciousness: Theory, Research, and Practice*, *5*(1), 46.

Mason, M. C., Zamparo, G., Marini, A., & Ameen, N. (2022). Glued to your phone? Generation Z's smartphone addiction and online compulsive buying. *Computers in Human Behavior*, *136*, 107404.

Mazumdar, A., Karak, G., & Sharma, S. (2020). Machine learning model for prediction of smartphone addiction. *Turkish Journal of Computer and Mathematics Education*, *11*(1), 545–550.

Mizuno, T., Ohnishi, T., & Watanabe, T. (2021). Visualizing social and behavior change due to the outbreak of COVID-19 using mobile phone location data. *New Generation Computing*, *39*(3), 453–468.

Moraglio, A., Chio, C. D., & Poli, R. (2007, April). Geometric particle swarm optimisation. In: *Genetic Programming: 10th European Conference, EuroGP 2007, Valencia, Spain, April 11-13, 2007. Proceedings 10* (pp. 125–136). Springer, Berlin, Heidelberg.

Shin, C., & Dey, A. K. (2013, September). Automatically detecting problematic use of smartphones. In: *Proceedings of the 2013 ACM International Joint Conference on Pervasive and Ubiquitous Computing (UbiComp '13)* (pp. 335–344). Association for Computing Machinery, New York. https://doi.org/10.1145/2493432.2493443

Tossell, C., Kortum, P., Shepard, C., Rahmati, A., & Zhong, L. (2015). Exploring smartphone addiction: Insights from long-term telemetric behavioral measures. *International Journal of Interactive Mobile Technologies*, *9*(2), 37–43.

Valderrama, J. A. (2014). *Development and validation of the problematic smartphone use scale*. Alliant International University, Alhambra.

Vernon, L., Modecki, K. L., & Barber, B. L. (2018). Mobile phones in the bedroom: Trajectories of sleep habits and subsequent adolescent psychosocial development. *Child Development*, *89*(1), 66–77.

Wang, J., Wang, P., Yang, X., Zhang, G., Wang, X., Zhao, F., Zhao, M., & Lei, L. (2019). Fear of missing out and procrastination as mediators between sensation seeking and adolescent smartphone addiction. *International Journal of Mental Health and Addiction*, *17*(4), 1049–1062.

3 Artificial Neural Network Model for Automated Medical Diagnosis

Shambhavi Mishra, Tanveer Ahmed, Mohd. Abuzar Sayeed, and Umesh Gupta
Bennett University

3.1 INTRODUCTION

The process of medical diagnosis is vital and demands a high level of accuracy and precision [1]. Applying machine learning methods in automating medical diagnosis has garnered significant attention. One such technique is the artificial neural network (ANN) model, which has shown promising results in many applications, including cancer diagnosis, cardiovascular disease prediction, and diabetic retinopathy detection [2,3]. These models can process large amounts of data quickly and accurately, making them ideal for use in automated medical diagnosis systems to improve the accuracy and efficiency of diagnosis. ANNs can also be trained to recognize patterns in different types of medical data, such as imaging, clinical, and genetic data, allowing for a more comprehensive analysis of patient health [4,5]. As technology improves and more medical data becomes available, ANNs will likely play an increasingly important role in automating medical diagnosis and improving patient outcomes.

Although ML algorithms do not need human intervention during the learning process, preparing the data that will be used by these algorithms, selecting the optimal algorithm, and adjusting it to achieve the best results all require professional data scientists. In addition to business expertise in processed data, these procedures depend on humans and require specialized computer science, programming, mathematics, and statistics knowledge. Because individuals with all these skills, referred to as "data science unicorns," are rare, many businesses hire data scientists and analysts to complete the complete ML processes. AutoML, a completely automated procedure that minimizes human interference, is one method for solving this problem. As Figure 3.1 shows, AutoML automates the main ML procedures. Data integration, transformation, cleaning, and reduction are all examples of data preparation, which occurs first in these processes. Most of a data engineer's effort is spent on data preparation because it is time-consuming [6]. In the subsequent stage, feature extraction and selection choose a subset of dataset characteristics that preserves the dataset's information while enhancing learning generalizations [7]. The algorithm selection process involves using a strategy to select the algorithm that will produce

DOI: 10.1201/9781003405368-3

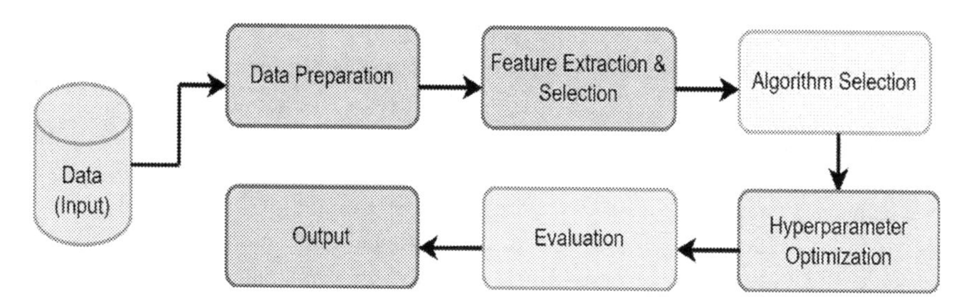

FIGURE 3.1 Main processes of the AutoML model.

the most accurate results. Hyperparameter optimization adjusts the algorithms' parameters to improve results [8]. AutoML systems employ several methodologies and optimization strategies to attain the desired accuracy and efficiency. Based on Bayes' theorem, a well-known probability theory, Bayesian optimization is a method for improving hyperparameters for machine learning algorithms [8–10]. Moreover, more straightforward methods like grid search and random search are used. Another technique for optimizing hyperparameters is dubbed "meta-learning," in which the AutoML system learns from its use of machine learning.

ANNs are a class of machine learning algorithms inspired by the structure and function of the human brain. ANNs can learn and generalize from large amounts of data, making them an ideal tool for automated medical diagnosis. Different types of ANN models are used in automated medical diagnosis. One example of a model used in this context is the convolutional neural network (CNN), often employed for image classification tasks. CNNs can learn features from medical images such as X-rays or computed tomography (CT) scans, enabling them to diagnose conditions such as cancer, pneumonia, or bone fractures [11]. They can also detect skin lesions, diabetic retinopathy, and eye diseases. Another type of ANN model utilized in automated medical diagnosis is the recurrent neural network (RNN) [12].

RNNs are specifically designed to handle sequential data, such as time-series data or clinical records, and can capture temporal dependencies within the data. These networks can be applied for predicting diseases, identifying patient risk factors for conditions such as heart disease or diabetes, or anticipating seizure onset in epilepsy patients. RNNs can also be employed for speech recognition tasks, analyzing patients' voices to detect symptoms or disease patterns like Parkinson's disease. Deep belief networks (DBNs) represent another type of ANN, consisting of multiple layers of nodes. DBNs extract features from medical data, including genetic or medical images, and can be applied to disease diagnosis and drug discovery [13,14]. The ANN model used in automated medical diagnosis autoencoder neural networks (AENs) is also used for unsupervised learning, where the ANN learns to reconstruct the input data. AENs have been used to detect anomalies in medical images and identify rare conditions [15,16]. Transfer learning is also one of the famous techniques where a pre-trained ANN model is used for a different task by reusing its knowledge [17,18]. Transfer learning can be used in medical diagnosis by training a pre-trained

ANN on a new medical dataset. This technique has been used for detecting cancer and other diseases. Different ANN models can be used in automated medical diagnosis, depending on the type of data available and the specific diagnosis task.

A complex layer structure used by deep neural networks to define inputs to outputs presents building blocks such as transformations and nonlinear functions [19]. Deep learning can now solve issues that traditional artificial intelligence (AI) has difficulty resolving [20]. Deep learning can use unlabeled data during training, making it suitable for dealing with diverse information and data to acquire information [21]. Deep learning applications could result in malicious behaviour. However, many more good things can be done with this technology. Deep learning has a clear path toward functioning with massive data sets, and as a result, its applications are expected to be more varied in the future [20]. This was highlighted way back in 2015 [20]. Other more recent research has emphasized the capabilities of cutting-edge deep learning techniques, including learning from difficult data [22,23], image recognition [24], text classification [25], and others. Deep learning is widely used in medical diagnosis [26,27]. Deep learning includes but is not limited to biomedicine [28], magnetic resonance imaging (MRI) analysis [29], and health informatics [30]. Segmentation, diagnosis, classification, prediction, and detection of different anatomical regions of interest are more focused applications of deep learning in the medical industry. Deep learning has many hidden layers that enable it to learn abstractions based on inputs. It is far superior to traditional machine learning because it can learn from raw data [31]. The neural networks' ability to learn from data using general-purpose learning procedures is the key to their deep learning capabilities. In this research paper, we investigate the use of ANNs for automated medical diagnosis (see Section 3.1). In Section 3.2, we discuss the related work. We focus on the ANN overview, its mathematical background, and network learning analysis in Section 3.3. The performance of automated medical diagnosis and comparing their traditional diagnostic methods with the application of ANNs to different medical conditions are discussed in Section 3.4. Section 3.5 addresses limitations. Finally, Section 3.6 provides concluding remarks and future directions.

3.2 RELATED WORK

Much research is being done on using ANNs for automated medical diagnosis. Several studies have proposed ANN-based approaches for medical diagnosis, such as using electronic health records (EHRs) or gene expression profiles for the diagnosis and survival prediction of different diseases, including colorectal cancer [32,33]. Furthermore, recent developments in deep learning, including ANNs, have shown promising results for medical diagnosis, particularly in medical imaging [34]. Some studies have reviewed using ANNs in medical diagnosis, focusing on cardiology, dermatology, neurology, breast cancer detection, and thyroid disease diagnosis [26]. Another study highlighted the challenges, opportunities, and future directions for medical diagnosis using deep learning, including ANNs. These studies emphasize the potential of ANNs in medical diagnosis and their ability to leverage vast amounts of medical data to improve patient outcomes. Various applications found the different automated machine and deep learning techniques useful. Reference [35] in this review article summarizes the recent advancement in deep learning–based methods for medical image analysis, including automated diagnosis. The authors discuss deep learning architectures and their applications in different medical imaging modalities, such as MRI and CT [36]. This systematic review summarizes the current state of the art in the automated diagnosis of melanoma using machine learning techniques, including CNN models. The authors highlight the potential of these models in improving the accuracy and efficiency of melanoma diagnosis. These instances emphasise the varied applications of ANN models in automated medical diagnosis, encompassing areas such as heart disease, melanoma, and medical image analysis. Deep learning–based techniques have demonstrated considerable potential in enhancing the accuracy and efficiency of medical diagnosis, but additional research is required to confirm their clinical value.

Another study comprehensively reviews various AI techniques, including ANNs, for disease diagnosis. The authors examine the applications of AI methods in diverse medical fields, such as neurology, cardiology, and oncology [37]. The study in Reference [38] focuses on the automated diagnosis of skin lesions, using a dataset of dermoscopic images and reporting high accuracy in categorizing skin lesions. Reference [39] explores an ANN-based method for automated breast cancer detection and classification, comparing the performance of ANNs and support vector machines (SVMs) and noting high accuracy for both techniques. Reference [40] introduces an ANN-based approach for medical diagnosis utilizing EHRs, reporting high accuracy in classifying patients into various diagnostic categories using the proposed method. Reference [41] presents an ANN-based approach for diagnosing and predicting the survival of colon cancer patients, using a dataset of gene expression profiles and achieving high accuracy in patient survival prediction. Another study reviews recent advancements in deep learning, including ANNs, for medical diagnosis. The authors discuss the challenges and opportunities associated with using deep learning in medical diagnosis and provide insights into future research directions [42].

We can use different new algorithms for diagnosing medical data using classification and prediction tasks. In this regard, the authors in Reference [43] propose a new machine learning algorithm called Kernel-Target Alignment based Fuzzy Lagrangian Twin Bounded Support Vector Machine (KTAF-LTBSVM) for classification tasks. The authors suggest that the proposed algorithm can be applied in various fields, such as image classification, bioinformatics, and financial analysis. Another research [44] described the twin bounded support vector machine (TB-SVM) algorithm, a popular machine learning algorithm for classification and regression tasks. However, the authors introduce bipolar fuzzy sets into the TB-SVM framework, allowing more flexibility in dealing with uncertain or ambiguous data. In one research [45], the authors employed a real-world dataset of individuals with autism spectrum disorder (ASD) and healthy controls to assess various demographic, clinical, and genetic traits. The authors reported that the randomization-based approaches help identify essential variables associated with ASD and can provide more reliable estimates of effect sizes than traditional statistical methods. In another study, authors compare the performance of least squares structural twin bounded support vector machine on class scatter (LS-STBSVM-CS) with several other machine learning algorithms on several real-world datasets, including imbalanced datasets. The results show that

LS-STBSVM-CS outperforms different algorithms regarding classification accuracy and other performance metrics [46]. The authors begin by discussing the challenges associated with early detection of brain tumours, which often require accurate segmentation of tumour regions from MRI scans. They then introduce an improved Otsu's thresholding method, a widely used image segmentation technique, and combine it with several supervised learning techniques, including SVMs and decision trees, to create a classification model for detecting brain tumours at an early stage [47].

ANNs have demonstrated high accuracy and sensitivity in various medical domains, indicating their potential for automated medical diagnosis. However, challenges related to the interpretability and generalization of ANNs still need to be addressed to enable their wider adoption in clinical practice.

3.3 OVERVIEW OF ARTIFICIAL NEURAL NETWORK

The "learning" and "generalization" abilities of the human brain are mathematically represented by an ANN. A subset of artificial neural networks is known as ANNs. Due to its ability to simulate highly nonlinear systems with ambiguous or complicated variable interactions, ANNs are commonly utilized in research. The basic kinds of neural networks are outlined in reference [48] and [49]. Natural language processing, anomaly detection, image and speech recognition, and predictive analytics are just a few of the uses for ANNs. They excel at processing massive volumes of data and frequently outperform conventional statistical models in challenging situations. However, they can also be computationally expensive and require significant data and computing power to train.

3.3.1 Mathematical Background

ANNs are inspired by the functioning of biological neurons in the brain and are constructed using mathematical models. The fundamental element of an ANN is a mathematical representation of a biological neuron, referred to as a perceptron [50]. A perceptron accepts multiple inputs, each multiplied by a corresponding weight. The weighted inputs are then combined and passed through an activation function to generate an output, which is the input for the next layer of neurons. Typically, the activation function is nonlinear, such as a sigmoid or rectified linear unit (ReLU), enabling the network to capture intricate relationships between inputs and outputs [51]. The following formula can represent the output of a perceptron:

$$O_n = f(w1I1 + w2I2 + \cdots + wnIn + b) \tag{3.1}$$

In this equation, O_n denotes the output; $I1$, $I2$ represent the inputs; $w1$, $w2$, wn are the corresponding weights; b is the bias term; and f stands for the activation function. ANNs comprise multiple layers of perceptrons, with each layer receiving input from the preceding layer and generating outputs to serve as inputs for the subsequent layer. The input layer accepts the input data, while the output layer delivers the network's ultimate output. The layers between the input and output are known as

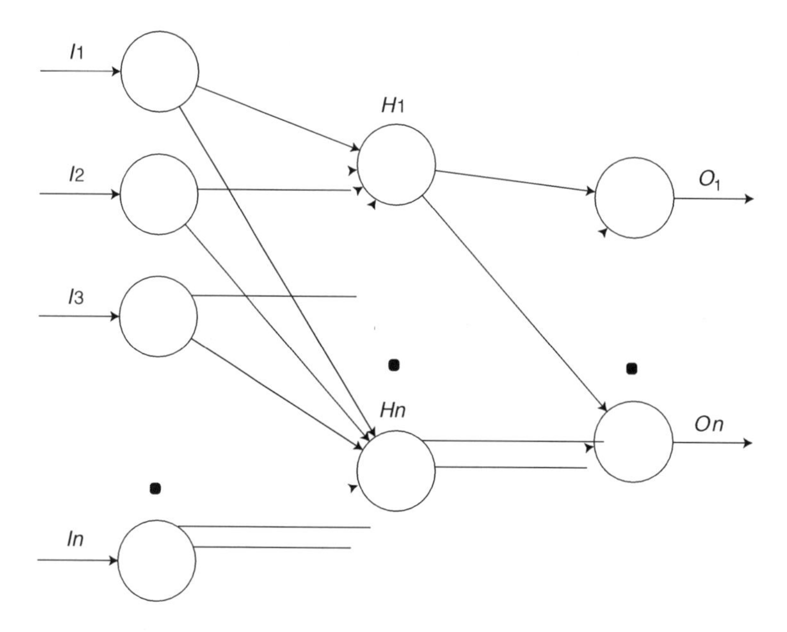

FIGURE 3.2 General architecture of the artificial neural network.

hidden layers. As illustrated in Figure 3.2, the general structure of an ANN involves adjusting the perceptron weights and biases during training to minimize the error between the predicted output and the actual output for a given dataset. This process is typically done using a form of "gradient descent optimization," where the weights and biases are updated in small increments to gradually improve the accuracy of the network. Overall, the mathematical background of ANNs involves linear algebra, calculus, and optimization, as well as knowledge of activation functions and their properties [52].

3.3.2 Network Learning of Artificial Neural Network

The training procedure of an ANN entails fine-tuning the network's weights and biases to minimize the discrepancy between predicted and actual outputs for a given set of training data. There are two primary learning approaches in ANNs: supervised learning and unsupervised learning [53]. Supervised learning is employed when the expected output is known for each input within the training data. In this learning method, the network is trained to associate inputs with outputs by modifying the weights and biases of the neurons. The goal is to minimize the discrepancy between the anticipated and predicted outputs for each input within the training data [54]. One of the most frequently utilized algorithms for supervised learning is backpropagation [55]. Backpropagation is a supervised learning algorithm that employs a variant of gradient descent optimization to adjust the weights and biases of neurons. It back-propagates the error through the network, starting at the output layer and moving toward the input layer. The error at each neuron is utilized to adjust the neuron's weights and biases, ultimately reducing the output error.

On the other hand, unsupervised learning is utilized when no expected output is available for each input in the training data. In this learning approach, the network is trained to detect patterns in the input data without any external guidance. Unsupervised learning is suitable for tasks such as clustering, where the objective is to group similar inputs. The Self-Organizing Map (SOM) is the most popular unsupervised learning algorithm. SOMs are neural networks that learn to project high-dimensional input data onto a two-dimensional grid [56,57]. They employ competitive learning, in which each neuron in the network vies with its neighbors for activation by the input data. The activated neurons then form clusters representing similar inputs. In summary, the training process of an ANN revolves around adjusting the weights and biases of neurons to minimize the error between the predicted output and the actual output, either with or without external guidance.

3.4 APPLICATIONS OF ARTIFICIAL NEURAL NETWORK IN AUTOMATED MEDICAL DIAGNOSIS

The use of ANNs has become widespread in the field of medical diagnosis, particularly in the realm of automated diagnostics [58]. ANNs are designed to identify patterns within data and can be employed to discover irregularities or estimate the probability of a disease. Figure 3.3 illustrates various applications of ANNs in automated medical diagnosis. This section will explore some of the most prevalent applications of ANNs in medical diagnosis.

3.4.1 Automated Image Analysis

The use of ANNs in automated image analysis for medical diagnosis is an expanding area of research. ANNs are employed to examine various medical images, such as X-rays, CT scans, and MRI scans, to assist in identifying diseases [59,60]. This analysis method offers advantages, including increased accuracy and reduced time

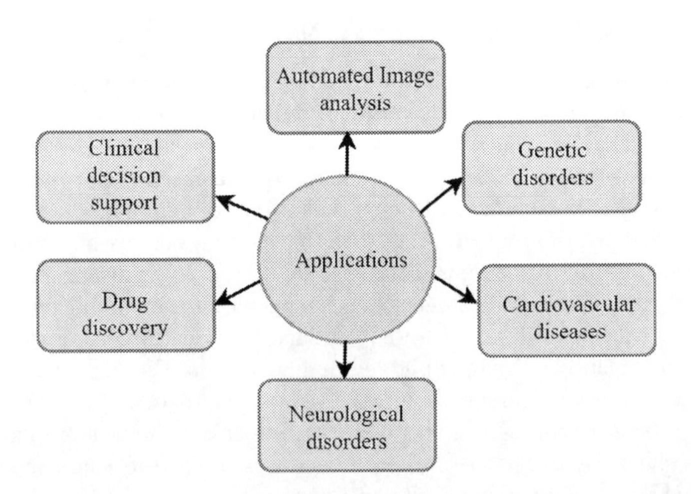

FIGURE 3.3 Applications of artificial neural network in automated medical diagnosis.

consumption compared to manual analysis. Moreover, ANNs can identify features in medical images that might be too subtle for human experts to notice. The image analysis process typically starts with gathering data from a medical imaging device, such as an X-ray or CT scan, which is subsequently input into the ANN. The ANN analyzes the data and generates a set of results that can be utilized for diagnosis. To enhance diagnostic accuracy, additional information, such as medical history or lab results, can be integrated into the results. ANNs can detect subtle features in medical images that may elude human experts. Furthermore, ANNs can track disease progression over time, offering valuable insights into treatment efficacy. Automated image analysis can also compare images from multiple patients, helping evaluate disease progression or treatment effectiveness.

- **Diagnosis of genetic disorders**: Diagnosing genetic disorders is complex because of the intricacies of these conditions and the challenges in accurately identifying which genetic variations are linked to specific disorders [61,62]. ANNs have gained increasing popularity in improving the diagnosis of genetic disorders. ANNs are computational algorithms capable of learning from data and recognizing patterns. By training these networks using extensive genetic and medical information datasets, they can be employed to identify genetic disorders more accurately in medical diagnosis. ANNs can discern patterns in genetic data, such as mutations or deletions, and subsequently associate them with corresponding medical conditions. By examining the data, ANNs can detect patterns that might be difficult for doctors to identify. They can also predict a patient's diagnostic outcome, assisting doctors in providing more accurate diagnoses and treatment plans. To ensure precision, ANNs must be adequately trained and tested on extensive datasets, which can be time-consuming and expensive. Moreover, the accuracy of the results may still rely on the quality of the data used. Therefore, it is crucial to guarantee that the data employed is accurate and comprehensive.
- **Diagnosis of cardiovascular diseases**: ANNs have demonstrated potential in diagnosing cardiovascular diseases because of their ability to learn from vast quantities of data and identify complex patterns within patient information [63 64]. One of the most prevalent applications of ANNs in cardiovascular diagnosis involves interpreting electrocardiogram (ECG) data.

  ANNs can be taught to recognize abnormalities in ECG signals, such as arrhythmias, ischemia, and other heart-related conditions. By examining extensive datasets of ECG signals, ANNs can learn to discern patterns indicative of specific heart issues. Another application of ANNs in cardiovascular diagnosis is interpreting cardiac imaging data, including echocardiograms and cardiac MRI scans. ANNs can be trained to identify and categorise abnormalities in these images based on particular heart conditions. ANNs have also been utilized in predicting the risk of cardiovascular events for patients. By analyzing patient data, such as medical history, lifestyle factors, and lab results, ANNs can learn to predict the risk of heart disease, stroke, and other cardiovascular incidents. This enables clinicians to identify high-risk patients and offer early intervention to prevent such events. Although ANNs exhibit promise in diagnosing cardiovascular disorders, challenges remain. One such challenge is the interpretability of ANNs, as it can be difficult to comprehend how the network reaches a specific diagnosis. Furthermore, ANNs necessitate extensive data for effective training, which can be a constraint in some healthcare environments where data might be limited. Nonetheless, ANNs hold the potential to significantly enhance the accuracy and speed of cardiovascular diagnosis, with ongoing research likely to yield further advancements in the field.

- **Diagnosis of neurological disorders**: ANNs have demonstrated potential in diagnosing neurological disorders because of their capacity to learn from vast datasets and identify intricate patterns within patient data [65,66]. ANNs can be trained to assess various data types, including medical imaging, clinical information, and patient histories, to pinpoint neurological disorders accurately. One prevalent application of ANNs in diagnosing neurological disorders involves interpreting medical imaging data, such as MRI and CT scans. ANNs can be taught to recognize anomalies in brain structures, such as tumors, lesions, and other pathologies. In addition, ANNs can analyze functional MRI data to detect brain regions activated during specific tasks, helping diagnose conditions such as epilepsy and Parkinson's disease. Another application of ANNs in diagnosing neurological disorders involves interpreting electroencephalogram (EEG) data. ANNs can be trained to identify abnormal patterns in EEG signals indicative of conditions such as epilepsy, sleep disorders, and traumatic brain injury. ANNs have also been employed to estimate the risk of developing neurological disorders by analyzing patient data, including family history, lifestyle factors, and genetic information. Although ANNs show promise in diagnosing neurological disorders, challenges remain. One such challenge is the interpretability of ANNs, as understanding how the network reaches a specific diagnosis can be difficult.

  Moreover, effective training of ANNs requires enormous amounts of data, which can be a constraint in specific healthcare environments where data might be limited. Nevertheless, ANNs hold the potential to significantly enhance the accuracy and speed of neurological diagnosis, with ongoing research likely to yield further advancements in the field.
- **Drug discovery**: In the realm of drug discovery, ANNs have gained popularity for their capacity to process extensive and intricate datasets related to molecular and pharmacological information [67]. ANNs assist researchers in pinpointing promising drug candidates, forecasting toxicity, and streamlining drug development. One critical application of ANNs in drug discovery involves the examination of molecular structures. ANNs can be trained to recognize patterns in molecular structures and predict the biological activity of a given compound. This can help identify potential drug candidates and optimize their structures for increased efficacy and reduced side effects. Another application of ANNs in drug discovery is in analyzing pharmacological data. ANNs can be trained to analyze drug–target interactions, predict drug efficacy, and identify potential drug–drug interactions. This

can help researchers optimize drug dosing, reduce toxicity, and improve patient outcomes. ANNS have also been used in the prediction of drug toxicity. By analyzing large datasets of toxicological data, ANNs can learn to predict the toxicity of a given drug compound and identify potential safety concerns. This can help researchers identify potential drug toxicity early in drug development and avoid costly clinical trial failures. ANNs have shown potential in the drug development process; however, issues remain to be solved. One difficulty is requiring extensive and varied datasets for efficient ANN training. The interpretability of ANNs can also provide a barrier in the drug development process because it might be challenging to understand how the network arrived at a specific prediction. However, ANNs have the potential to enhance the drug development process considerably, and continued research in this area will probably result in future advancements in the field [68].

- **Clinical decision support**: Clinical Decision Support Systems (CDSS) are computer-based tools designed to assist healthcare professionals in making clinical decisions by providing patient-specific recommendations or alerts based on relevant clinical information [69,70]. ANNs are CDSSs that have gained popularity in medical diagnosis because of their ability to learn complex relationships between clinical variables and make accurate predictions. ANNs are computer algorithms inspired by the human brain's structure and function. They consist of interconnected nodes, or" neurons," that process input data and produce output signals. ANNs are trained using a dataset of input–output pairs, and they learn to map inputs to outputs by adjusting the strength of connections between neurons. In medical diagnosis, ANNs have been used to diagnose various conditions, such as cancer, heart disease, and diabetes. ANNs have several advantages over traditional diagnostic tools, including handling complex, multi-dimensional data, learning from experience, and adapting to new situations. Nevertheless, ANNs come with certain drawbacks.

  Their interpretability can be challenging, making it unclear how they reached a specific diagnosis. Moreover, ANNs demand substantial data for efficient training, with the input data quality directly impacting the results' quality. In general, ANNs have demonstrated potential as a medical diagnosis tool, but their efficiency relies on data quality and the precise context of their application. Healthcare professionals must collaborate closely with data scientists and specialists to guarantee the effective use of ANNs in clinical settings.

3.5 LIMITATIONS OF ARTIFICIAL NEURAL NETWORK MODEL FOR AUTOMATED MEDICAL DIAGNOSIS

Because of their ability to learn and generalize from data, ANN models are often employed in automated medical diagnosis [49,71]. Figure 3.4 illustrates the various constraints of ANNs for automated medical diagnosis. Nevertheless, there are multiple challenges associated with their application in this domain, which include:

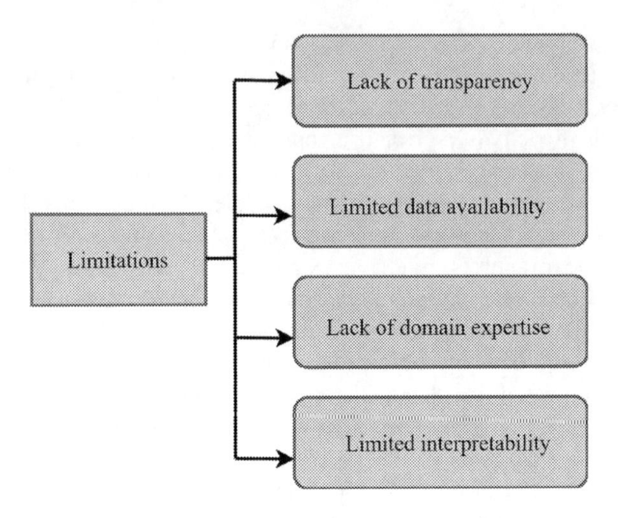

FIGURE 3.4 Limitations of the artificial neural network (ANN) in automated medical diagnosis.

- **Lack of transparency**:
  - Limited transparency in ANN models for automated medical diagnosis is a notable constraint affecting these models' precision, dependability, and credibility [72,73]. ANNs function as a "black box," making comprehending the reasoning behind their diagnoses challenging. This lack of transparency can be especially problematic in healthcare, where professionals must make crucial treatment decisions based on the diagnosis.
  - The issue of transparency in ANN models stems from their complex nature. ANNs consist of interconnected nodes that learn to recognize patterns in the input data to predict outcomes for new data. However, humans do not easily understand the specific decision-making process within these models. This lack of interpretability implies that it can be tough to determine the rationale behind a model's diagnosis, and it may be difficult to identify errors or biases in the decision-making process.
  - Limited transparency in ANN models can also restrict their practical applications in healthcare. For instance, in some cases, ANNs may offer accurate diagnoses but cannot justify their decisions. This can make it hard for healthcare professionals to explain the diagnosis to patients, undermining trust in the diagnosis and the healthcare system.
  - To tackle this constraint, researchers are striving to develop more interpretable ANN models that can offer greater transparency in decision-making. This includes creating methods for visualizing the inner workings of ANNs, such as pinpointing the input features most crucial in making a diagnosis. Furthermore, approaches such as explainable AI and model-agnostic methods can assist in understanding the decision-making processes of ANN models [74,75].

- Consequently, addressing the constraint of limited transparency in ANN models for automated medical diagnosis is essential to enhance these models' accuracy, reliability, and trustworthiness. To increase the adoption of ANNs in medical diagnostics, developing more interpretable ANN models and methods for examining their internal operations is crucial.

- **Limited data availability**:
  - Limited data availability is a substantial constraint of ANN models in automated medical diagnosis. ANNs necessitate extensive, high-quality data for effective training. However, acquiring high-quality data in medical diagnosis can be difficult because of patient privacy, data protection, and the infrequency of certain diseases [76,77]. Consequently, ANN models for medical diagnosis might be restricted by the volume and quality of accessible data.
  - Limited data availability can result in several issues with ANN models for medical diagnosis. With limited data, ANNs might struggle to learn the intricate patterns and relationships required for precise diagnoses, leading to reduced accuracy and unreliable diagnoses.
  - Secondly, data shortage could cause overfitting, where the ANN excessively focuses on recognizing specific patterns in the training data and may not perform well with new data [78]. Overfitting can lead to poor generalization performance, implying that the ANN will not function effectively on previously unseen data.
  - Lastly, limited data availability can also introduce biases in the training data. If the available data do not represent the entire population, the ANN model might be biased toward specific patterns or features, resulting in inaccurate diagnoses. For instance, if a dataset used to train an ANN for detecting heart disease includes only patients from a particular demographic group, the ANN might not perform well on patients from other groups.
  - To tackle the constraint of limited data availability, researchers are investigating methods to enhance data collection and augmentation techniques that can help generate additional data [79,80]. Transfer learning, which involves pre-training ANN models on large datasets and subsequently fine-tuning the models on smaller datasets, can help overcome limited data accessibility in medical diagnosis [81,82].
  - In conclusion, limited data availability is a substantial constraint of ANN models for automated medical diagnosis, potentially leading to reduced accuracy, overfitting, and biases in the training data. Developing methods to enhance data collection and augmentation and employ transfer learning techniques can help overcome this constraint and improve the accuracy and dependability of ANN models for medical diagnosis.
- **Lack of domain expertise**:
  - Insufficient domain knowledge is a considerable constraint of ANN models in automated medical diagnosis [83,84]. ANNs depend on high-quality input data to identify patterns and connections, enabling

accurate diagnoses. Nevertheless, guaranteeing that the input data are pertinent, precise, and meaningful for the specific medical diagnosis task becomes difficult without domain knowledge.

- In medical diagnosis, domain knowledge is essential for comprehending the intricacies of the diseases and conditions being diagnosed, ensuring that the input data accurately represents these complexities. Without domain knowledge, the input data might be incomplete or misinterpreted, resulting in inaccurate diagnoses and decreased reliability of the ANN model.
- A lack of domain knowledge can also contribute to biases in the ANN model. For instance, if the input data doesn't represent the entire population, the ANN might favour specific patterns or features, causing inaccurate diagnoses. Moreover, correctly interpreting the ANN model's results can be challenging without domain knowledge, reducing trust in the diagnosis and limiting its practical applications.
- To mitigate the constraint of insufficient domain knowledge, it is crucial to involve domain experts, such as medical professionals and researchers, in developing and implementing ANN models for medical diagnosis [85,86]. Domain experts can offer valuable insights into the complexity of the diseases and conditions being diagnosed, ensuring that the input data are accurate, relevant, and significant. Furthermore, engaging domain experts in interpreting ANN model results can help guarantee that the diagnoses are accurate and trustworthy.
- In conclusion, insufficient domain knowledge is a notable constraint of ANN models for automated medical diagnosis, potentially leading to inaccurate diagnoses, model biases, and diminished trust in the diagnosis. Including domain experts in developing, implementing, and interpreting ANN models for medical diagnosis is vital to ensure these models' accuracy, dependability, and practical applications.

- **Limited interpretability**:
  - Limited interpretability is a notable drawback of ANN models utilized in automated medical diagnosis. ANNs are highly intricate algorithms capable of discerning patterns and relationships in input data that may not be apparent to human analysts. This complexity can make it challenging to comprehend how the ANN reached its diagnosis or pinpoint the specific attributes or patterns that contributed to the diagnosis [87,88].
  - Limited interpretability can pose challenges in medical diagnosis. Comprehending the diagnostic process and identifying potential errors or biases in the ANN model is vital. Furthermore, restricted interpretability may hinder the practical applications of ANN models in medical diagnosis, as healthcare professionals could be reluctant to depend on a diagnosis they cannot understand or justify.
  - To tackle the issue of limited interpretability, researchers are investigating techniques to enhance the interpretability of ANN models in

medical diagnosis. One method involves developing visualization tools to display the ANN's internal representations, simplifying the identification of features or patterns leading to the diagnosis. Another approach includes post-hoc interpretability methods, such as local interpretable model-agnostic explanations (LIME) or Shapley additive explanations (SHAP), which can determine the most critical features for the ANN's diagnosis [89,90].

- Moreover, several researchers are examining the use of hybrid models that merge ANNs with other machine learning models, such as decision trees or rule-based systems, which are easier for human users to comprehend [91]. These hybrid models can offer the accuracy and complexity of ANNs while facilitating more transparent interpretation and detection of potential errors or biases.
- In conclusion, limited interpretability is a significant drawback of ANN models for automated medical diagnosis, potentially limiting their practical applications and diminishing trust in their diagnoses. Developing techniques to enhance the interpretability of ANN models, such as visualization methods or hybrid models, can help overcome this limitation and improve the accuracy, dependability, and practical applications of ANN models in medical diagnosis.

Although ANNs have demonstrated potential in automated medical diagnosis, they also possess considerable limitations to address before achieving widespread adoption in clinical settings.

3.6 CONCLUSION AND FUTURE DIRECTION

In summary, the advancement of ANN models in automating medical diagnoses holds immense potential to transform healthcare services. By merging machine learning with medical expertise, these models exhibit considerable capability in identifying various medical conditions and enhancing patient recovery. The application of ANNs in medical diagnosis is a swiftly progressing field that has displayed significant success in identifying diverse health issues. Previous research findings illustrate the high accuracy of ANNs in diagnosing medical conditions, establishing them as a vital instrument in medical diagnostics. Furthermore, ANN models can assist medical professionals in detecting health issues earlier and more swiftly, leading to prompt treatment and better patient results. Besides minimizing medical mistakes, increasing efficiency, and lowering healthcare costs, ANNs offer additional advantages. In summary, implementing ANNs in automated medical diagnosis represents a thrilling research area with considerable potential for enhancing healthcare outcomes. It is crucial to continue the development of these models to fine-tune their performance, broaden their applicability in diagnosing various medical conditions, and guarantee their ethical and responsible use in healthcare environments.

Concerning future research directions, several potential pathways exist for further exploration in this domain. Firstly, additional research is required to refine the performance of ANNs for specific health conditions and enhance the models' interpretability. Secondly, efforts should be made to design more advanced ANN frameworks that can incorporate a broader array of medical data sources, such as genomics and EHRs.

Finally, research into the ethical implications of using ANN models in healthcare should be explored to ensure the technology is deployed responsibly and transparently. Overall, using ANNs in medical diagnosis is a rapidly evolving field with significant potential to improve patient outcomes and revolutionize healthcare delivery. With continued research and development, ANNs have the potential to become an essential tool in medical diagnosis, providing healthcare professionals with accurate, fast, and efficient diagnostic tools. However, there are several avenues for future research and development in this area to improve further the accuracy and reliability of these models and their practical applications in healthcare. Figure 3.5 shows the different future directions. Some of the future directions for research on ANNs in medical diagnosis are as follows:

- **Incorporating more complex data sources**: ANNs are primarily trained on structured data such as medical imaging and patient records. More complex data sources such as genomics, proteomics, and metabolomics could be integrated into ANN models to improve diagnostic accuracy and precision.
- **Improving the interpretability of ANN models**: Although ANNs have excellent performance in diagnosing medical conditions, their black-box nature can make interpreting how they arrive at their conclusions difficult. Further research is required to improve the interpretability of these models to help healthcare professionals understand and trust their results.
- **Addressing ethical and legal issues**: Several ethical and legal issues, including data privacy, security, and the effect of automation on healthcare workers, are brought up by using ANNs in medical diagnosis. Future research should address these concerns to guarantee that ANNs are utilized morally and sensibly.
- **Expanding the use of ANNs in low-resource settings**: ANNs can significantly improve healthcare outcomes in low-resource settings, where access to specialized medical personnel and diagnostic tools is limited. To increase access to healthcare, future research should concentrate on creating ANNs that can be used in resource-constrained settings, including rural or isolated places.

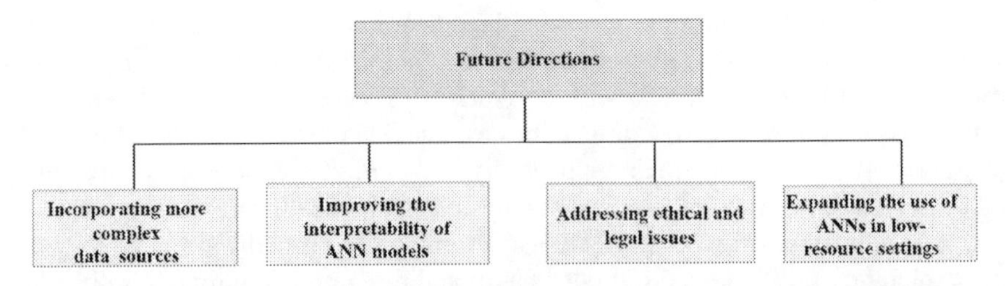

FIGURE 3.5 Limitations of the artificial neural network in automated medical diagnosis.

Applying ANNs in medical diagnosis is a swiftly advancing domain with considerable promise for enhancing healthcare results. Additional investigation is required to fine-tune these models for various health conditions, boost their comprehensibility, tackle ethical and legal concerns, and broaden their utilization in low-resource environments.

REFERENCES

1. Euan A Ashley. Towards precision medicine. *Nature Reviews Genetics*, 17(9):507–522, 2016.
2. Huma Naz and Sachin Ahuja. Deep learning approach for diabetes prediction using Pima Indian dataset. *Journal of Diabetes & Metabolic Disorders*, 19:391–403, 2020.
3. Junaid Rashid, Saba Batool, Jungeun Kim, Muhammad Wasif Nisar, Amir Husain, Sapna Juneja, and Riti Kushwaha. An augmented artificial intelligence approach for chronic diseases prediction. *Frontiers in Public Health*, 10:559, 2022.
4. AN Ramesh, Chandra Kambhampati, John RT Monson, and PJ Drew. Artificial intelligence in medicine. *Annals of the Royal College of Surgeons of England*, 86(5):334, 2004.
5. Miao Cui and David Y Zhang. Artificial intelligence and computational pathology. *Laboratory Investigation*, 101(4):412–422, 2021.
6. Shichao Zhang, Chengqi Zhang, and Qiang Yang. Data preparation for data mining. *Applied Artificial Intelligence*, 17(5–6):375–381, 2003.
7. Samina Khalid, Tehmina Khalil, and Shamila Nasreen. A survey of feature selection and feature extraction techniques in machine learning. In: *2014 Science and Information Conference*, pp. 372–378. IEEE, New York, 2014.
8. Chengrun Yang, Yuji Akimoto, Dae Won Kim, and Madeleine Udell. Oboe: Collaborative filtering for automl model selection. In: *Proceedings of the 25th ACM SIGKDD International Conference on Knowledge Discovery and Data Mining*, pp. 1173–1183, 2019. https://doi.org/10.1145/3292500.3330909
9. Matthias Feurer, Aaron Klein, Katharina Eggensperger, Jost Springenberg, Manuel Blum, and Frank Hutter. Efficient and robust automated machine learning. *Advances in Neural Information Processing Systems*, 28:2962–2970, 2015.
10. Chris Thornton, Frank Hutter, Holger H Hoos, and Kevin Leyton-Brown. Autoweka: Combined selection and hyperparameter optimization of classification algorithms. In: *Proceedings of the 19th ACM SIGKDD International Conference on Knowledge Discovery and Data Mining*, pp. 847–855, 2013. https://doi.org/10.1145/2487575.2487629
11. Hang Yu, Laurence T Yang, Qingchen Zhang, David Armstrong, and M Jamal Deen. Convolutional neural networks for medical image analysis: State-of-the-art, comparisons, improvement and perspectives. *Neurocomputing*, 444:92–110, 2021.
12. Elif Derya U. Beyli. Implementing automated diagnostic systems for breast cancer detection. *Expert Systems with Applications*, 33(4):1054–1062, 2007.
13. Yogesh Kumar and Manish Mahajan. 5. Recent advancement of machine learning and deep learning in the field of healthcare system. *Computational Intelligence for Machine Learning and Healthcare Informatics*, 1:77, 2020.
14. Justin Ker, Lipo Wang, Jai Rao, and Tchoyoson Lim. Deep learning applications in medical image analysis. *Ieee Access*, 6:9375–9389, 2017.
15. Christoph Baur, Stefan Denner, Benedikt Wiestler, Nassir Navab, and Shadi Albarqouni. Autoencoders for unsupervised anomaly segmentation in brain mr images: A comparative study. *Medical Image Analysis*, 69:101952, 2021.
16. Haibo Zhang, Wenping Guo, Shiqing Zhang, Hongsheng Lu, and Xiaoming Zhao. Unsupervised deep anomaly detection for medical images using an improved adversarial autoencoder. *Journal of Digital Imaging*, 35(2):153–161, 2022.

17. Sajja Tulasi Krishna and Hemantha Kumar Kalluri. Deep learning and transfer learning approaches for image classification. *International Journal of Recent Technology and Engineering (IJRTE)*, 7(5S4):427–432, 2019.
18. Nermeen Abou Baker, Nico Zengeler, and Uwe Handmann. A transfer learning evaluation of deep neural networks for image classification. *Machine Learning and Knowledge Extraction*, 4(1):22–41, 2022.
19. Martin Abadi, Andy Chu, Ian Goodfellow, H Brendan McMahan, Ilya Mironov, Kunal Talwar, and Li Zhang. Deep learning with differential privacy. In: *Proceedings of the 2016 ACM SIGSAC Conference on Computer and Communications Security*, pp. 308–318, 2016. https://doi.org/10.1145/2976749.2978318
20. Yann LeCun, Yoshua Bengio, and Geoffrey Hinton. Deep learning. *Nature*, 521(7553):436–444, 2015.
21. Xue-Wen Chen and Xiaotong Lin. Big data deep learning: Challenges and perspectives. *IEEE Access*, 2:514 525, 2014.
22. Riccardo Miotto, Fei Wang, Shuang Wang, Xiaoqian Jiang, and Joel T Dudley. Deep learning for healthcare: Review, opportunities and challenges. *Briefings in Bioinformatics*, 19(6):1236–1246, 2018.
23. Jian Wei, Jianhua He, Kai Chen, Yi Zhou, and Zuoyin Tang. Collaborative filtering and deep learning based recommendation system for cold start items. *Expert Systems with Applications*, 69:29–39, 2017.
24. Hoo-Chang Shin, Holger R Roth, Mingchen Gao, Le Lu, Ziyue Xu, Isabella Nogues, Jianhua Yao, Daniel Mollura, and Ronald M Summers. Deep convolutional neural networks for computer-aided detection: Cnn architectures, dataset characteristics and transfer learning. *IEEE Transactions on Medical Imaging*, 35(5):1285–1298, 2016.
25. Jia Song, Sijun Qin, and Pengzhou Zhang. Chinese text categorization based on deep belief networks. In: *2016 IEEE/ACIS 15th International Conference on Computer and Information Science (ICIS)*, pp. 1–5. IEEE, New York, 2016.
26. June-Goo Lee, Sanghoon Jun, Young-Won Cho, Hyunna Lee, Guk Bae Kim, Joon Beom Seo, and Namkug Kim. Deep learning in medical imaging: General overview. *Korean Journal of Radiology*, 18(4):570–584, 2017.
27. Kenji Suzuki. Overview of deep learning in medical imaging. *Radiological Physics and Technology*, 10(3):257–273, 2017.
28. Polina Mamoshina, Armando Vieira, Evgeny Putin, and Alex Zhavoronkov. Applications of deep learning in biomedicine. *Molecular Pharmaceutics*, 13(5):1445–1454, 2016.
29. Jin Liu, Yi Pan, Min Li, Ziyue Chen, Lu Tang, Chengqian Lu, and Jianxin Wang, Applications of deep learning to MRI images: A survey. *Big Data Mining and Analytics*, 1(1):1–18, 2018.
30. Daniele Rav'ı, Charence Wong, Fani Deligianni, Melissa Berthelot, Javier Andreu-Perez, Benny Lo, and Guang-Zhong Yang. Deep learning for health informatics. *IEEE Journal of Biomedical and Health Informatics*, 21(1):4–21, 2016.
31. Miotto Riccardo, Wang Fei, Wang Shuang, Jiang Xiaoqian, and T Dudley Joel. Deep learning for healthcare: Review, opportunities and challenges. *Briefings in Bioinformatics*, 19(6):1236–1246, 2017.
32. Hanyang Li, Lijie Lai, and Jun Shen. Development of a susceptibility gene based novel predictive model for the diagnosis of ulcerative colitis using random forest and artificial neural network. *Aging (Albany NY)*, 12(20):20471, 2020.
33. Muhammad Javed Iqbal, Zeeshan Javed, Haleema Sadia, Ijaz A Qureshi, Asma Irshad, Rais Ahmed, Kausar Malik, Shahid Raza, Asif Abbas, Raffaele Pezzani, et al. Clinical applications of artificial intelligence and machine learning in cancer diagnosis: Looking into the future. *Cancer Cell International*, 21(1):1–11, 2021.
34. Jianmin Jiang, P Trundle, and Jinchang Ren. Medical image analysis with artifi cial neural networks. *Computerized Medical Imaging and Graphics*, 34(8):617–631, 2010.

35. Geert Litjens, Thijs Kooi, Babak Ehteshami Bejnordi, Arnaud Arindra Adiyoso Setio, Francesco Ciompi, Mohsen Ghafoorian, Jeroen Awm Van Der Laak, Bram Van Ginneken, and Clara I Sánchez. A survey on deep learning in medical image analysis. *Medical Image Analysis*, 42:60–88, 2017.
36. Ravi Manne, Snigdha Kantheti, and Sneha Kantheti. Classification of skin cancer using deep learning, convolutionalneural networks-opportunities and vulnerabilities-a systematic review. *International Journal for Modern Trends in Science and Technology, ISSN*, 6:2455–3778, 2020.
37. S, tefan Busnatu, Adelina-Gabriela Niculescu, Alexandra Bolocan, George ED Pe-trescu, Dan Nicolae Păduraru, Iulian Năstasă, Mircea Lupus, oru, Marius Geantă, Octavian Andronic, Alexandru Mihai Grumezescu, et al. Clinical applications of artificial intelligence-an updated overview. *Journal of Clinical Medicine*, 11(8):2265, 2022.
38. Delia-Maria Filimon and Adriana Albu. Skin diseases diagnosis using artificial neural networks. In: *2014 IEEE 9th IEEE International Symposium on Applied Computational Intelligence and Informatics (SACI)*, pp. 189–194. IEEE, New York, 2014.
39. Anju Yadav, Vivek K Verma, Vipin Pal, Vanshika Jain, and Vanshika Garg. Automated detection and classification of breast cancer tumour cells using machine learning and deep learning on histopathological images. In: *2021 6th International Conference for Convergence in Technology (I2CT)*, pages pp. 1–6. IEEE, New York, 2021.
40. Shamsul Huda, John Yearwood, Herbert F Jelinek, Mohammad Mehedi Hassan, Giancarlo Fortino, and Michael Buckland. A hybrid feature selection with ensemble classification for imbalanced healthcare data: A case study for brain tumor diagnosis. *IEEE Access*, 4:9145–9154, 2016.
41. Farid E Ahmed. Artificial neural networks for diagnosis and survival prediction in colon cancer. *Molecular Cancer*, 4(1):1–12, 2005.
42. Ahsan Bin Tufail, Yong-Kui Ma, Mohammed KA Kaabar, Francisco Martínez, AR Junejo, Inam Ullah, Rahim Khan, et al. Deep learning in cancer diagnosis and prognosis prediction: A minireview on challenges, recent trends, and future directions. *Computational and Mathematical Methods in Medicine*, 2021, 2021.
43. Umesh Gupta and Deepak Gupta. Kernel-target alignment based fuzzy lagrangian twin bounded support vector machine. *International Journal of Uncertainty, Fuzziness and Knowledge-Based Systems*, 29(05):677–707, 2021.
44. Umesh Gupta and Deepak Gupta. Bipolar fuzzy based least squares twin bounded support vector machine. *Fuzzy Sets and Systems*, 449:120–161, 2022.
45. Umesh Gupta, Deepak Gupta, and Umang Agarwal. Analysis of randomization based approaches for autism spectrum disorder. In: Gupta, D., Goswami, R.S., Banerjee, S., Tanveer, M., Pachori, R.B. (eds) *Pattern Recognition and Data Analysis with Applications*. Lecture Notes in Electrical Engineering, pp. 701–713. Springer, New York, 2022.
46. Umesh Gupta and Deepak Gupta. Least squares structural twin bounded support vector machine on class scatter. *Applied Intelligence*, 53:1–31, 2022.
47. Madhuri Gupta, Divya Srivastava, Deepika Pantola, and Umesh Gupta. Brain tumor detection using improved otsu's thresholding method and supervised learning techniques at early stage. In: Noor, A., Saroha, K., Pricop, E., Sen, A., Trivedi, G. (eds) *Proceedings of Emerging Trends and Technologies on Intelligent Systems*. Advances in Intelligent Systems and Computing, pp. 271–281. Springer, New York, 2022.
48. Giles M Foody. Land cover classification by an artificial neural network with ancillary information. *International Journal of Geographical Information Systems*, 9(5):527–542, 1995.
49. Filippo Amato, Alberto López, Eladia María Peña-Méndez, Petr Vaňhara, Aleš Hampl, and Josef Havel. Artificial neural networks in medical diagnosis, *Journal of Applied Biomedicine*, 11(2):47–58, 2013.

50. Neha Gupta et al. Artificial neural network. *Network and Complex Systems*, 3(1):24–28, 2013.
51. Dabal Pedamonti. Comparison of non-linear activation functions for deep neural networks on mnist classification task. arXiv preprint arXiv:1804.02763, 2018. https://doi.org/10.48550/arXiv.1804.02763
52. Stanley Cohen. The basics of machine learning: Strategies and techniques. In: *Artificial Intelligence and Deep Learning in Pathology*, pp. 13–40. Elsevier, Amsterdam, 2021.
53. Christian Janiesch, Patrick Zschech, and Kai Heinrich. Machine learning and deep learning. *Electronic Markets*, 31(3):685–695, 2021.
54. Taiwo Oladipupo Ayodele. Types of machine learning algorithms. *New Advances in Machine Learning*, 3:19–48, 2010.
55. Raul Rojas and Raúl Rojas. The backpropagation algorithm. In: *Neural Networks: A Systematic Introduction*, pp. 149–182. Springer, Berlin, Heidelberg, 1996. https://doi.org/10.1007/978-3-642-61068-4_7
56. Rémi Domingues, Maurizio Filippone, Pietro Michiardi, and Jihane Zouaoui. A comparative evaluation of outlier detection algorithms: Experiments and analyses. *Pattern Recognition*, 74:406–421, 2018.
57. Christian Lopez, Scott Tucker, Tarik Salameh, and Conrad Tucker. An unsupervised machine learning method for discovering patient clusters based on genetic signatures. *Journal of Biomedical Informatics*, 85:30–39, 2018.
58. Yu-chen Wu and Jun-wen Feng. Development and application of artificial neural network. *Wireless Personal Communications*, 102:1645–1656, 2018.
59. Shouvik Chakraborty, Sankhadeep Chatterjee, Amira S Ashour, Kalyani Mali, and Nilanjan Dey. Intelligent computing in medical imaging: A study. In: *Advancements in Applied Metaheuristic Computing*, pp. 143–163. IGI global, Philadelphia PA, 2018. doi: 10.4018/978-1-5225-4151-6.ch006
60. Hafsa Khalid, Muzammil Hussain, Mohammed A Al Ghamdi, Tayyaba Khalid, Khadija Khalid, Muhammad Adnan Khan, Kalsoom Fatima, Khalid Masood, Sultan H Almotiri, Muhammad Shoaib Farooq, et al. A comparative systematic literature review on knee bone reports from mri, x-rays and ct scans using deep learning and machine learning methodologies. *Diagnostics*, 10(8):518, 2020.
61. Oleg Yu Atkov, Svetlana G Gorokhova, Alexandr G Sboev, Eduard V Generozov, Elena V Muraseyeva, Svetlana Y Moroshkina, and Nadezhda N Cherniy. Coronary heart disease diagnosis by artificial neural networks including genetic polymorphisms and clinical parameters. *Journal of Cardiology*, 59(2):190–194, 2012.
62. B Samanta. Gear fault detection using artificial neural networks and support vector machines with genetic algorithms. *Mechanical Systems and Signal Processing*, 18(3):625–644, 2004.
63. Zeeshan Ahmed, Khalid Mohamed, Saman Zeeshan, and XinQi Dong. Artificial intelligence with multi-functional machine learning platform development for better healthcare and precision medicine. *Database*, 2020:baaa010, 2020.
64. Barry J Maron, Richard A Friedman, Paul Kligfield, Benjamin D Levine, Sami Viskin, Bernard R Chaitman, Peter M Okin, J Philip Saul, Lisa Salberg, George F Van Hare, et al. Assessment of the 12-lead electrocardiogram as a screening test for detection of cardiovascular disease in healthy general populations of young people (12-25 years of age) a scientific statement from the american heart association and the american college of cardiology. *Journal of the American College of Cardiology*, 64(14):1479–1514, 2014.
65. Jigneshkumar L Patel and Ramesh K Goyal. Applications of artificial neural networks in medical science. *Current Clinical Pharmacology*, 2(3):217–226, 2007.
66. Lakhmi C Jain and Norman M Martin. *Fusion of Neural Networks, Fuzzy Systems and Genetic Algorithms: Industrial Applications*, Vo. 4. CRC Press, Boca Raton, FL, 1998.

67. Vineetha Mandlik, Pruthvi Raj Bejugam, and Shailza Singh. Application of artificial neural networks in modern drug discovery. In: *Artificial Neural Network for Drug Design, Delivery and Disposition*, pp. 123–139. Elsevier, Academic Press, Amsterdam, 2016. https://doi.org/10.1016/B978-0-12-801559-9.00006-5
68. Erik Gawehn, Jan A Hiss, and Gisbert Schneider. Deep learning in drug discovery. *Molecular Informatics*, 35(1):3–14, 2016.
69. Ida Sim, Paul Gorman, Robert A Greenes, R Brian Haynes, Bonnie Kaplan, Harold Lehmann, and Paul C Tang. Clinical decision support systems for the practice of evidence-based medicine. *Journal of the American Medical Informatics Association*, 8(6):527–534, 2001.
70. ATM Wasylewicz and AMJW Scheepers-Hoeks. Clinical decision support systems. In: *Fundamentals of Clinical Data Science*, pp. 153–169, 2019. https://doi.org/10.1007/978-3-319-99713-1_11
71. Turgay Ayer, Jagpreet Chhatwal, Oguzhan Alagoz, Charles E Kahn Jr, Ryan W Woods, and Elizabeth S Burnside. Comparison of logistic regression and artificial neural network models in breast cancer risk estimation. *Radiographics*, 30(1):13–22, 2010.
72. Paulo JG Lisboa. A review of evidence of health benefit from artificial neural networks in medical intervention. *Neural Networks*, 15(1):11–39, 2002.
73. Nathan Radakovich, Matthew Nagy, and Aziz Nazha. Machine learning in haematological malignancies. *The Lancet Haematology*, 7(7):e541–e550, 2020.
74. Christoph Molnar, Gunnar König, Julia Herbinger, Timo Freiesleben, Susanne Dandl, Christian A Scholbeck, Giuseppe Casalicchio, Moritz Grosse-Wentrup, and Bernd Bischl. General pitfalls of model-agnostic interpretation methods for machine learning models. In: *XXAI: Extending Explainable AI Beyond Deep Models and Classifiers, ICML 2020 Workshop*, July 18, 2020, Vienna, Austria, *Revised and Extended Papers*, pp. 39–68. Springer, New York, 2022.
75. Jirayus Jiarpakdee, Chakkrit Kla Tantithamthavorn, Hoa Khanh Dam, and John Grundy. An empirical study of model-agnostic techniques for defect prediction models. *IEEE Transactions on Software Engineering*, 48(1):166–185, 2020.
76. Simon Kern, Sascha Liehr, Lukas Wander, Martin Bornemann-Pfeiffer, Simon Müller, Michael Maiwald, and Stefan Kowarik. Artificial neural networks for quantitative online nmr spectroscopy. *Analytical and Bioanalytical Chemistry*, 412:4447–4459, 2020.
77. Igor Pantic, Jovana Paunovic, Jelena Cumic, Svetlana Valjarevic, Georg A Petroianu, and Peter R Corridon. Artificial neural networks in contemporary toxicology research. *Chemico-Biological Interactions*, 369:110269, 2022.
78. Laith Alzubaidi, Jinglan Zhang, Amjad J Humaidi, Ayad Al-Dujaili, Ye Duan, Omran Al-Shamma, José Santamaría, Mohammed A Fadhel, Muthana Al-Amidie, and Laith Farhan. Review of deep learning: Concepts, cnn architectures, challenges, applications, future directions. *Journal of Big Data*, 8:1–74, 2021.
79. Jason Wei and Kai Zou. Eda: Easy data augmentation techniques for boosting performance on text classification tasks. arXiv preprint arXiv:1901.11196, 2019. https://doi.org/10.48550/arXiv.1901.11196
80. Maayan Frid-Adar, Idit Diamant, Eyal Klang, Michal Amitai, Jacob Goldberger, and Hayit Greenspan. Gan-based synthetic medical image augmentation for increased cnn performance in liver lesion classification. *Neurocomputing*, 321:321–331, 2018.
81. Malliga Subramanian, Kogilavani Shanmugavadivel, and PS Nandhini. On finetuning deep learning models using transfer learning and hyper-parameters optimization for disease identification in maize leaves. *Neural Computing and Applications*, 34(16):13951–13968, 2022.
82. Pronnoy Dutta, Pradumn Upadhyay, Madhurima De, and RG Khalkar. Medical image analysis using deep convolutional neural networks: CNN architectures and transfer learning. In: *2020 International Conference on Inventive Computation Technologies (ICICT)*, pp. 175–180. IEEE, 2020.

83. Jyoti Soni, Ujma Ansari, Dipesh Sharma, Sunita Soni, et al. Predictive data mining for medical diagnosis: An overview of heart disease prediction. *International Journal of Computer Applications*, 17(8):43–48, 2011.
84. Yidong Chai, Hongyan Liu, and Jie Xu. Glaucoma diagnosis based on both hidden features and domain knowledge through deep learning models. *Knowledge-Based Systems*, 161:147–156, 2018.
85. Vimla L Patel, David R Kaufman, and Jose F Arocha. Emerging paradigms of cognition in medical decision-making. *Journal of Biomedical Informatics*, 35(1):52–75, 2002.
86. Kai Siang Chan and Nabil Zary. Applications and challenges of implementing artificial intelligence in medical education: Integrative review. *JMIR Medical Education*, 5(1):e13930, 2019.
87. Nida Shahid, Tim Rappon, and Whitney Berta. Applications of artificial neural networks in health care organizational decision-making: A scoping review. *PloS ONE*, 14(2):e0212356, 2019.
88. Alfredo Vellido. The importance of interpretability and visualization in machine learning for applications in medicine and health care. *Neural Computing and Applications*, 32(24):18069–18083, 2020.
89. W James Murdoch, Chandan Singh, Karl Kumbier, Reza Abbasi-Asl, and Bin Yu. Definitions, methods, and applications in interpretable machine learning. *Proceedings of the National Academy of Sciences*, 116(44):22071–22080, 2019.
90. Pantelis Linardatos, Vasilis Papastefanopoulos, and Sotiris Kotsiantis. Explainable ai: A review of machine learning interpretability methods. *Entropy*, 23(1):18, 2020.
91. Zili Meng, Minhu Wang, Jiasong Bai, Mingwei Xu, Hongzi Mao, and Hongxin Hu. Interpreting deep learning-based networking systems. In: *Proceedings of the Annual Conference of the ACM Special Interest Group on Data Communication on the Applications, Technologies, Architectures, and Protocols for Computer Communication*, pp. 154–171, 2020. https://doi.org/10.1145/3387514.3405859

4 Analyzing of Heterogeneous Perceptions of a Mutually Dependent Health Ecosystem System Survey

Manish Bhardwaj
KIET Group of Institutions

Sumit Kumar Sharma
Ajay Kumar Garg Engineering College

Jyoti Sharma
KIET Group of Institutions

Vivek Kumar
ABES Engineering College

4.1 INTRODUCTION

As the cost of knowledge acquisition has plummeted in the digital era, innovation has shifted from centralized institutions to decentralized value chains, networks, and even business ecosystems. The healthcare industry as a whole is benefiting from information and communications technology (ICT) development, just as many other tech-heavy industries [1]. There is an encouraging parallel trend when these developments are compared to those of eHealth and mHealth services. It has been found that healthcare innovations and improvements frequently occur beyond organizational boundaries in the ecosystemic environment, where boundaries are porous and various stakeholders are engaged through tangled of value aggregation [2].

It is vital to comprehend the significance of inventions that necessitate substantial investments of time and financial resources. It has been claimed that there is a drastic shift in world demographics, with an accompanying rise in the senior population's share of the total. There will be two billion persons over the age of 60 by 2050 (a jump from 12% to 22% of the world's population) according to the World Health Organization (WHO). The rate at which the population is aging is increasing.

DOI: 10.1201/9781003405368-4

As the global population ages, chronic diseases are also on the rise [3]. This rapidly increasing senior population calls into question the viability of current health and social care models because they will undoubtedly lead to increased healthcare needs and expenses.

This study focuses on the business ecosystems in the healthcare domain and the business models of participating stakeholders to explain "how" multimedia-based healthcare companies might work together to tackle the problems of the future. Focusing on the new phenomena of ecosystemic collaboration in healthcare, this article explores its implications (i.e., Connected Health) [4]. Business ecosystems are defined as "a network of interconnected business models." As a boundary-crossing unit of analysis, business models are thought to link high-level strategies with concrete execution.

Newer, more sustainable, and more integrated healthcare models, like those suggested by, call for a number of different approaches, such as improvements in technology, care processes, and care delivery. All of these need more research. Healthcare systems around the world are undergoing transformations as a result of new models designed to reduce costs without compromising quality of treatment.

"Associated health" is an umbrella word for a variety of healthcare management concepts, including telemedicine, telehealth, telecare, mHealth, and eHealth. Chronic diseases and the worldwide aging population are the primary foci of extensive Connected Health research [5]. Preventive medicine, the pharmacist's role, and patient–professional electronic communication are just a few examples of how linked health can save healthcare spending. Healthcare professionals' digitalized workflows will be a key factor in the enhancement of healthcare services in the future. In the future healthcare industry, ecosystemic business development is both necessary and promising, as highlighted [6].

The healthcare industry relies heavily on public funding. But as pointed out, more and more private companies are entering the healthcare industry with an eye on both income and service provision [7]. This dissertation examines the empirical background of linked health as a business environment, particularly in light of the proliferation of service-providing businesses in recent years [8]. In particular, this study focuses on the interconnected health business ecosystems, where different business models are always interacting to better discover and seize opportunities, create and capture value, and find and utilize competitive advantages.

Cultural and societal trends, combined with economic concerns, have spurred the need for us to reevaluate how we deliver health and social care in our community. Healthcare exerts substantial cost constraints on both public budget and consumer lending [9]. In addition, societal changes have increased the need for care to be provided in a uniquely tailored manner, with an emphasis on individualized, high-tech "smart" solutions. The field of "Connected Health" is a new and fast expanding one, with the potential to improve the quality, security, and cost-effectiveness of healthcare delivery systems.

Integrated Health is employed in a wide variety of sectors despite being primarily a healthcare industry phenomenon (for example, healthcare, social care, and the wellness sector). Thus, many definitions exist with differing emphasis put on healthcare, business, technology and support service providers, or any combination of these [10]. Connected Health is not well defined and is still an area of debate among researchers.

Connected Health is "the overarching descriptor covering digital health, eHealth, mHealth, telecare, telehealth, and telemedicine," as the ECHAlliance (2014) group puts it. Integrated care is "a process framework for healthcare system where equipment, facilities, or treatments are developed around the patient's needs, and associated health data is shared so that the patient can receive care in the most proactive and efficient manner possible." The interdependence and enabling nature of technological solutions for healthcare are central to this discussion. As the Food and Drug Administration notes (2014), "Connected Health" refers to "electronic ways of health care delivery that allow users to deliver and receive care outside of traditional health care venues." Examples include mobile medical apps, medical device data systems, software, and wireless technology. Therefore, there is a renewed interest in researching ICT to foster the growth of Connected Health as innovative solutions strive to enable new healthcare ties and partnerships [11]. Connected Health has been defined as "patient-centred care resulting from process-driven health care delivery conducted by healthcare providers, patients and/or carers who are facilitated by the use of advanced technologies (software and/or hardware)." Therefore, Integrated Health can be viewed as a socio-technical healthcare model that provides healthcare beyond traditional healthcare settings. The word "ecosystem" encompasses this concept [12].

To achieve a healthy Connected Health Ecosystem, we must find a way to accommodate the competing needs of the diverse groups of stakeholders in today's medical system. Primary care, secondary care, payers, policy makers, pharmacies, clinicians, patients, families, creative thinkers, state officials, patient groups, intellectuals, and entrepreneurs are all examples of stakeholders who can work together to test, develop, and explore innovative protocols and assessments for integrated primary healthcare solutions [13].

There is a rising interest in investigating how ICT allows integrated products and solutions as innovative solutions strive to facilitate such connectedness across different providers. Patients can be put in harm's way if healthcare technology is ill-conceived at any stage, whether in its conception, creation, implementation, maintenance, or use [14]. This means that throughout the service lifecycle, it is essential to have an assessment process in place that is always running. Evidence-based evaluation of the role of ICT in bolstering healthcare services is lacking however, and this is the case with healthcare technology like Connected Health.

This study provides an introduction to several prominent e-health and IS evaluation frameworks and explores how they may be applied to the evaluation of Connected Health. We propose a Connected Health Evaluation Framework (CHEF) to unify these initiatives [15]. To fully understand the possible effects of Connected Health technology, it is necessary to examine all of the 'ingredients,' which is where CHEF comes into play.

4.2 SCOPE OF STUDY

As a current design for clinical governance, this research takes the summary of Integrated Health as its starting point. Connecting technologies for data integration for various stakeholders allows for the sharing of information between patient populations, guardians, and medical researchers in a way that facilitates quicker and more accurate judgment call for patient wellness and health [16].

It's becoming increasingly common for ICT-heavy workplaces to have an ecosystemic approach, where digital and nondigital elements of the workplace work together. Organizational activities (innovation, product creation, etc.) in an important ecosystem approach extend beyond the borders of individual organizations and into the territories of their collaborating stakeholders [17]. It has also been noted in the field of Connected Health. Multi-party, interdependent, and porous corporate ecosystems are often conceptualized through the lens of an "ecosystem" [18]. Therefore, it is arguable that a business ecosystems viewpoint can add to the literature on how various stakeholders in the linked health business ecosystem might increase collaboration at various stages of the business ecosystem lifecycle from a practical and sustainable standpoint.

The purpose of this research is to go into the topic of business ecosystems as they pertain to the field of linked health [19]. The steps or series of events that happen during experience-to-experience transformation are often hard to deal with. Management researchers, in order to better comprehend changing companies, have turned to a wide variety of metaphors and theories from other fields. A business ecosystem is a metaphor that has been used for a long time to describe the interconnectedness, diversity, complexity, and evolution of a company's various parts and the relationships between them. The idea of a corporate ecosystem can be traced back to ecological systems in nature [20].

This research extends the work that viewed business ecosystems as a network of business models by identifying business models as the "strategic vehicle" that translates high-level organizational strategies into actionable plans in real-world contexts [21–24]. Corporations' business ecosystems, rivals, complementors, buyers, individuals, and society are all linked through economic models, making up the larger ecosystem. In order to investigate and comprehend the dynamics and interdependencies inside the intricate smart healthcare modern enterprise, business models are used as the boundary-spanning unit of analysis in this study.

Purposeful inflows and outflows of knowledge are essential for fostering innovation and having an impact on the corporate environment. And, to improve innovation process as a cluster, stress the importance of entrepreneurs and other stakeholders working together to share and apply relevant information [25]. Engagement between partners can be encouraged in an ecosystemic framework through the use of concepts, such as purposeful information sharing, cooperation, and co-creation, all of which contribute to long-term performance.

"An ecosystem" as a metaphorical idea has permeated numerous disciplines in academics and in reality. First, the notion of healthcare communities displays numerous involved players and their relationship levels. Provider–payer interactions, practitioner partnerships, contacts between device makers and clinicians, employer–payer partnerships, and consumer–provider partnerships are among the most common [32]. Multi-stakeholder environments have always been at the center of healthcare ecosystems, but now consumer/patient-centricity and significant technology use are rapidly gaining ground [26].

Second, the concept of digital ecosystems is relevant to digital business and technological evolution. In theory, digital environments are made up of various online resources (often in the form of digital platforms or cloud services) that

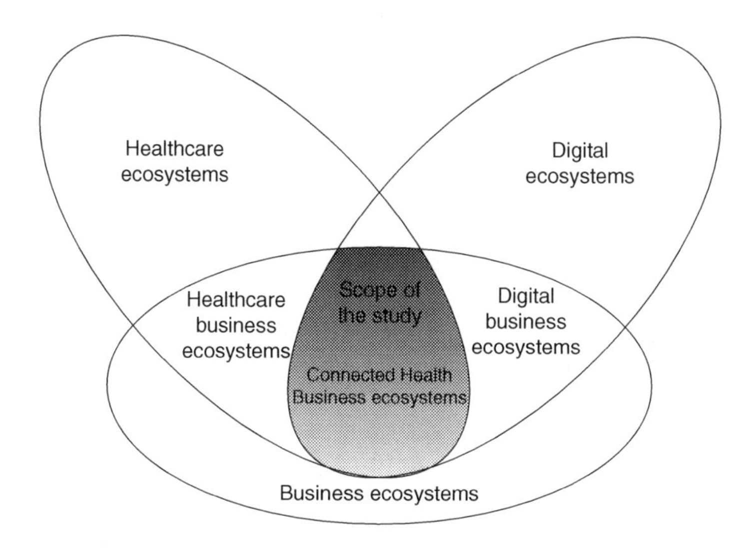

FIGURE 4.1 Systematic view of healthcare ecosystem in business ecosystem.

promote greater interaction and cooperation between people and companies [27]. In reality, digital ecosystems have enabled most of the digital disruptions in most businesses. Furthermore, business ecosystems are viewed as interdependent multi-stakeholder business environments that enable co-evolution and mutual development of stakeholders.

Based on what has been said so far, the scope of this study is shown in Figure 4.1. A Venn diagram depicting the intersection of the healthcare ecosystem, the digital environment, and the business ecosystem is presented in the figure. In the overlapping sectors, new concepts can be recognized, which are also discussed in the literature, such as digital business ecosystems, healthcare business ecosystems, and digital/connected healthcare ecosystems. However, the area of this research is where these three notions overlap. The area where the two circles meet is what I call the "smart healthcare Business Environment." Connected Health is a new and more holistic approach to digital transformation that focuses on assistance for carrying out the necessary judgments using state-of-the-art technology (the collection of sensor-based data, artificial intelligence, etc.). A fundamental inclination based on the examined literature is that business settings in Connected Health tend to adopt a business ecosystem approach, and there can be numerous smart healthcare business environments in action at a given point in time. In order to define the many aspects of the corporate ecosystem, authors analyze specific ecosystems in a variety of first-hand sources [28].

4.3 RESEARCH GAP

Publicly, the importance of the medical industry can't be denied, but studies show that it's only just starting to become clear on a business level. Academic works are tasked with proving the healthcare industry's worth to the economy and laying forth concrete plans to boost that worth. Future linked medical services must be designed to be more accessible and scalable. In order to build and culture market demand, this idea suggests using ecosystemic methodologies and adopting open marketing strategies that involve training potential client segments. Increasing collaboration in the linked health domain necessitates the creation of frameworks for contextualization and coordination of choices.

The goal of linked healthcare resources is the creation of individualized services, stressing the need of ecosystems for fostering cooperation rather than concentrating on individual solutions. Understanding the connected heath business ecosystem through the lens of business models provides a fresh perspective on how to foster long-term cooperation [29].

For linked health to flourish in the long run, a number of different approaches that incorporate a multi-stakeholder perspective are required, as pointed out [30]. In addition, the importance of investigating the supporting framework for Connected Health is highlighted. Business infrastructure for sustainability and feasibility is essential, especially in a multi-stakeholder and complex setting, and technology infrastructure is essential for successful implementation.

At this point in time, in the midst of the information technology revolution, many healthcare organizations use collaborative approaches and open innovation strategies to boost efficiency and effectiveness. Despite the obvious synergy between open innovation tactics and ecosystemic approaches, the intricacy of these methods prevents them from gaining widespread traction. As pointed out, the complexity of business ecosystems derives from the fact that they have not been thoroughly dissected and explained in the healthcare context from a business process viewpoint. Managed inflows and outflows of knowledge across organizational boundaries are at the heart of open innovation, which is a form of distributed business and innovation processes. Discussions of business models and business ecosystems in the field of linked health must take into account the complexities and interdependencies of the various business processes involved.

It is becoming clear that Connected Health holds great promise for enhancing the accessibility, safety, and results of healthcare delivery.

Wireless, digital, electronic, mobile, and telehealth are all included under the umbrella term "Integrated Health," which describes a theoretical framework for the administration of healthcare centered on the individual requirements of the patient as the center of attention.

Because Connected Health is still in its infancy, there have been limited attempts to create evaluation frameworks that may be used to help determine how to study the effects of Connected Health tools.

To fill this void, we propose the following study:

How can evaluations of Connected Health solutions be facilitated by existing technology assessment models?

To delve deeper into this subject, we conducted a literature analysis, focusing on works that evaluated information systems (IS) and healthcare information systems (HIS).

4.4 IS AND HIS EVALUATION MODELS

There are a few main purposes that evaluation is meant to achieve. Understanding the complex nature of technology and its potential to enhance clinical performance, patient care, and service operations requires an assessment of the influence of IS in the healthcare setting. For better process, care, economics, and patient happiness in the future, evaluation provides us with the opportunity to learn from the past and present achievement [31].

The emergence and evidence-base of Connected Health innovation are supported by identifying numerous assessment methodologies throughout the IS research, which allows us to capitalize on modern evidence and discover strategies for improving medical systems. Following in these footsteps, we take a generic approach to evaluate use on innovation [32].

For the purpose of creating a CHEF, we can look at a number of widely used modeling techniques from the IS and service industry. There are many perspectives from which to conduct an IS review, and these include the technical, social, economic, human, and organizational perspectives. HIS assessment is also a key component of a number of existing frameworks [33].

Author used a study for modeling techniques of IS that takes into account numerous assessment viewpoints as a primary criterion for making our selections. Author found that many of the models only dealt with one small facet of systems engineering and hence were inappropriate for the broad scope of Connected Health. The following is a synopsis of these points of view:

- Medical: knowing the processes that govern the production, distribution, and consumption of goods and services that affect healthcare;
- Technical: the framework of equipment and software equipment to communicate quality healthcare operations more efficiently;
- Human: how healthcare workers are educated and regulated; how staff members think and feel about their jobs; how comfortable and safe their workplaces are for patients.

This can also explore the changes in social conduct and maturation through the involvement of both personal (e.g., attitudes, mood, or health history) and contextual variables (e.g., service accessibility or finance of care);

- Organizational: how a healthcare organization is structured, its culture, and its politics all play a role in the evaluation;
- Regulatory: how a public body exercises authority over activities held in high esteem by the healthcare sector.

4.1.1 IS Success Model

This study also looked at a number of often referenced studies on the development from the discipline of IS.

The IS success model is one such framework that does just that, classifying IS achievements into six distinct types based on the vantage point from which they were initially analyzed.

Multidimensionality is used in the model to quantify interdependencies among categories (Figure 4.2).

- Information;
- Quality of systems and services;
- Use;
- Satisfaction of users.

It is obvious that the success of the IS is influenced by the following factors, which point to a link between the six categories (i.e., net benefits).

Users' happiness and engagement with an IT system are affected by its net advantages.

1. Technology Acceptance Model (TAM)
    - Users' attitudes toward technology adoption are studied through the TAM, which takes into account a wide range of relevant criteria [43].
    - The perceived usefulness (U) and the perceived ease of use (E) of the technology are two of these aspects (Figure 4.3).
    - According to TAM, these elements are what ultimately decide whether or not people will adopt a new technology. Although TAM is a great tool for analyzing how people respond to new technologies, it has limitations when it comes to explaining the "worth" of these innovations.
2. Success Model of Search Engine
    - To reach this goal, looks at how hard it is to figure out how search engine algorithms affect individual users and builds the search engine success model as an extension of the IS success model.

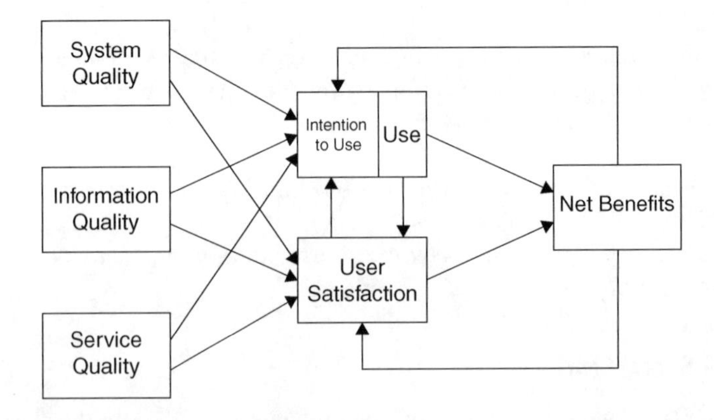

FIGURE 4.2 Block diagram of the success model of information systems (IS).

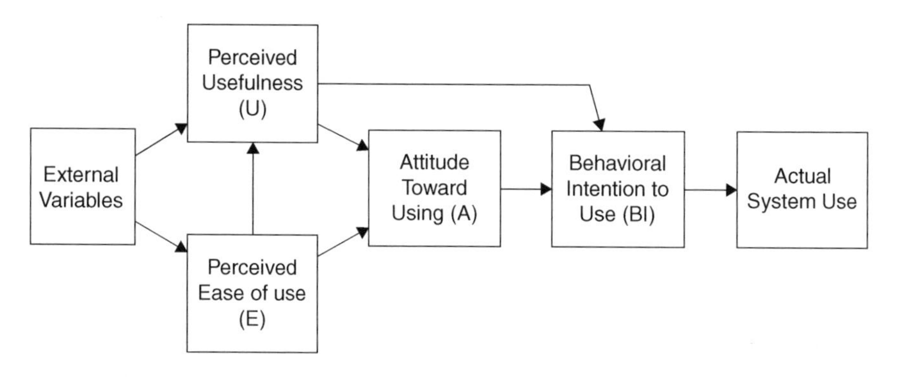

FIGURE 4.3 Model of technology acceptance.

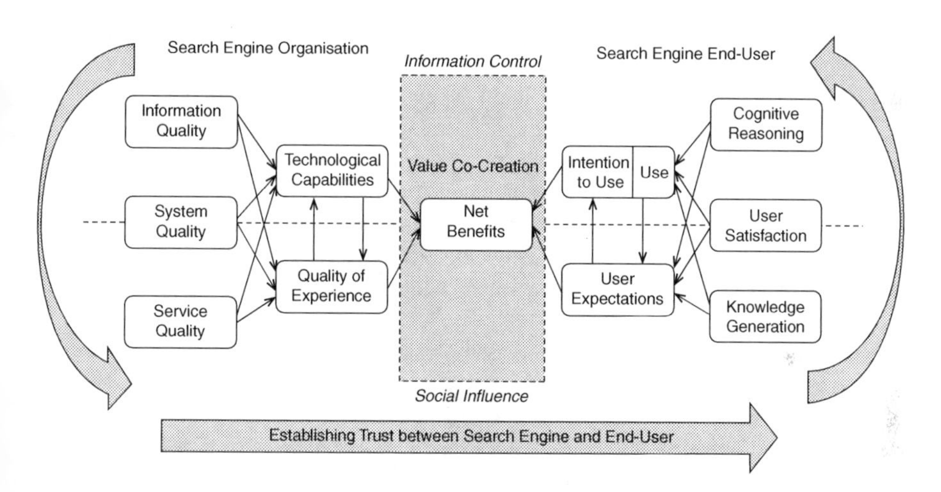

FIGURE 4.4 Block diagram of success model of search engine.

- The interdependencies among the parts expand upon Delone and McLean's IS Success Model by providing a more holistic perspective, especially in the post connection between the organization and the end-user [34].
- When viewed through the lens of Connected Health, this approach exemplifies the recursive nature of building trust in order to produce and maintain net benefits.
- In order to quantify interdependencies across the model's categories, uses an effect, which is shown in Figure 4.4, to measure how the model's different categories rely on each other.
  - Learning synthesis,
  - Intellectual processing,
  - Net advantages through a co-creation relationship.

4.5 THE CONTEXT OF CONNECTED HEALTH

Note that the development of ICT has consistently influenced medical improvement. Midway through the 1990s saw the dawn of the Internet, the early 2000s saw the rise of mobile phone technology, and the later part of the last decade saw the emergence of the social web paradigm, which has had a profound effect on the way in which information is shared and processed around the world. Because of the rapid development of ICT, a wide range of health and wellness-related services, including eHealth, mHealth, telehealth, and telemedicine, have mushroomed [35,36].

According to the WHO, there has been a dramatic shift in world demographics across a variety of dimensions, with the share of the aged population growing. According to the WHO, by 2050, the world's senior population will make up 22% of the world population, up from 12% in 2015. An aging population poses unanticipated challenges for the healthcare system, such as an increase in the prevalence of chronic diseases [37]. This impending demographic shift raises questions about the sufficiency of current health and social care models in light of the inevitable increase in healthcare requirements and associated expenses. To effectively manage resources and provide quality care in the face of such threats, more inclusive models have been called for in scholarly literature.

Existing research has already included popular terms, such as "ehealth," "mhealth," "telehealth," and "telemedicine," into the creation of future healthcare models that are robust and inclusive. The concept of "Connected Health" is one such future healthcare paradigm that has received a lot of attention from researchers, medical practitioners, and corporations.

Health informatics is an all-encompassing paradigm that has grown from its roots in the more established medical technology revolution models (eHealth, mHealth, telemedicine, etc.). Using a variety of technological solutions, smart healthcare bridges the gap between the care giver and the care receiver as well as many stakeholders throughout the care delivery chain. While integrating current modern digital healthcare approaches and tapping into the promise of data collected using sensors and analyzed with the aid of artificial intelligence, this paper explores how healthcare delivery in the future might look [38].

According to the goal of linked health, which is to get the right information to the right person at the right time, putting integrated care models into place rests on how well that goal is met [39]. By facilitating framework collaboration and data collection, "Connected Health" promotes timely health and wellness-related decision-making.

Connected Health is predicated on developing healthcare technologies that promote optimal Internet connectivity, sharing, analysis, and utilization. Better health and care decisions can be made at all stakeholder levels by a wider range of actors (citizens, patients, professionals, and policymakers). Connected Health makes use of everything from simple text messages to cutting-edge technologies (artificial intelligence, machine learning, chatbots, etc.) to provide individualized, patient-centered care. Among the many benefits of Connected Health, they highlight the availability of constant "feedback loops" that allow for more informed and faster decision-making [40]. More people are involved and there is a broader societal impact thanks to the use of data and feedback loops.

There are many ways in which linked health might contribute to better healthcare delivery. Connected solutions will first enable real-time, two-way communication between patients and their care teams through a variety of services. Secondly, emergency care and routine care are both enhanced through sensor data collecting and virtual reality services. Third, HD video conferencing tailored to individual need facilitates the provision of a range of virtual care services to geographically dispersed populations [41]. Although concepts along these lines have been proposed before, until now, their full potential has not been realized because of a lack of appropriate technical and commercial underpinnings [42–43]. Fourth, the model is strengthened by Connected Health's decision-support features and feedback loop qualities.

Finally, the concept of developing a patient-centered Connected Health model that makes effective use of patient data by means of digital analytics holds great promise [44].

4.6 APPLICATION AREA AND SCOPE OF CONNECTED HEALTH

In their article, "Connected Health: Definition and Utilization," highlights three primary domains where this concept can be used. There are three main areas that need to be addressed: (1) degenerative illness and comorbidity management; (2) resident and patients' behavior, avoidance, and recovery; and (3) infirmity and contextual assisted living. In this regard, the "digitally enabled hospitals" are identified as a fundamental intended application of linked health that will aid in facilitating a wide range of digitally intensive services within the confines of hospitals.

As the world's population ages, chronic diseases have taken the spotlight as the leading cause of hospitalization. Connected Health projects have always prioritized those suffering from long-term conditions. It can benefit people with chronic diseases by: (1) allowing for disease prevention at residence and simplifying the fine-tuning of therapy, (2) avoiding hospitalization through timely decisions, and (3) teaching the patient for self-care and compliance.

Some studies have found that implementing Connected Health drastically increases patient involvement and active self-management among those with chronic diseases. Although Connected Health's usefulness in managing chronic diseases has been praised, its limited utility in preventing such conditions has been noted. Nonetheless, more lifestyle and rehabilitation-focused linked health services are entering the market, and these provide preventative measures in addition to their other services.

Despite healthcare professionals' occasional tendency to view lifestyle management as a distinct topic, it's clear that the two go hand in hand.

Illness avoidance has been proven to work in the medical journals. More precise digital health projects are being implemented for orthopedic treatment and regeneration cases, wholesome nutrition, the fight against obesity, good sleep sanitation and efficiency, and the maintenance of adequate levels of physical activity as part of the larger framework of disease prevention. The breadth of lifestyle and prevention-related areas covered by Connected Health is shown in the aforementioned list. Connected Health solutions can be used in a variety of situations, including those where they might aid in the recovery process for people who have been injured.

When it pertains to in-home care for older people, the ambience supportive housing solutions prioritize independence and quality of life by addressing the unique demands of the senior population. Vital signs and activity tracking, psychological and physical coaching, self-evaluation, and social protection are all part of this comprehensive approach. As a result of testing, smart home technologies are making strides.

To further advance staff efficiency, hospital procedures, clinical governance, and patient satisfaction, the next generation of hospitals will be equipped with a wide range of cutting-edge technologies, including essential medical devices, efficient content systems, and online communications tools. There will be millions of connected devices in use by 2020, the vast majority of which will be generation communications devices, cordless wearables, and thousands of Internet of things devices for various sorts of assessments.

Future hospitals and surgical equipment will be intelligent and connected, i.e., they will provide more helpful services by leveraging smart technology, processors, data collecting, software algorithms, and interfaces to aid surgeons in their decision-making. A prime example of this type of change is the advent of robotic-assisted surgery (RAS). Noninvasive surgery's (MIS) widespread use and the updated information revolution have altered the traditional doctor–patient dynamic while also enhancing surgical outcomes and the quality of life for patients.

Hospitals will become "smart," or equipped with Internet of things–based technologies, as a result of advancements in medical technology and the expansion of healthcare services. In the future, hospitals will be empowered to extrapolate information gleaned from systems that are digitally and mechanically automated from start to finish. To do this, there will be a need for a mechanism of surveillance and transmission between healthcare providers, devices, patients, and other stakeholders.

In the not-too-distant future, medical care will include virtual hospitals and tailor-made drugs. It is anticipated that patients will soon be able to attend e-clinics or simulated hospitals for medicinal practice, without the participation of a healthcare practitioner, with treatments and interaction controlled remotely, thanks to network access and digital health technologies.

4.7 CHEF

The Connected Health analysis approach is described here (CHEF). Both the benefits of existing HIS/IS models and their drawbacks, as revealed by the literature analysis, impacted the creation of CHEF, as shown in Figure 4.5. Moreover, although economics and legislation are significant influences on innovation, they are essentially disregarded by the majority of the modeling techniques we found. Healthcare net benefits are the foundation of the CHEF framework. In order to ascertain how the many components of Connected Health contribute to one another's value, CHEF is structured around four distinct but interrelated layers: healthcare, business, people, and platforms. The medical positive externalities are generated through the many stages in the service lifespan that each category supports.

Here's an illustration:

In this stage of developing a quality healthcare model, the emphasis is on expanding the business by increasing the company's market share and facilitating

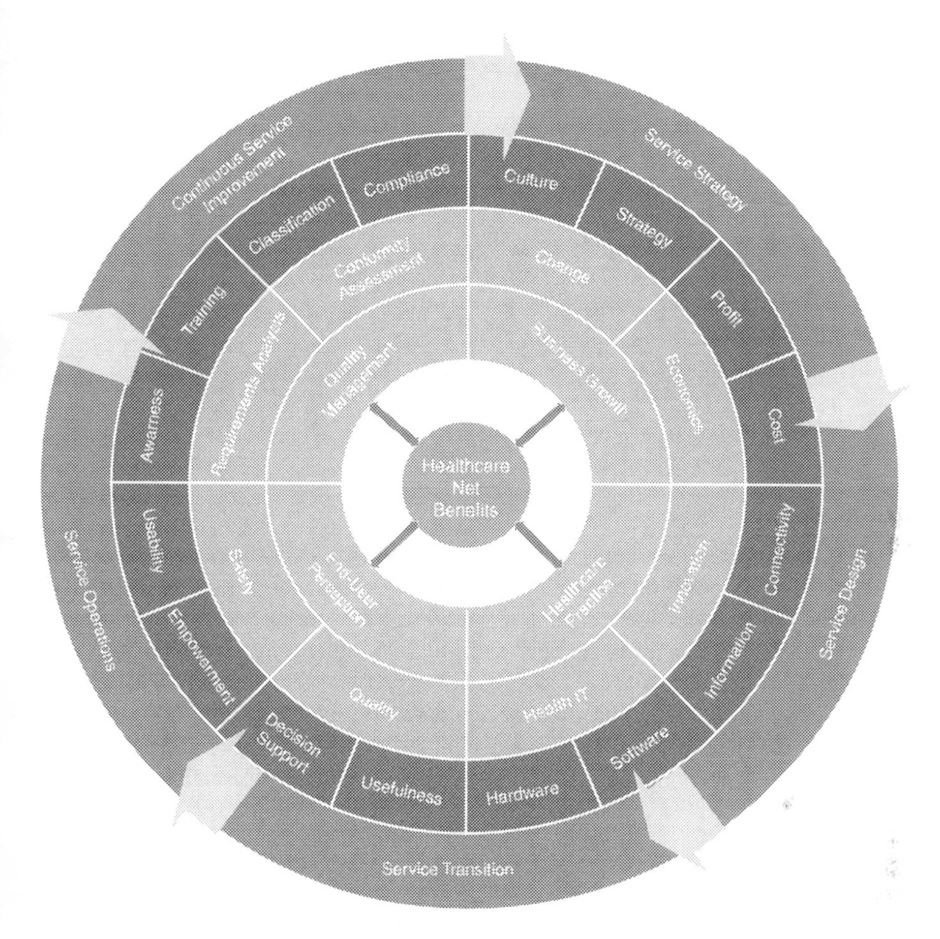

FIGURE 4.5 Various field structures of connected health evaluation framework (CHEF).

positive social and financial improvement in the field of healthcare. The study places special emphasis on the shift in mindset and approach required to successfully implement Connected Health advancements. It is important to conduct an economic review of the benefits and costs of implementing Connected Health development before doing so.

Healthcare Practice: This area addresses the impact of health IT and innovation on practice/clinical pathways during both the program planning and migration phases of healthcare delivery. The technological feasibility of providing a Connected Health solution, in terms of both equipment and software, is assessed. Also considered are socio-technical and ethnographic perspectives on the innovativeness of different approaches to healthcare delivery.

End-user Perception: As part of both the quality healthcare migration and management stages, this centers on the safety and effectiveness of patient care from a user's perspective (e.g., patients).

Physician, patient, or caregiver: In this test, the reliability and effectiveness of Connected Health services are measured. In order to strike a good balance between

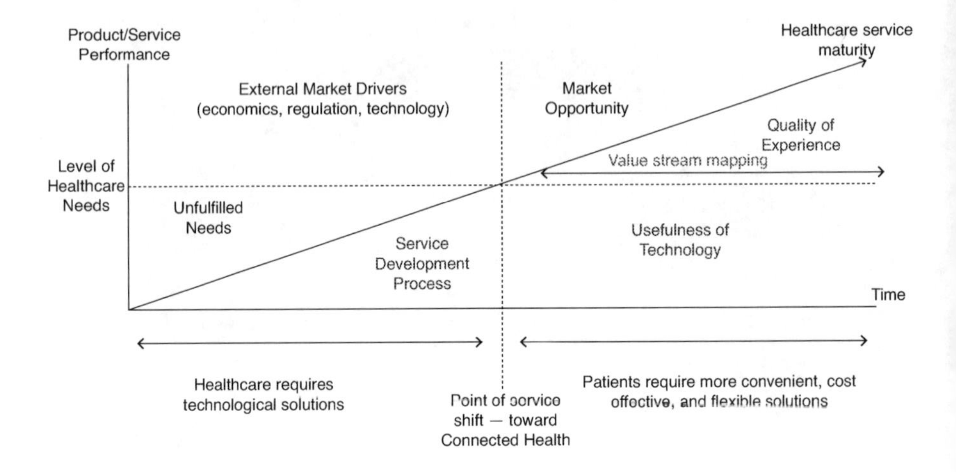

FIGURE 4.6 Environment of Integrated Health.

empowerment and safety, it may be necessary to conduct an evaluation of the usability and level of empowerment a solution may provide. From a standards perspective, we can determine whether the implementation of Connected Health technologies has boosted the quality of healthcare decision-making and increased the value of technological breakthroughs.

Careful attention to technical and regulatory requirements and conformity evaluation is at the heart of quality management, which is an integral aspect of both the healthcare service operations and continuous service improvement phases. In this stage, we can assess the needs of healthcare stakeholders in order to boost user engagement with Connected Health innovations and develop more effective training strategies. The study may also look at how the company rates in terms of technological classification and compliance with medical device laws. With this in mind, businesses may better adjust their service strategies, and the service lifecycle itself is perpetually updated by means of a philosophy of continuous enhancement.

This study will then determine the most important metrics for judging the success of Connected Health products in each of these subfields. To further the development of Connected Health, we will be working to establish deployable performance factors for each segment and its constituent parts. The outside environment of CHEF is made up of different operating stages of the lifecycle, and each of these stages stresses the need to find period following.

When properly aligned, the business project plan and market prospects can flourish in tandem thanks to the service lifecycle phases (Figure 4.6). The Connected Health ecosystem meets the demands of healthcare IT to improve the quality of healthcare delivery. External influences and different demographic drivers may leave certain unmet healthcare demands, and these gaps may be filled by Connected Health. A number of these factors are also creating fresh business prospects for Integrated Health technology, allowing for higher levels of healthcare service maturity and higher improvement in customer effectiveness. Connected Health technologies add value because they enhance the efficiency and effectiveness of healthcare delivery through the use of technology.

Despite the fact that technology can aid in healthcare delivery, it is essential to ask fundamental questions throughout the service lifecycle, such as "What problem does information solve?" and "To what problem is this technology a solution?" As a fundamental evaluative tool, the issue posed by Postman is equally applicable to the growing subject of Connected Health.

To build on this, it is essential to simulate the present medical system, including entertainer engagement, product life cycle management, resource exchange, service constraints, workflows, organizational structures, and mapping the leading healthcare industry outlook. In the context of Connected Health, it is necessary to establish a process for verifying the authenticity of data before it can be deemed reliable

CHEF provides a framework to direct the assessment procedure. Consequently, it is crucial to consider the following two key factors as we progress in the field of Interactive Diagnosis and treatment alternatives: (1) guaranteeing that the mechanisms, devices, and offerings meet the health and social needs of users through evidence-based research and (2) providing new patient-centered innovative approaches to motivate people to effectively maintain the well-being at home and within the community.

It is also important to evaluate the effects of Connected Health across the full range of care services. There are many diverse stakeholders, and CHEF recognizes this fact. By utilizing an assessment process, CHEF will be able to provide the following for evaluators: An overarching perspective on a healthcare system; tailored analysis of the healthcare service lifecycle; monitoring on service organizations and customer analytics; cheat sheet and key performance indicator tools to assess national healthcare innovation integrations, healthcare interventions, and medical providers.

These will also be incorporated into our research plan as we work to create a CHEF, and they will then be used to analyze different healthcare offerings and identify essential indicators for quality. The Connected Health features and the overall positive effects on healthcare are directly correlated. To meet the explosive expansion of healthcare IT solutions, CHEF will be further verified by ongoing industry involvement and Connected Health technologies.

In addition to encouraging organizations to take into account the intricate socio-technical ecology in which medical applications are designed, implemented, and used, CHEF can help spur innovation by leading evaluation at each stage of the health IT product lifecycle. Data management, patient privacy, exposure to healthcare data, communicating and interacting, and knowledge management are all important aspects of Connected Health that deserve special attention from the quality systems that regulate it. From a health and safety standpoint, legislation and standardization are a great help to the technology evaluation procedures. We anticipate that CHEF will assist institutions in assessing the dangers associated with Connected Health features and weighing them against the opportunities presented by these features, as in the creation of a benefit–risk profile. In addition, conformity assessment will determine whether or not medical devices that pass regulatory muster pose any difficulties for Connected Health advancement. To increase patient safety, CHEF advocates for the adoption of quality management concepts, the adoption of continually improved standards, and the harnessing of a learning and perpetual improvement environment.

Using CHEF, businesses will be able to spot subpar healthcare solutions, evaluate system performance, track end-user engagement, and spot holes in their overall business plan. Furthermore, by assessing best practice and by suggesting solutions and possibilities for enhancement based on the CHEF examination and knowledge acquired, CHEF provides a first step toward applying evaluation to broaden the evidence-based basis for Connected Health.

4.8 CONCLUSION

Healthcare costs will rise proportionally as the population ages and life expectancy improves. Governments often strive to cut these rising expenses by decreasing healthcare overhead such as personnel, and improve the communication time, discussion, and management. Healthcare performance and patient safety might be compromised as a result of service bottlenecks caused by this.

There is mounting data suggesting a paradigm shift toward encouraging individuals to take greater responsibility for their health. Improved service efficiencies and efficacy, as well as financial support for these societal transitions, are made possible by technological advancements that are both enabling and aligned with these trends in healthcare. There is great promise in Connected Health as a means of reimagining healthcare delivery as a whole. The true value or advantages (healthcare, care quality, profitability, etc.) connected with digitalization in healthcare delivery systems are essential for the success of Connected Health evaluation methodologies. This chapter presents a concise overview of how prior studies have identified the existence of e-health and telecoms equipment, and how this knowledge can be utilized to guide future assessments in the field of Connected Health.

Our proposed CHEF will serve as a bridge between these initiatives, and we plan to put it to use through collaboration with business. It became clear to us during the course of our study that people's perspectives on Connected Health innovations are heavily influenced by factors such as their location, socioeconomic standing, and level of technological expertise. Therefore, technology plays a significant role in developing healthcare partnerships by helping healthcare stakeholders feel more connected to one another. Evaluating the effect of IT development on a healthcare ecosystem is made easier if we have a firmer grasp of the architecture of the Integrated Health network.

CHEF is the first step toward evaluating healthcare technology developments and providing a comprehensive view of Connected Health. Our future objectives encompass the sustenance of a robust collaborative alliance with members of ARCH – Scientific Science for Integrated Digital Health Centre, spanning across the realms of business and academia. Our diverse research team will continue our work and validate CHEF with a wide range of healthcare consumers and IT service providers.

REFERENCES

[1] Adner, R. (2012). *The Wide Lens: A New Strategt for Innovation*. Penguin, London.
[2] Adner, R., & Kapoor, R. (2010). Value creation in innovation ecosystems: How the structure of technological interdependence affects firm performance in new technology generations. *Strategic Management Journal*, 31, 306–333.

[3] Agboola, S. O., Ball, M., Kvedar, J. C., & Jethwani, K. (2013). The future of connected health in preventive medicine. *Qjm*, 106(9), 791–794. doi:10.1093/qjmed/hct088.
[4] Lakkadi, S., Mishra, A., & Bhardwaj, M. (2015). Security in ad hoc networks. *American Journal of Networks and Communications*, 4(3–1), 27–34.
[5] Jain, I., & Bhardwaj, M. (2022). A survey analysis of COVID-19 pandemic using machine learning. In: *Proceedings of the Advancement in Electronics & Communication Engineering 2022*, doi:10.2139/ssrn.4159523.
[6] Ågerfalk, P., & Fitzgerald, B. (2008). Outsourcing to an unknown workforce: Exploring opensourcing as a global sourcing strategy. *California Management Review*, 32(2), 385–409.
[7] Ahokangas, P., Boter, H., & Iivari, M. (2018). Ecosystems perspective on entrepreneurship. In: *The Palgrave Handbook of Multidisciplinary Perspectives on Entrepreneurship* (pp. 387–407). doi:10.1007/978-3-319-91611-8_18
[8] Sharma, A., Tyagi, A., & Bhardwaj, M. (2022). Analysis of techniques and attacking pattern in cyber security approach: A survey. *International Journal of Health Sciences*, 6(S2), 13779–13798. doi:10.53730/ijhs.v6nS2.8625
[9] Tyagi, A., Sharma, A., & Bhardwaj, M. (2022). Future of bioinformatics in India: A survey. *International Journal of Health Sciences*, 6(S2), 13767–13778. doi:10.53730/ijhs.v6nS2.8624.
[10] Chauhan, P., & Bhardwaj, M. (2017). Analysis the performance of interconnection network topology C2 torus based on two dimensional torus. *International Journal of Emerging Research in Management &Technology*, 6(6), 169–173.
[11] Ahokangas, P., Juntunen, M., & Myllykoski, J. (2014). Cloud computing and transformation of international e-business models. *Research in Competence-Based Management*, 7, 3–28.
[12] Ahokangas, P., Perälä-Heape, M., & Jämsä, T. (2015). Alternative futures for individualized connected health. In S. Gurtner & K. Soyez (Eds.), *Challenges and Opportunities in Health Care Management* (pp. 61–74). Springer, Cham.
[13] Alt, R., & Zimmermann, H.-D. (2001). Preface: Introduction to special section: Business models. *Electronic Markets*. doi:10.1080/713765630
[14] Alvesson, M., & Sköldberg, K. (2017). *Reflexive methodology: New Vistas for Qualitative Research* (2nd edn). Sage Publication, London. doi:10.1080/ 13642531003746857
[15] Amit, R., & Zott, C. (2001). Value creation in e-business. *Strategic Management Journal*. doi:10.1002/smj.187
[16] Anderson, G., & Arsenault, N. (1998). *Fundamentals of Educational Research*. Routledge, London.
[17] Atkova, I. (2018). From opportunity to business model : An entrepreneurial action perspective. ACTA Universitatis Ouluensis, Oulu.
[18] Atzori, L., Iera, A., & Morabito, G. (2010). The internet of things: A survey. *Computer Networks*, 54(15), 2787–2805. doi:10.1016/j.comnet.2010.05.010
[19] Pourush, N. S., & Bhardwaj, M. (2015). Enhanced privacy-preserving multi-keyword ranked search over encrypted cloud data. *American Journal of Networks and Communications*, 4(3), 25–31.
[20] Wu, J., Haider, S. A., Bhardwaj, M., Sharma, A., & Singhal, P. (2022). Blockchain-based data audit mechanism for integrity over big data environments. *Security and Communication Networks*, 2022.
[21] Bhardwaja, M., & Ahlawat, A. (2019). Evaluation of maximum lifetime power efficient routing in ad hoc network using magnetic resonance concept. *Recent Patents on Engineering*, 13(3), 256–260.
[22] Bahrami, H., & Evans, S. (1995). Flexible re-cycling and high-technology entrepreneurship. *California Management Review*. doi:10.2307/41165799

[23] Balandin, S., Balandina, E., Koucheryavy, Y., Kramar, V., & Medvedev, O. (2013). Main trends in mhealth use scenarios. *Journal on Selected Topics in Nano Electronics and Computing*, 1(1), 64–70. doi:10.15393/j8.art.2013.3022

[24] Baldwin, C. Y. (2012). Organization Design for Business Ecosystems. *Journal of Organization Design*, 1(1), 20–23. doi:10.7146/jod.6334

[25] Ballon, P. (2007). Business Modelling Revisited: The Configuration of Control and Value (August 2007). Available at http://dx.doi.org/10.2139/ssrn.1331554.

[26] Barr, P. J., McElnay, J. C., & Hughes, C. M. (2012). Connected health care: The future of health care and the role of the pharmacist. *Journal of Evaluation in Clinical Practice*, 18(1), 56–62. doi:10.1111/j.1365-2753.2010.01522.x

[27] Bhardwaj, M., & Ahalawat, A. (2019). Improvement of lifespan of ad hoc network with congestion control and magnetic resonance concept. In: Bhattacharyya, S., Hassanien, A., Gupta, D., Khanna, A., Pan, I. (eds) *International Conference on Innovative Computing and Communications*. Lecture Notes in Networks and Systems (pp. 123–133). Springer, Singapore.

[28] Bhardwaj, M., & Ahlawat, A. (2017). Optimization of network lifetime with extreme lifetime control proficient steering algorithm and remote power transfer. *DEStech Transactions on Computer Science and Engineering.*

[29] Bhardwaj, M., & Ahlawat, A. (2017). Enhance Lifespan of WSN Using Power Proficient Data Gathering Algorithm and WPT. *DEStech Transactions on Computer Science and Engineering.*

[30] Basole, R. C. (2009). Visualization of interfirm relations in a converging mobile ecosystem. *Journal of Information Technology*. doi:10.1057/jit.2008.34

[31] Berglund, H., & Sandström, C. (2013). Business model innovation from an open systems perspective: Structural challenges and managerial solutions. *International Journal of Product Development*, 18(2–4), 274–285.

[32] BHR. (2014). The new healthcare ecosystem: 5 emerging relationships. Retrieved 6 January 2019, from https://www.beckershospitalreview.com/hospital-management-administration/the-new-healthcare-ecosystem-5-emerging-relationships.html

[33] BMZ. (2015). Business model innovation. Retrieved 2 January 2019, from https://businessmodelzoo.com/resources/business-model-innovation

[34] Brown, A., Fleetwood, S., & Roberts, J. M. (2002). Critical realism and marxism. *Critical Realism and Marxism*. doi:10.4324/9780203299227

[35] Carroll, N. (2014). In search we trust: Exploring how search engines are shaping society. *International Journal of Knowledge Society Research (IJKSR)*, 5(1), 12–27.

[36] Caulfield, B. M., and Donnelly, S. C. (2013). What is connected health and why will it change your practice?. *QJM*, 106, hct114.

[37] Sharma, M., Rohilla, S., & Bhardwaj, M. (2015). Efficient routing with reduced routing overhead and retransmission of manet. *American Journal of Networks and Communications*, 4(3–1), 22–26. doi:10.11648/j.ajnc.s.2015040301.15

[38] Bhardwaj, M., & Ahlawat, A. (2018). Wireless power transmission with short and long range using inductive coil. *Wireless Engineering and Technology*, 9, 1–9. doi:10.4236/wet.2018.91001.

[39] Kaur, K. D., & Bhardwaj, M. (2015). Effective energy constraint routing with on-demand routing protocols in MANET, *American Journal of Networks and Communications*, 4(2), 21–24. doi:10.11648/j.ajnc.20150402.12

[40] Christensen, C. M., Bohmer, R., & Kenagy, J. (2000). Will disruptive innovations cure health care? *Harvard Business Review*, 78(5), 102–112.

[41] Dansky, K.H., Palmer, L., Shea, D., Bowles, K.H. (2001). Cost analysis of telehomecare. *Telemedicine Journal and e-Health*, September, 225–232.

[42] Dávalos, M. E., French, M. T., Burdick, A. E., & Simmons, S.C. (2009). Economic evaluation of telemedicine: Review of the literature and research guidelines for benefit-cost analysis. *Telemedicine and e-Health*, December, 933–948.
[43] Davis, F. D. (1989). Perceived usefulness, perceived ease of use, and user acceptance of information technology. *MIS Quarterly*, 13, 319–340.
[44] Delbanco, T., Walker, J., Bell, S. K., Darer, J. D., Elmore, J. G., Farag, N., Feldman, H.J., Mejilla, R., Ngo, L., Ralston, J.D., Ross, S.E. Trivedi, N., Vodicka, E., & Leveille, S.G. (2012). Inviting patients to read their doctors' notes: A quasi-experimental study and a look ahead. *Annals of Internal Medicine*, 157(7), 461–470.

5 Intuitionistic Fuzzy-Based Technique Ordered Preference by Similarity to the Ideal Solution (TOPSIS) Method

An MCDM Approach for the Medical Decision Making of Diseases

Vijay Kumar
Manav Rachna International Institute of Research and Studies

H. D. Arora
Amity University

Kiran Pal
Delhi Institute of Tool Engineering

5.1 INTRODUCTION

Since the inception of computers, many scientific as well as decision support tools have been developed that helps decision makers to easily take decisions under unfavorable circumstances. The development of many soft computing techniques such as fuzzy theory and its generalization provide handheld support to the decision theory and allow the discipline to grow exponentially in every sphere of the universe. Fuzzy set (FS) and its generalizations have applications across domains and contribute to solving real-life problems with uncertain and imprecise information. Zadeh (1965) introduced the concept of fuzzy set theory, over the generalization of classical set

DOI: 10.1201/9781003405368-5

theory, which has the inbuilt capability to represent incomplete information. For the past five decades, fuzzy theory has grown many folds and many generalized versions of fuzzy theory have emerged as a potential area of interdisciplinary research. Among this, Intuitionistic fuzzy set (IFS) theory introduced by Atanassov (1986) is one such generalization, which is characterized by membership as well as nonmembership grades, respectively. It is known that IFSs describe the fuzzy character more comprehensively in some special situations and have a variety of applications in decision theory. Zadeh (1969) anticipated that fuzzy theory handles the problems of medical diagnosis very well, as it is a computational tool for dealing with imprecision and uncertainty of human reasoning. In any diagnostic process, medical knowledge is usually imprecise and uncertain and is a relationship between the symptom and the disease. Sanchez (1979) proposed the fuzzy-based model, which represents the knowledge base by establishing a fuzzy max–min relation between the symptom and the disease. Real-world problems involve decision-making to obtain the best solution of the problem on hand. Multi-criteria decision-making (MCDM) techniques are very much popular in dealing with the situations ranging from the procurement of best suited land to medical decision-making, where uncertainty is involved. TOPSIS method is a multi-criteria assessment technique developed by Hwang and Yoon (1981) to improve the Zenley's concept (1974), in which compromise solution is closest to the positive ideal solution (PIS). TOPSIS method found the distance closest to the PIS and farthest away from the negative ideal solution (NIS) and considers that each criterion is monotonic, increasing or decreasing by conditions such as:

i. For the benefit criterion, both the performance value and the preference value would be larger;
ii. For the cost criterion, the performance value would be smaller.

The optimality of all the criteria have been taken care by the ideal solution, in which worst values of all the criteria have been taken up by a negative-ideal solution and the selection of alternatives has been done by the Euclidean distance. The ranking of alternatives for the problem on hand has been done easily by the implemented TOPSIS method. In decision-making problems, to get the most reliable alternative among the available ones, MCDM techniques are used by the researchers. Many Researchers, such as Ananda and Herath (2005), Adeel et al. (2019c), Akram and Adeel (2019), Akram et al. (2018), Bai (2013), Lai et al. (1994), Boran et al. (2009), Chen (2000), Chen and Tsao (2008), Chu (2002, 2002a), Gao et al. (2008), Hung and Chen (2009), Krohling and Campanharo (2011), Li and Nan (2011), Mahdaviet al. (2009), Nadaban et al. (2016), Kakushadze et al. (2017), Balioti et al. (2018), Vahdani et al. (2011) and many more, have applied the TOPSIS method in various problems of decision-making, such as supplier selection, selection of land, robotics, medical diagnosis, ranking of water quality, and many other real-life situations flavored with fuzzy sets and generalized fuzzy sets.

In this chapter, a decision support system using IF-based TOPSIS method has been proposed for the medical diagnosis of diseases. For this purpose, a case study for the diagnosis of the type of yellow fever has been considered to discuss the proposed technique.

5.2 SOME PRELIMINARIES

A crisp set is a conventional bivalent set in which the element is either a member of the set or not.

Zadeh (Lotfi Asker Zadeh is a mathematician and computer scientist, and a professor of computer science at the University of California, 1965). An FS is the extension of the classical set in which elements have a degree of membership and is defined as

> In mathematics, a set can be thought of as any collection of distinct things considered as a whole…

Let $X = \{x_1, x_2, \ldots x_n\}$ be a discrete universe of discourse, the fuzzy set A defined is given as: $A = \left\{ < x, \mu_A(x) > \; : \; x \in X \right\}$

where $\mu_A : X \rightarrow [0, 1]$ and $\mu_A(x)$ is called the grade of membership of $x \in X$.

IFSs introduced by Atanassov (1986), which are quite useful and applicable, are defined as:

An IFS A in $X = \{x_1, x_2, \ldots x_n\}$ is given as: $A = \left\{ < x, \mu_A(x), \nu_A(x) > | \, x \in X \right\}$

where $\mu_A(x) : X \rightarrow [0, 1]$ and $\nu_A : X \rightarrow [0, 1]$ are the membership function and non-membership function of the element $x \in X$. The function $\pi_A(x) = 1 - \mu_A(x) - \nu_A(x)$ is the hesitation index of x.

In limiting case, if $\pi_A(x) = 0$, IFS reduces automatically to the fuzzy set.

Let $A \in R$ be a fuzzy set, the fuzzy number satisfies the following conditions as:

a. $\lambda x_1 + (1 - \lambda) x_1 \in A$.
b. $\max \mu_A(x) = 1$.
c. Membership function $\mu_A(x)$ is piecewise continuous over $A \subseteq R$.
d. It is defined in a real number.

Xia and Xu (2012) introduced entropy-based SIFWA operator in MCDM problems and is defined as:

Let $A_i = (\mu_{A_i}, \nu_{A_i})$ be the collection of IFNs with weight vector as $w = (w_1, w_2, \ldots, w_n)^T$ such that $w_i \in [0, 1]$ and $\sum_{i=1}^{n} w_i = 1$.

If $SIFWA : V^n \rightarrow V$ and $SIFWA(A_1, A_2, \ldots, A_n) = w_1 A_1, w_2 A_2, \ldots, w_n A_n$

$$SIFWA(A_1, A_2, \ldots, A_n) = \left(\frac{\prod_{i=1}^{n} \mu_{A_i}^{w_i}}{\prod_{i=1}^{n} \mu_{A_i}^{w_i} + \prod_{i=1}^{n} (1 - \mu_{A_i})^{w_i}}, \frac{\prod_{i=1}^{n} \nu_{A_i}^{w_i}}{\prod_{i=1}^{n} \nu_{A_i}^{w_i} + \prod_{i=1}^{n} (1 - \nu_{A_i})^{w_i}} \right)$$

5.3 TOPSIS DECISION-MAKING MODEL

Hwang and Yoon proposed the TOPSIS method in 1981 to deal with ranking the alternatives and decision-making problems. The IF-based TOPSIS method for medical decision-making problems has been discussed as:

Let $F_i = \left\{\left\{S_j\left\langle, \mu_{F_i}\left(S_j\right), v_{F_i}\left(S_j\right)\right\rangle \mid S_j \in S\right\}\right.$ be the IF representation of the alternatives (diseases) and $F = \{F_1,\ldots,F_m\}$ and $S = \{S_1,\ldots,S_n\}$ be the set of criteria (symptoms) of the disease with criteria weight (CW) vector as $W = \{w_1,\ldots,w_n\}$ such that $w_j \geq 0, \sum_{j=1}^{n} w_j = 1.$

The decision-making algorithm is given as follows:

Step 2.1. Construction of IFDM:

To get the collective opinion of the medical experts ME_k decision-making problem, the average of an individual's opinion of medical experts is taken and aggregated into a collective IFDM.

Consider that r_{ij}^k be the IFN assigned by medical experts ME_k for the evaluation of $F_i = \left\{\left\{S_j\left\langle, \mu_{F_i}\left(S_j\right), v_{F_i}\left(S_j\right)\right\rangle \mid S_j \in S\right\}\right.$ against S_i. Each of the medical experts ME_k is allocated a weight $\lambda_k > 0$ with $\sum_{k=1}^{d} \lambda_k = 1$. The aggregated IF rating $r_{ij}^k = \left(\mu_{ij}^k, v_{ij}^k\right)$ of diseases with regard to each symptom can be evaluated with the help of the SIFWA operator as follows:

$$r_{ij} = SIFWA\left(r_{ij}^{(1)},\ldots,r_{ij}^{(d)}\right) = \sum_{k=1}^{d} \lambda_k r_{ij}^k$$

$$= \left(\frac{\prod_{k=1}^{d}\left(\mu_{ij}^k\right)^{\lambda_k}}{\prod_{k=1}^{d}\left(\mu_{ij}^k\right)^{\lambda_k} + \prod_{k=1}^{d}\left(1-\mu_{ij}^k\right)^{\lambda_k}}, \frac{\prod_{k=1}^{d}\left(v_{ij}^k\right)^{\lambda_k}}{\prod_{k=1}^{d}\left(v_{ij}^k\right)^{\lambda_k} + \prod_{k=1}^{d}\left(1-v_{ij}^k\right)^{\lambda_k}} \right)$$

The decision matrix is defined as:

$$D = \begin{pmatrix} x_{11} & x_{12}\cdots & x_{1n} \\ x_{21} & x_{22}\cdots & x_{2n} \\ \vdots & \vdots & \vdots \\ x_{m1} & x_{m2}\cdots & x_{mn} \end{pmatrix}$$

Step 2.2. Assigning the CWs:

Entropy-based method has been utilized to assess the CWs with the entropy introduced by Vlachos and Sergiadis (2007) as:

$$E_{VS}^{IFS}\left(S_j\right) = -\frac{1}{n\ln 2}\sum_{i=1}^{n}[\mu_F(x_i)\ln \mu_F(x_i) + \nu_F(x_i)\ln \nu_F(x_i) -$$

$$\left(1-\pi_F(x_i)\right)\ln\left(1-\pi_F(x_i)\right) - \pi_F(x_i)\ln 2]$$

where $\frac{1}{n\ln 2}$ is the constant such that $0 \le E_{VS}^{IFS}\left(S_j\right) \le 1$

Degree of divergence

$$\delta_j = 1 - E_{VS}^{IFS}\left(S_j\right)$$

Entropy Weight

$$w_j = \frac{\delta_j}{\sum_{j=1}^{n}\delta_j}$$

Step 2.3. Create the Weighted IFDM (R):

The IFDM is defined by Atanassov (1986) to determine the aggregated IFDM:

$$R = W^T \otimes D = W^T \otimes \left[x_{ij}\right]_{m\times n} = \left[x_{ij}\right]_{m\times n}$$

where R = aggregated decision matrix

$$x_{ij} = \left\langle \mu_{ij}, \nu_{ij} \right\rangle = \left\langle 1-\left(1-\mu_{ij}\right)^{w_j}, \nu_{ij}^{w_j} \right\rangle, \mathrm{w}_j > 0$$

Step 2.4. IF Derivations IFPIS (F^+) **and IFNIS** (F^-)**:**

For benefit criteria (B) and cost criteria (G), the *IFPIS* (F_i^+) and *IFNIS* (F_i^-) are represented by Zeleny (1974) as:

$$F_i^+ = \left\{\left\langle \begin{array}{l} S_j, \left(\max_i \mu_{ij}\left(S_j\right) \middle| j \in G\right), \left(\min_i \mu_{ij}\left(S_j\right) \middle| j \in B\right), \left(\min_i \nu_{ij}\left(S_j\right) \middle| j \in G\right), \\ \left(\max_i \nu_{ij}\left(S_j\right) \middle| j \in B\right) \end{array} \right\rangle i \in m\right\}$$

$$F_i^- = \left\{\left\langle \begin{array}{l} S_j, \left(\min_i \mu_{ij}\left(S_j\right) \middle| j \in G\right), \left(\max_i \mu_{ij}\left(S_j\right) \middle| j \in B\right), \left(\max_i \nu_{ij}\left(S_j\right) \middle| j \in G\right), \\ \left(\min_i \nu_{ij}\left(S_j\right) \middle| j \in B\right) \end{array} \right\rangle i \in m\right\}$$

Step 2.5. Distance Measures of F_i from IFPIS (F_i^+) and IFNIS (F_i^-):

The distance measure is expressed by Szmidt and Kacprzyk (2001, 2002) as:

$$\delta_{IFS}\left(F_i, F_i^+\right) = \sqrt{\sum_{j=1}^{n}\left[\left(\mu_{F_i}(S_j) - \mu_{F_i^+}(S_j)\right)^2 + \left(v_{F_{i}^+}(S_j) - v_{F_i^+}(S_j)\right)^2 + \left(\pi_{F_i}(S_j) - \pi_{F_i^+}(S_j)\right)^2\right]}$$

$$\delta_{IFS}\left(F_i, F_i^-\right) = \sqrt{\sum_{j=1}^{n}\left[\left(\mu_{F_i}(S_j) - \mu_{F_i^-}(S_j)\right)^2 + \left(v_{F_i}(S_j) - v_{F_i^-}(S_j)\right)^2 + \left(\pi_{F_i}(S_j) - \pi_{F_i^-}(S_j)\right)^2\right]}$$

Step 2.6. Closeness Coefficient CC_i:

$$CC_i = \frac{\delta_{IFS}\left(F_i, F_i^-\right)}{\delta_{IFS}\left(F_i, F_i^+\right) + \delta_{IFS}\left(F_i, F_i^-\right)}$$

Step 2.7. Ranking of Alternatives:

Ranking is in the descending order as per the value of CC_i. Highest CC_i is the decision value.

5.4 CASE STUDY

The diagnosis of three types of vector-borne diseases has been taken on the basis of data collected from the health care center situated in Delhi and is explained in the following paragraphs.

5.4.1 Representation of Collected Data

MCDM models do not have mathematical equations but have criteria in the form of verbal/linguistic variables, which are used under critical or complex situations. In this chapter, the rating of alternatives and their CWs are assumed to be IFNs, which are defined on the basis of qualitative as well as quantitative data and are given in Table 5.1.

TABLE 5.1
Verbal (Linguistic) Terms Used in the Data Collection

Linguistic terms	Low (L)	Satisfactory (Sa)	Hard (H)	Very hard (VH)

The data of four patients with the help of three medical experts is given in Tables 5.2–5.6 as:

TABLE 5.2
Medical Experts Opinion—Vector-Borne Diseases Symptoms

	Chikungunya (F^C)			Dengue (F^D)			Malaria (F^M)		
Symptoms	ME_1	ME_2	ME_3	ME_1	ME_2	ME_3	ME_1	ME_2	ME_3
S_1	H	H	H	VH	VH	VH	H	H	H
S_2	VH	VH	VH	H	H	H	L	L	L
S_3	L	L	L	L	L	L	VH	VH	VH
S_4	H	Sa	H	H	H	H	L	L	L
S_5	L	L	L	VH	VH	VH	L	L	L
S_6	L	L	L	H	H	H	F	L	F
S_7	L	L	F	L	L	F	L	Sa	H

TABLE 5.3
Post Examination Medical Experts' Opinion for the Patient "p1001"

		Diseases		
Symptoms (S)	**Medical Expert (ME)**	F^C **(Chikungunya)**	F^D **(Dengue)**	F^M **(Malaria)**
S_1	ME_1	Sa	Sa	VH
	ME_2	L	Sa	VH
	ME_3	L	L	H
S_2	ME_1	L	L	L
	ME_2	L	Sa	L
	ME_3	Sa	Sa	L
S_3	ME_1	L	L	VH
	ME_2	L	L	VH
	ME_3	Sa	L	H
S_4	ME_1	L	L	VH
	ME_2	L	L	VH
	ME_3	Sa	L	VH
S_5	ME_1	H	L	H
	ME_2	H	L	H
	ME_3	VH	Sa	VH
S_6	ME_1	L	Sa	H
	ME_2	Sa	H	Sa
	ME_3	VH	VH	Sa
S_7	ME_1	L	L	L
	ME_2	H	Sa	Sa
	ME_3	H	Sa	H

TABLE 5.4
Post Examination Medical Experts' Opinion for the Patient "p1002"

Symptoms (S)	Medical Expert (D)	Diseases		
		F^C (Chikungunya)	F^D (Dengue)	F^M (Malaria)
S_1	ME_1	VH	Sa	L
	ME_2	H	L	L
	ME_3	H	H	Sa
S_2	ME_1	VH	Sa	L
	ME_2	VH	Sa	L
	ME_3	H	L	Sa
S_3	ME_1	Sa	Sa	Sa
	ME_2	L	Sa	L
	ME_3	L	L	H
S_4	ME_1	VH	Sa	Sa
	ME_2	VH	Sa	L
	ME_3	VH	h	L
S_5	ME_1	L	Sa	L
	ME_2	L	Sa	L
	ME_3	L	H	Sa
S_6	ME_1	H	Sa	L
	ME_2	H	Sa	L
	ME_3	H	Sa	Sa
S_7	ME_1	Sa	H	Sa
	ME_2	Sa	H	L
	ME_3	L	H	L

TABLE 5.5
Post Examination Medical Experts' Opinion for the Patient "p1003"

Symptoms (S)	Medical Expert (D)	Diseases		
		F^C (Chikungunya)	F^D (Dengue)	F^M (Malaria)
S_1	ME_1	H	VS	Sa
	ME_2	Sa	H	Sa
	ME_3	H	VS	H
S_2	ME_1	H	H	L
	ME_2	VH	H	L
	ME_3	Sa	VH	Sa
S_3	ME_1	L	VH	L
	ME_2	L	VH	L
	ME_3	L	VH	L

(Continued)

TABLE 5.5 (*Continued*)
Post Examination Medical Experts' Opinion for the Patient "p1003"

		Diseases		
Symptoms (S)	Medical Expert (D)	F^C (Chikungunya)	F^D (Dengue)	F^M (Malaria)
S_4	ME_1	H	VH	L
	ME_2	VH	VH	L
	ME_3	H	H	L
S_5	ME_1	Sa	VH	L
	ME_2	Sa	VH	L
	ME_3	Sa	H	L
S_6	ME_1	Sa	Sa	L
	ME_2	Sa	VH	Sa
	ME_3	L	VH	L
S_7	ME_1	L	Sa	Sa
	ME_2	L	Sa	Sa
	ME_3	L	H	Sa

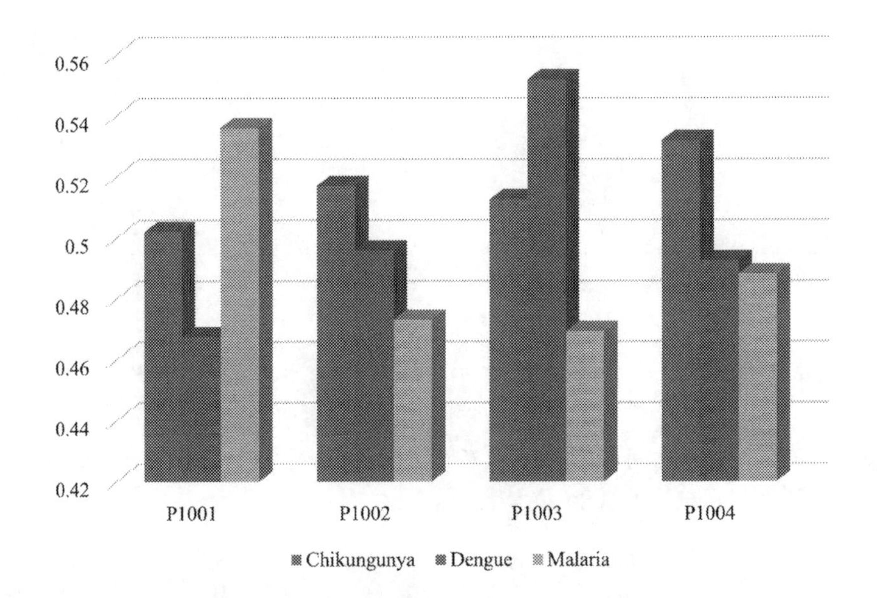

FIGURE 5.1 Ranking of diseases for each patient.

5.4.2 Evaluation of Case Study

Let {$p1001$, $p1002$, $p1003$, $p1004$} be the suspected patients in a hospital, which may come across with a disease caused by mosquito bite. Medical experts examined the patients by means of related tests and symptoms present in patients to prescribe the suitable treatment. In the disease detection procedure, specified weight is allocated to three medical experts as $\lambda_1 = 0.20$, $\lambda_2 = 0.35$, $\lambda_3 = 0.45$ on the basis of distinctive domains such as knowledge, background and expertise. Let $\{F^C, F^D, F^M\}$ be the given diseases as chikungunya, dengue, and malaria, respectively, and called as alternatives. Let $\{S_1, \ldots S_7\}$ be the set of associated symptoms of the vector-borne diseases as rashes, pain in joints and muscle, orbital headache, fever, and vomiting, respectively.

On the basis of the given data, the computational procedure is discussed as:

Step 3.1: Using step (2.1), construct the collective opinion of experts in aggregated IFDM using the SIFWA operator.

Step 3.2: CWs, entropy values, and degree of divergence for each criterion has been presented in Table 5.7.

Step 3.3: Using step (2.3), the value of R is obtained and the weighted IFDM is presented in Table 5.8.

Step 3.4: The IFPIS (F_i^+) and IFNIS (F_i^-) for each alternative are presented in Tables 5.9 and 5.10.

Step 3.5: Closeness coefficient CC_i and ranking is presented in Table 5.11.

Higher the value of closeness coefficient CC_i, more likely is the alternative. The alternative is nearer to IFPIS and farthest from IFNIS and the ranking is given as per the descending order of CC_i and is presented in Table 5.12.

TABLE 5.6
Post Examination Medical Experts' Opinion for the Patient "p1004"

		Diseases		
Symptoms (S)	Medical Expert (D)	(F^C Chikungunya)	(F^D Dengue)	F^M (Malaria)
S_1	ME_1	VH	Sa	H
	ME_2	H	Sa	L
	ME_3	H	L	Sa
S_2	ME_1	VH	L	Sa
	ME_2	VH	Sa	L
	ME_3	Sa	Sa	Sa
S_3	ME_1	VH	L	L
	ME_2	H	H	Sa
	ME_3	VH	Sa	L
S_4	ME_1	H	L	Sa
	ME_2	H	L	Sa
	ME_3	VH	L	Sa
S_5	ME_1	Sa	L	L
	ME_2	Sa	L	Sa
	ME_3	H	Sa	Sa

(Continued)

TABLE 5.6 (*Continued*)
Post Examination Medical Experts' Opinion for the Patient "p1004"

Symptoms (S)	Medical Expert (D)	Diseases		
		(F^C Chikungunya)	(F^D Dengue)	F^M (Malaria)
S_6	ME_1	H	Sa	Sa
	ME_2	H	Sa	Sa
	ME_3	Sa	H	L
S_7	ME_1	Sa	Sa	L
	ME_2	Sa	Sa	L
	ME_3	H	L	Sa

TABLE 5.7
Collective Weights, Entropy, and Degree of Divergence

Patient	w_1	w_2	w_3	w_4	w_5	w_6	w_7
				Weights (W_i)			
p1001	0.1378	0.1609	0.1955	0.1955	0.2350	0.0653	0.0101
p1002	0.1257	0.1519	0.1022	0.1805	0.1509	0.1552	0.1336
p1003	0.1070	0.1244	0.2239	0.1976	0.1386	0.1260	0.0825
p1004	0.1356	0.1186	0.1285	0.2700	0.1444	0.0898	0.1132
Patient	E_1	E_2	E_3	E_4	E_5	E_6	E_7
				Entropy (E_i)			
p1001	0.6388	0.5782	0.4873	0.4873	0.3839	0.8288	0.9736
p1002	0.6993	0.6366	0.7554	0.5680	0.6390	0.6285	0.6803
p1003	0.6812	0.6292	0.3326	0.4110	0.5870	0.6245	0.7541
p1004	0.7463	0.7781	0.7596	0.4947	0.7298	0.8320	0.7882
Patient	δ_1	δ_2	δ_3	δ_4	δ_5	δ_6	δ_7
				Degree of Divergence (δ_i)			
p1001	0.3612	0.4218	0.5127	0.5127	0.6161	0.1721	0.0264
p1002	0.3007	0.3634	0.2446	0.4320	0.3610	0.3715	0.3192
p1003	0.3188	0.3708	0.6674	0.5890	0.4130	0.3755	0.2459
p1004	0.2537	0.2219	0.2404	0.5053	0.2702	0.1680	0.2118

5.5 RESULT AND DISCUSSION

The results are displayed in graphical form in Figure 5.1. Based on the given information, the ranking of alternatives obtained by the application of the TOPSIS method is given in Table 5.12. According to this, patient p1001 was diagnosed with Malaria, patient p1003 was diagnosed with Dengue, and patients p1002 and p1004 were diagnosed with Chikungunya. Moreover, the diagnosis parameters used by the doctors were the same as that of findings obtained from the proposed technique.

TABLE 5.8
Weighted IF Decision Matrix (*R*)

	S_1	S_2	S_3	S_4	S_5	S_6	S_7
				P1001			
F^C	(0.9779, 0.7177)	(0.7520, 0.9588)	(0.7072, 0.9502)	(0.7072, 0.9502)	(0.9625, 0.5353)	(0.9672, 0.9223)	(0.9932, 0.9900)
F^D	(0.9089, 0.9089)	(0.7520, 0.9588)	(0.5567, 0.9796)	(0.5567, 0.9796)	(0.4947,0.9755)	(0.9863, 0.8658)	(0.9898, 0.9950)
F^M	(0.9794, 0.6933)	(0.9832, 0.5946)	(0.9709, 0.5946)	(0.9709, 0.5946)	(0.9704,0.5163)	(0.9557, 0.9557)	(0.9932, 0.9900)
				P1002			
F^C	(0.9754, 0.7389)	(0.9774, 0.6677)	(0.7818, 0.9835)	(0.9812, 0.5823)	(0.6364, 0.9842)	(0.966, 0.6994)	(0.9116, 0.9116)
F^D	(0.9473, 0.8483)	(0.7890, 0.9533)	(0.8526, 0.9683)	(0.9252, 0.7895)	(0.9371,0.8207)	(0.8980, 0.8980)	(0.9706, 0.7352)
F^M	(0.8004, 0.9677)	(0.7640, 0.9611)	(0.9106,0.9216)	(0.6474, 0.9711)	(0.7654,0.9613)	(0.6281, 0.9838)	(0.7249, 0.9785)
				P1003			
F^C	(0.9640, 0.8372)	(0.9632, 0.8133)	(0.5113, 0.9767)	(0.9661, 0.6071)	(0.9084, 0.9084)	(0.8215, 0.9611)	(0.7810, 0.9913)
F^D	(0.9852, 0.7401)	(0.9800, 0.7183)	(0.9767, 0.5113)	(0.9706, 0.5913)	(0.9793, 0.6918)	(0.9797, 0.7383)	(0.8976, 0.9519)
F^M	(0.9550, 0.8693)	(0.8021, 0.9680)	(0.5113, 0.9767)	(0.5532, 0.9794)	(0.6603, 0.9855)	(0.7733, 0.9738)	(0.9443, 0.9444)
				P1004			
F^C	(0.9735, 0.7215)	(0.9823, 0.7295)	(0.9717, 0.7616)	(0.9631, 0.4678)	(0.9397, 0.8277)	(0.9660, 0.8764)	(0.9524, 0.8623)
F^D	(0.8093, 0.9582)	(0.8859, 0.9431)	(0.9171, 0.8801)	(0.5219, 0.9571)	(0.7742, 0.9630)	(0.9621, 0.8891)	(0.8381, 0.9650)
F^M	(0.8569, 0.9288)	(0.8859, 0.9431)	(0.7964, 0.9670)	(0.5764, 0.9447)	(0.7448, 0.9700)	(0.8692, 0.9721)	(0.8183, 0.9709)

TABLE 5.9
IFPIS (F_i^+) of Each Alternative

Patient	f_1^+	f_2^+	f_3^+	f_4^+	f_5^+	f_6^+	f_7^+
p1001	(0.9794, 0.6933)	(0.7520, 0.9588)	(0.9709, 0.5946)	(0.9709, 0.5946)	(0.9704, 0.5163)	(0.9863, 0.8658)	(0.9898, 0.9950)
p1002	(0.9754, 0.7389)	(0.9774, 0.6677)	(0.9106, 0.9216)	(0.9812, 0.5823)	(0.9371, 0.8207)	(0.9660, 0.6994)	(0.7249, 0.9785)
p1003	(0.9852, 0.7401)	(0.9800, 0.7183)	(0.9767, 0.5113)	(0.9706, 0.5913)	(0.9793, 0.6918)	(0.9797, 0.7383)	(0.7810, 0.9913)
p1004	(0.9735, 0.7215)	(0.9823, 0.7295)	(0.9717, 0.7616)	(0.9631, 0.4678)	(0.9397, 0.8277)	(0.9660, 0.8764)	(0.8183, 0.9709)

TABLE 5.10
IFPIS (F_i^-) of Each Alternative

Patient	f_1^-	f_2^-	f_3^-	f_4^-	f_5^-	f_6^-	f_7^-
p1001	(0.9089, 0.9089)	(0.6176, 0.9832)	(0.5567, 0.9796)	(0.5567, 0.9796)	(0.4947, 0.9755)	(0.9557, 0.9557)	(0.9932, 0.9900)
p1002	(0.8004, 0.9677	(0.7640, 0.9611)	(0.7818, 0.9835)	(0.6474, 0.9711)	(0.6364, 0.9842)	(0.6281, 0.9838)	(0.9706, 0.7352)
p1003	(0.9550, 0.8693)	(0.8021, 0.9680)	(0.5113, 0.9767)	(0.5532, 0.9794)	(0.6803, 0.9855)	(0.7733, 0.9738)	(0.9443, 0.9443)
p1004	(0.8093, 0.9582)	(0.8859, 0.9431)	(0.7964, 0.9670)	(0.5219, 0.9571)	(0.7448, 0.9700)	(0.8692, 0.9721)	(0.9524, 0.8623)

TABLE 5.11
Closeness Coefficient CC_i and Ranking

Alternatives	$\delta_{IFS}\left(F_i, F_i^{+}\right)$	$\delta_{IFS}\left(F_i, F_i^{-}\right)$	CC_i	Rank
		P1001		
F^C	2.7921	2.8147	0.5020	2
F^D	2.9423	2.5817	0.4674	3
F^M	2.6201	3.0283	0.5361	1
		P1002		
F^C	2.7033	2.8935	0.5170	1
F^D	2.6836	2.6392	0.4958	2
F^M	2.9604	2.6577	0.4731	3
		P1003		
F^C	2.6665	2.8027	0.5125	2
F^D	2.5881	3.1872	0.5511	1
F^M	2.9170	2.5780	0.4692	3
		P1004		
F^C	2.7175	3.0893	0.5320	1
F^D	2.8454	2.7611	0.4925	2
F^M	2.8783	2.7443	0.4881	3

TABLE 5.12
Ranking and Disease Diagonosis for Each Patient

Patients	Ranking	Diagnosis of Disease
P1001	$F^M \succ F^C \succ F^D$	Malaria
P1002	$F_C \succ F_D \succ F_M$	Chikungunya
P1003	$F_D \succ F_C \succ F_M$	Dengue
P1004	$F^C \succ F^D \succ F^M$	Chikungunya

5.6 CONCLUSION

The TOPSIS approach under IF environment has been discussed for the diagnosis of given vector-borne diseases. Diagnosis of diseases was performed on the basis of data of four patients under patient ID p1001–p1004. The original data are given in qualitative terms, and further defined in IFNs. The ranking of diseases were presented with the help of the given method. The given method is cost effective, reduces laboratory footprints to the environment, and saves the money of patients. Thus, it is very much effective for an initial start of the treatment and suitable for places where laboratory facility is not available, and is indeed a decision support tool.

REFERENCES

Adeel, A., Akram, M. and Koam, A.N.A. (2019). Group decision making based on mm-polar fuzzy linguistic TOPSIS method. *Symmetry*. 11 (6): 735.

Akram, M. and Adeel, A. (2019). TOPSIS approach for MAGDM based on interval-valued hesitant fuzzy NN-soft environment. *International Journal of Fuzzy Systems*. 21 (3): 993–1009.

Akram, M., Shumaiza and Smarandache, F. (2018). Decision making with bipolar neutrosophic TOPSIS and bipolar neutrosophic. *ELECTRE-I. Axioms*. 7 (2): 33.

Ananda, J. and Herath, G. (2005). Evaluating public risk preferences in forest land-use choices using multi-attribute utility theory. *Ecological Economics*. 55 (3): 408–419.

Atanassov, K. (1986). Intuitionistic fuzzy sets. *Fuzzy Sets and Systems*. 20: 87–96.

Bai, Z. (2013). An interval-valued intuitionistic fuzzy TOPSIS method based on an Improved score function. *Scientific World Journal*. 1–6.

Balioti, V., Tzimopoulos, C. and Evangelides, C. (2018). Multi-criteria decision making using TOPSIS method under fuzzy environment: Application in spillway selection. *Proceeding*. 2: 637.

Boran, F. E., Genc, S., Kurt, M. and Akay, D. (2009). A multi-criteria intuitionistic fuzzy group decision making for supplier selection with TOPSIS method. *Expert Systems with Applications*. 36: 11363–11368.

Chen, C.T. (2000). Extensions of the TOPSIS for group decision making under fuzzy environment. *Fuzzy Sets and Systems*. 1 (114): 1–9.

Chen, T.Y. and Tsao, C.Y. (2008). The interval-valued fuzzy TOPSIS method and experimental analysis. *Fuzzy Sets and Systems*. 159 (11): 1410–1428.

Chu, T. C. (2002a). Selecting plant location via a fuzzy TOPSIS approach. *International Journal of Advance Manufacturing Technology*. 20 (11): 859–864.

Chu, T.C. (2002b). Facility location selection using fuzzy TOPSIS under group decisions. *International Journal of Uncertainty Fuzziness and Knowledge-Based Systems*. 10 (6): 687–701.

Gao, P., Feng, J. and Yang, L. (2008). Fuzzy TOPSIS algorithm for multiple criteria decision making with an application in information systems project selection. In: *4th International Conference on Wireless Communications, Networking and Mobile Computing*, Dalian, 2008, 1–4.

Hung, C.C. and Chen, L.H. (2009). A multiple criteria group decision making model with entropy weight in an intuitionistic fuzzy environment. In: Huang, X., Ao Si. and Castillo, O. (eds) *Intelligent Automation and Computer Engineering. Lecture Notes in Electrical Engineering*. Springer, Dordrecht.

Hwang, C. L. and Yoon, K. (1981). *Multiple Objective Decision Making Methods and Applications: A state-of-the-art Survey. Lecture Notes in Economics and Mathematical Systems*. Springer, New York, NY.

Kakushadze, Z., Raghubanshi, R. and Yu, W. (2017). Estimating cost savings from early cancer diagnosis. *Data*. 2: 1–16.

Krohling, R.A. and Campanharo, V.C. (2011). Fuzzy TOPSIS for group decision making: A case study for accidents with oil spill in the sea. *Expert Systems with Applications*. 38 (4): 4190–4197.

Lai, Y.J., Liu, T.Y. and Hwang, C.L. (1994). TOPSIS for MODM. *European Journal of Operational Research*. 76 (3): 486–500.

Li, D.F. and Nan J.X. (2011). Extension of the TOPSIS for multi-attribute group decision making under Atanassov IFS environments. *International Journal of Fuzzy System Applications*. 1 (4): 47–61.

Mahdavi, I., Heidarzade, A., Sadeghpour-Gildeh, B. and Mahdavi-Amiri, N. (2009). A general fuzzy TOPSIS model in multiple criteria decision making. *International Journal of Advanced Manufacturing Technology*. 45: 406–420.

Nadaban, S., Dzitac, S. and Dzitac, I. (2016). Fuzzy TOPSIS: A general view. *Procedia Computer Science*. 91: 823–831.

Szmidt, E and Kacprzyk, J. (2001). Entropy for intuitionistic fuzzy sets. *Fuzzy Sets and Systems*. 118: 467–477.

Szmidt, E. and Kacprzyk, J. (2002). Using intuitionistic fuzzy sets in group decision making. *Control and Cybernetics*. 31: 1037–1053.

Vahdani, B., Mousavi, S.M. and Tavakkoli, M.R. (2011). Group decision making based on novel fuzzy modified TOPSIS method. *Applied Mathematical Modelling*. 35: 4257–4269.

Vlachos, I.K. and Sergiadis, G.D. (2007). Intuitionistic fuzzy information-Applications to pattern recognition. *Pattern Recognition Letters*. 28: 197–206.

Xia, M.M. and Xu, Z.S. (2012). Entropy/cross entropy-based group decision making under intuitionistic fuzzy environment. *Information Fusion*. 13: 31–47.

Zadeh, L. A. (1965). Fuzzy sets. *Information and Control*. 8: 338–356.

Zadeh, L. A. (1969). Biological applicationof the theory of fuzzy sets and systems. In: Proctor, L.D. (ed.), The *Proceedings of an International Sysposiumon Biocybernatics of central Nervous System*. Little, Brown and Company, Boston, 199–206.

Zeleny, M. (1974). A concept of compromise solutions and the method of the displaced ideal. *Computer and Operation Research*. 1: 479–496.

6 Design of a Heuristic IoT-Based Approach as a Solution to a Self-Aware Social Distancing Paradigm

Amit Kumar Bhuyan and Hrishikesh Dutta
Michigan State University

6.1 INTRODUCTION

The recent surge in unstable scenarios because of pandemics, war-like environments, etc., has led to limited physical mobility. Although mobility is inevitable for the functioning of a locality, navigation in such environments has surfaced as a founded challenge. For example, because of the increase in the number of COVID variants, an escalation in safety norms is of surmounting importance. Similarly, political instabilities in Ukraine have led to rescue and relief operations that need stealth. Although these problems are ubiquitous, they compound for people with pre-existing impairments. As, for example, maneuvering a visually impaired person through a crowd while considering his/her own safety and following the safety protocols is a challenging task. The deliberation about safety to avoid mishaps is not limited to the above problems; rather, it has found its place in technology as well, like automatic self-driving cars, warehouse maintenance robots, household contraptions, etc. Though the safety norms are being implemented based on pre-described standards by every organization or institution, the effectiveness of these norms is not quantifiable. This poses a threat in mitigating dire situations such as healthcare and war, along with increasing cost ineffectiveness in case of technological equipment.

The uncertainty at the cusp of this transition from a world with no boundaries to a realm of limitations in movement makes it imperative for us to find a solution for assisting people with pre-existing medical conditions to navigate a crowded area while following social distancing norms. The movement of a subject to a pre-decided destination by following the social distancing protocols has been imperative and is modeled in this work as a "Heading Problem." The "Heading Problem" focuses on the choice of direction while maneuvering in a crowded area (Figure 6.1). The solution to the Heading Problem, thus, represents the path the user should follow to maneuver a crowd while following social distancing norms. With the advent of smart, intelligent, and

DOI: 10.1201/9781003405368-6

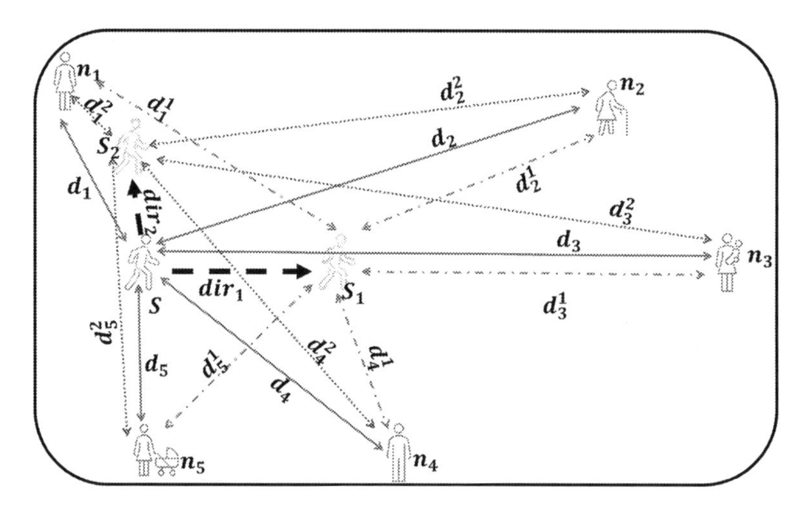

FIGURE 6.1 Heading problem.

perceptive systems at our disposal, decision-making by handheld devices isn't inconceivable. We propose a method that can calculate the heading of a subject if information about a potential threat is available to the smart user device. We propose a mobile Internet-of-things (IoT)-based framework to solve the "Heading Problem" mentioned above. A deployable architecture using IoT is developed, where each handheld user device acts as a mobile IoT node. The IoT node, possessing soft computing capabilities, makes decisions about the user's whereabouts and navigation prospects. Three different mechanisms were used for decision-making: a novel look-ahead (LA) approach, a differential evolution (DE)-based candidate selection technique, and a leader-based swarm optimization method. The proposed framework allows the user to decide the direction of his/her next step to avoid potential contact. The only requirement of the underlying hardware support of these handheld devices for this framework is the capability of computing distances between them. This functionality can be easily implemented using some readily available low-cost, low-power, low-complexity sensors, for example, distance computing ultrasonic sensors [11–13], Global Positioning System, etc. The devices also require a communication interface to share their current locations with their neighbors within the communication interface. The proposed solution approach requires location information within an 18-foot radius of the user. With this requirement, any short-range communication interface, such as Bluetooth, WiFi, Zigbee, etc., can be a viable and efficient communication interface among the devices.

While designing this, there were many perspectives that surfaced and many of them were addressed. To make the simulation more pragmatic, the following aspects were considered. First, to capture the dynamic nature of the environment, the environment was made stochastic. Second, to check the validity of the approaches and the adeptness of the algorithm, pre-decided destinations were provided. This ensures the universality of the methods used, by not making the solutions unique for a particular environment. Third, the density of the environment, where a subject (namely, a person) is maneuvering toward its target, is such that it replicates a realistic population density, e.g., the population density of Michigan. Fourth, experiments were also conducted with changing destination before convergence to the previous destination to check the adaptiveness of the methods proposed in this work. Though this work focuses on the aforesaid problem, where a subject maneuvers toward its destination with certain constraints, the proposed methods can be mapped to a variety of problems, namely, obstacle avoidance, collision detection in communication network, traffic monitoring systems, scaled autonomous environments such as self-driving vehicles, drone trajectory, etc.

The contribution of this work is as follows. First, a mobile-IoT-based short-range communication framework is developed for navigation assistance for persons with pre-existing impaired conditions. Second, three soft-computing-based decision-making mechanisms are used for solving the aforesaid problem algorithmically: (1) a novel heuristic-based LA technique is proposed, which tackles the said problem by modulating the locality of the decisions; (2) a DE algorithm is used to check the potency of mating-based evolutionary techniques to solve the given problem; and (3) a leader-based particle swarm optimizer (PSO) is employed to ameliorate the decision-making given the constraints and dynamic nature of the problem. The performances of these techniques are compared based on the readiness of the algorithms and the precision of the decisions. Third, an analytical study is provided comparing the effectiveness of each of these techniques in relation to the problem using extensive simulation experiments.

The usability of this proposal ranges across a wide roster of applications and its scope is boundless. Their application domains can be broadly classified as follows.

Medical Usage: Given the relevance with the present pandemic scenario, this can be a deliberate solution to cope with social distancing norms set by the Center for Disease Control and Prevention and other government/international medical institutions. Although maintaining social distance is an easily comprehensible concept, it is difficult to bring it into effect and ascertain its operability. It is even more difficult to apply in people with impaired physical abilities, such as those with vision impairment and challenged cognitive capabilities. Its applicability is also relevant in children and aged individuals who need monitoring and supervision. For the abovementioned categories, the decision to maneuver through a public domain geography while maintaining distancing norms can be a challenge. The proposed methodology can help as a guidance system for these people while acting as an alarming system for others in the vicinity.

Technological Usage: With the upsurge in the demand for smart homes and unmanned automobiles, the proposed solution to the Heading Problem finds its place as an adaptive obstacle avoidance and route determination tool. Ranging from intelligent wheelchairs to robotic Roomba vacuum cleaners, development and improvement of object avoidance methodologies has been of imperative importance. Similarly, unmanned cars and other vehicles use route planning and crash avoidance software to ensure safety and traffic control. Although most of these avoidance capabilities work adeptly, they use data acquisition sensors, such as camera, ultrasound sensors, and so on. The proposed method takes a leap toward solving this through ubiquitous interfaces such as Bluetooth.

The applicability of the proposed methodology isn't limited to the aforesaid application domains. It can be used as a generic guidance tool on handheld devices like mobile phones to alert while maneuvering niche geographies such as groceries, hospitals, offices, etc.

The structure of this chapter is as follows. Section 6.2 provides a detailed description of the overall system architecture. In Section 6.3, we provide a formal definition of the problem and show a connection of the Heading Problem with the classical knapsack problem. A nonevolutionary computation approach "LA" is proposed and discussed in Section 6.4. Sections 6.5 and 6.6 provide details on two evolutionary techniques, namely, DE and particle swarm optimization, used in this work. The experimental set-up and the parameters used in simulation are provided in Section 6.7. The results and performance of all the approaches are analyzed in Section 6.8. The work is summarized in Section 6.9, and finally possible future extensions of this work are discussed in Section 6.10.

6.2 SYSTEM MODEL

We propose a deployable application for handheld mobile devices carried by users in need of mobility assistance. These devices are interconnected with each other, through a network interface, when they are in communication range. This can be posed as a mobile IoT network where each node is a decision-making device capable of soft-computing. These IoT nodes can be low complexity, low power, and low-cost embedded devices, such as Arduino Nano, Raspberry Pi, etc. [14,15]. Also, the IoT nodes can be mobile phones where the proposed mobility assistance technique is installed as a software application or an Operating System update. As shown in Figure 6.2, each user node is connected with its neighboring user nodes through short-range communication

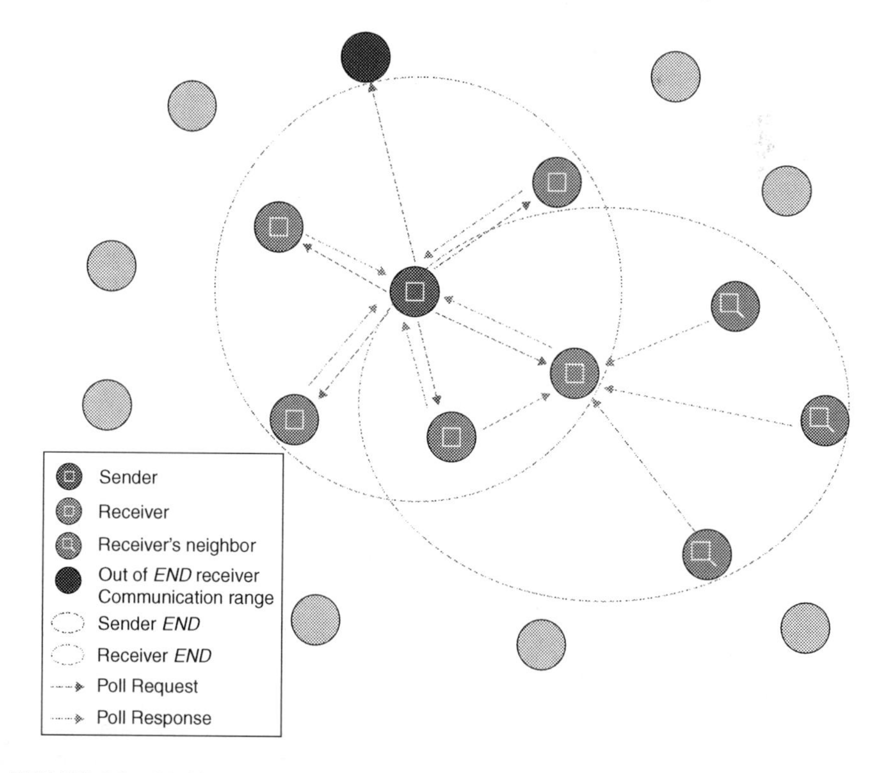

FIGURE 6.2 Mobile Internet-of-things (IoT) network model for mobility guidance.

Control Information	*END*	Latitude of i	Longitude of i

(a) PDU of Rq packets

Control Information	Latitude of i	Longitude of i	Latitude of i's neighbors	Longitude of i's neighbors

(b) PDU of Rs packets

FIGURE 6.3 The packet data unit (PDU) structure of Rq and Rs packets.

interfaces, such as Bluetooth, WiFi, Zigbee, etc. The user nodes communicate with each other to acquire the necessary geo-temporal information of their respective neighboring user nodes. Such information is then used for decision-making using the mobility assistance mechanism explained in Sections 6.4–6.6.

When a user is in a location where mobility assistance is required, it broadcasts Poll Request Packets (Rq) to all accessible IoT nodes in its vicinity. The "Rq" contains the location coordinates of the sender node. When a neighboring IoT node receives an "Rq," first it decides if the "Rq" is for itself. If the "Rq" is not for the receiving node, it ignores and doesn't respond to the request. Otherwise, it responds with a Poll Response Packet (Rs). An "Rs" contains the location of the receiver node and the receiver node's previously received neighborhood location information through Rq and Rs. The Packet Data Unit (PDU), i.e., Rq and Rs, structure is shown in Figure 6.3.

As can be seen from Figure 6.3, the request PDU has its latitude and longitude information along with its Expected Neighborhood Diameter (END). The "END" is used by the receiver of the Rq to decide if it lies within the specified "END." If so, it responds to the request packets with a response packet, which has information of its own geolocation (latitude and longitude) and its "k" nearest neighboring nodes' geolocation. For the experiments in this work, "k" is chosen to be 5. Considering the latitude, longitude, and END to be 64 bits double data types, the Rq and Rs PDUs are of size 192 and 256 bits, respectively. Ascribed to the small packets size of the PDUs, the communication cost is low. This makes it suitable for low bandwidth, low-cost communication channels.

6.3 "HEADING PROBLEM": A CLASSICAL KNAPSACK PERSPECTIVE

The design objective of the mobility assistance system is posed as a *Heading Problem*, as mentioned in Section 6.1. We propose a solution to the *Heading Problem* by representing it as a classical Knapsack Problem [1–4]. Here, items are selected to be added to a knapsack to maximize the profit while satisfying the weight constraints. Every item has a profit and a weight associated with it. The inclusion of the item depends on the maximization of cumulative profit while staying within the total weight limit of the knapsack. The analogy of the *Heading Problem* to a Knapsack Problem is given below.

Here, $d_i \in D = \{d_1, d_2, \ldots, d_n\}$ where, n is the number of neighbors to a subject within a certain range in a public place and d_i is the distance between the subject and neighbor i. Let $DIR = \{dir_1, dir_2, dir_3, \ldots, dir_m\}$ be the set of all possible directions the subject can head. For example, in Figure 6.1, dir_1 and dir_2 represent two of the possible directions the subject can head. Let $\Delta d_i^k = d_i - d_i^k$ represent the change in distance from neighbor n_i while heading in direction dir_k. Now, Δd_i^k can be used to compute the benefit associated with neighbor n_i along heading dir_k as $b_i^k = f\left(\Delta d_i^k\right)$, where the function $f(x)$ is proportional to x for $x > 0$ and zero otherwise. With respect to the situation explained in Figure 6.1, the benefit can be computed as follows. If $\Delta d_i^1 = d_i - d_i^1$ and $\Delta d_i^2 = d_i - d_i^2$, then benefit associated with each neighbor, $b_i^1 = f\left(\Delta d_i^1\right)$ and $b_i^2 = f\left(\Delta d_i^2\right)$, can be used to calculate the total benefit for dir_1 and dir_2, respectively. Along with benefit b_i^k, there is a penalty term p_i^k associated with each neighbor for each direction dir_k the subject takes. Once the subject heads toward a specific direction, the benefit and penalty for each neighbor i are computed as follows:

$$b_i = \begin{cases} \Delta d_i \text{ for } \Delta d_i > 0 \\ 0 \text{ otherwise} \end{cases} \tag{6.1}$$

$$p_i = \begin{cases} 1 \text{ for } d_i < M \\ 0 \text{ otherwise} \end{cases} \tag{6.2}$$

The benefit term in equation (6.1) favors the positive change in distance of the subject from the neighbor and the penalty term in equation (6.2) is used to avoid being within M ft from any neighbor. According to the present prevailing distancing norms, $M = 6$ can be used for constraint formulation. The benefits and penalties are associated with any direction of movement of the subject. Therefore, the aim is to maximize the total benefit from a heading and constrain the maximum accumulated penalty because of the heading. This is given below in equation (6.3).

$$\text{maximize} \sum_{i=1}^{n} b_i d_i \tag{6.3}$$

$$\text{subject to} \sum_{i=1}^{n} p_i < R$$

where

R is the maximum allowable cumulative penalty. The benefit b_i and p_i are analogous to the profit and weight of an item d_i in a classical knapsack problem.

6.4 LOOK-AHEAD TECHNIQUE: NON-EC APPROACH

In order to explore the validity of the approach and to create a baseline for comparison of evolutionary algorithmic approaches, a LA technique is used to solve the problem. In this approach, initially D directions are chosen stochastically uniformly from the initial position of the subject (refer Figure 6.1). Then the benefit parameter is computed along all D directions as follows. For any direction d_i, the distance from the nearest kth neighbor is computed (s_k^0). Now, the benefit value along direction d_i for lth LA is computed iteratively over Lth LA as shown in equations (6.4) and (6.5):

$$b^l(d_i) = w_i^l + \sum_{x=1}^{D} b^{l+1}(d_x), L-1 \geq l \geq 0,\ 1 \leq i \leq D \tag{6.4}$$

$$b^L(d_i) = \sum_{y=1}^{K} \Delta s_y^L \tag{6.5}$$

Here Δs_y^L is the change distance from yth nearest neighbors in Lth LA, and w_i^l represents the weight associated along direction d_i for LA l. This weight coefficient is chosen in order to give priority to the subject's preferred direction. It should be noted that LA technique considers plausible future steps to decide the next most beneficial step. The number of future steps for decision-making is represented by 'L'.

Similarly, there is a penalty term associated with each direction d_i for LA l, which takes care of the fact that the subject should avoid being within distance M from any neighbor. The penalty along d_i with LA l is computed in equations (6.6) and (6.7).

$$p^l(d_i) = \sum_{x=1}^{D} p^{l+1}(d_x), L-1 \geq l \geq 0,\ 1 \leq i \leq D \tag{6.6}$$

$$p^L(d_i) = \begin{cases} P \times \sum_{\forall y \in N,\ y<K} \delta\left(s_y^l < M\right) \\ 0 \text{ otherwise} \end{cases} \tag{6.7}$$

Here $\delta(C)$ is the delta function such that

$$\delta(C) = \begin{cases} \text{if} & \begin{matrix} 1 & C \text{ is true} \\ 0 & \text{otherwise} \end{matrix} \end{cases} \tag{6.8}$$

The subject decides to take the next step along the direction of maximum benefit and minimum penalty. An episode of the algorithm ends when the subject reaches the assigned destination within a set tolerance value. In addition to the benefit (profit in a classical Knapsack problem), the LA mechanism also considers a penalty associated with s_y^L. The benefit w_i^l, which favors the intended direction, biases the choice of selecting an intermediate destination, namely, the next step. One of the important advantages of using a LA mechanism is its ability to address the dynamic environment. Also, an intermediate decision to take a step doesn't just take the instantaneous benefit rather it considers the future benefits it may accumulate by making a choice.

This mechanism is motivated by the concept of future observable states in Markov Decision Process and Reinforcement Learning [5,6].

It is to be noted that this approach is a form of greedy search mechanism. Moreover, another drawback is that it searches for the preferred direction iteratively by looking ahead over multiple steps, which is computationally daunting. If the population size per LA is N_s, there are K neighbors, and the number of LA is L, then the computation increases by $(N_s \times K)^L$. Third, the intermediate solutions are selected based on a uniform distribution, which doesn't ensure the next solution near the previous solutions. Finally, the use of popular gradient-based methods was not considered because of the changing dynamics of the search space. The goal is to address these drawbacks and to show the performance improvement by using an evolutionary algorithm.

6.5 DIFFERENTIAL EVOLUTION FOR CANDIDATE SELECTION

For the application of an evolutionary strategy, we chose DE [7,8] to solve the problem by emphasizing on the problem locally. Looking at the problem locally entails focusing on deciding the favorable next step based on current observation. Every candidate solution is still weighted by change in distance and penalized by its vicinity from the nearest neighbor. The importance of the preferred direction is preserved as well. Its difference from the LA algorithms comes from the process it follows to produce new candidate solutions and update the candidate solutions over pre-decided number of generations. Mating in DE is between multiple chromosomes, as opposed to the most popular idea of mating between two chromosomes. Here there is flexibility in the choice of the number of solutions selected for mating because it takes difference vectors of a pair of solutions for mutation to produce a token solution, as given in equation (6.9).

$$v_i = x_{r_1} + F\left(x_{r_2} + x_{r_3}\right) \tag{6.9}$$

Here, x_i is a candidate solution and F denotes the mutation factor. Based on the type of problem and the validity of DE, more than one difference vector can be considered for mutation as shown in equation (6.10):

$$v_i = x_{r_1} + F_1\left(x_{r_2} + x_{r_3}\right) + F_2\left(x_{r_4} + x_{r_5}\right) \tag{6.10}$$

Then a target solution is selected for recombination with the token solution, which generates a candidate solution.

$$u_j = C\left(x_j, v_j\right) \tag{6.11}$$

u_j from equation (6.11) is selected as a solution for the next generation if the fitness function value for it is better than the parent solutions.

The focus while using DE is on the convergence, i.e., how soon it reaches the destination and on the accumulated penalty. The results are found to be comparable with the LA mechanism described before.

6.6 PARTICLE SWARM OPTIMIZER FOR CANDIDATE SELECTION

Given that the benefits and penalties for the intermediate candidate solutions are designed such that every solution can be considered as a unique solution for a knapsack problem, the idea of using algorithms that replicate social behavior has validity. Here, any solution, irrespective of its optimality, is influenced by its locality as well as the globally optimal solution. Therefore, this work considers PSO [9,10] as a possible solution for the posed problem. This technique is a leader-based evolutionary method, which is guided not only by its local and global bests but also by inertia, which captures the effect of past solutions on the new solutions. Updating the particles is based on the inertia of a particle, and best local and global decisions. Every particle changes position in the search space based on its velocity. The new velocity is calculated based on equation (6.12).

$$V^i(t+1) = wV^i(t) + c_1\left(pbest^i - X^i(t)\right) + c_2\left(gbest - X^i(t)\right) \tag{6.12}$$

After the new velocity is derived, it is used to generate a new prospective solution using the existing solutions. This is shown in equation (6.13).

$$X^i(t+1) = X^i(t) + V^i(t+1) \tag{6.13}$$

Here, $pbest^i$, X^i, and V^i are the best function value, present solution, and velocity of the particle i. The global best solution among all the solutions from the particles is $gbest$. Its implementation and effects as compared to the LA method and mating based evolutionary methods are detailed in the later sections.

6.7 EXPERIMENTAL SETUP

For the experimental setup of the LA mechanism and DE and PSO, U users are considered. To depict the mobile users in an enclosed area, a space of $M_1 \times M_2$ has been simulated. Here, each point is a user's temporal location, as shown in Figure 6.1. The idea of social distancing is implemented by considering that a user must be D_{sd} distance away from its adjacent neighbors. One step (intermediate solution) is constrained within Δc in both axes from the previous solution. The stopping criteria are set according to proximity of a solution to the destination with a tolerance of T.

For the LA mechanism, the number of solutions per LA is set to 10. For the calculation of benefit and penalty, the number of neighbors considered is 5. Neighbors are selected based on their proximity from the previous solution. To achieve the results using LA, the mechanism number of "LA's" L is set to 3. All the experimental parameters are tabulated in Table 6.1. One unit in the simulation is mapped to 5 ft in real-time scenarios.

For DE, all the above scenarios are the same except for the fitness function and the operators used to generate population. The fitness function is based on a combination of accumulated benefits and penalties for every solution. The function value is computed by using the following equation:

$$func = w_1 b + w_2 p \tag{6.14}$$

TABLE 6.1
Experimental Parameters

Parameter	Value
Population position	Random Uniform
Population density	176.8 per sq. ft (MI) 186,012 per sq. ft (Cambridge, MA)
Physical distancing (for penalty)	6 ft (1.8 m)
Each step	3.5 ft (1.06 m)
Number of neighbors	5
Number of solutions per look-ahead	10
M_1	1,000 feet
M_2	1,000 feet
D_{sd}	6 feet
Δc	0.7 feet
T	4 feet

Here, w_1 and w_2 are the weights associated with the benefit b and penalty p, respectively. The value of w_1 and w_2 are experimental hyperparameters that are chosen empirically, as discussed in Section 6.8. The best solution is decided based on the maximum function value. For the evolutionary parameters, mutation rate is set between 0.5 and 1 where dithering is used to select the mutation constant, whereas the recombination rate is set to 0.7, which is used to create a new population. For the initial setup, the maximum number of allowed generations is set to 1,000.

While using PSO to calculate candidate solutions, the population size is set to 10. A higher population (number of particles) will be beneficial, but a lower value is chosen for computational ease. This is because our intention is to explore the effect of leader-based solution strategies to solve such problems rather than finding an adept solution. This experiment delves into different combinations of hyperparameters to analyze the movement of the swarm. The details of hyperparameters for the experiments are provided in Section 6.8. The results comparing the performance of LA, DE, and PSO are stated in the next section.

6.8 RESULTS AND ANALYSIS

As stated before, all the aforementioned methodologies have been compared based on convergence duration, namely, the number of steps it takes to reach the destination and accumulated benefit along with accumulated penalty while avoiding being in the vicinity of neighbors. Figure 6.4 shows two example runs using the LA method. Here, the subject starts walking from the center of the simulation space and intends to reach one of the two equidistant destinations located at 1,414 ft from the center. Two different destinations are chosen for these experiments to ensure generalization of the aforesaid proposed mechanisms.

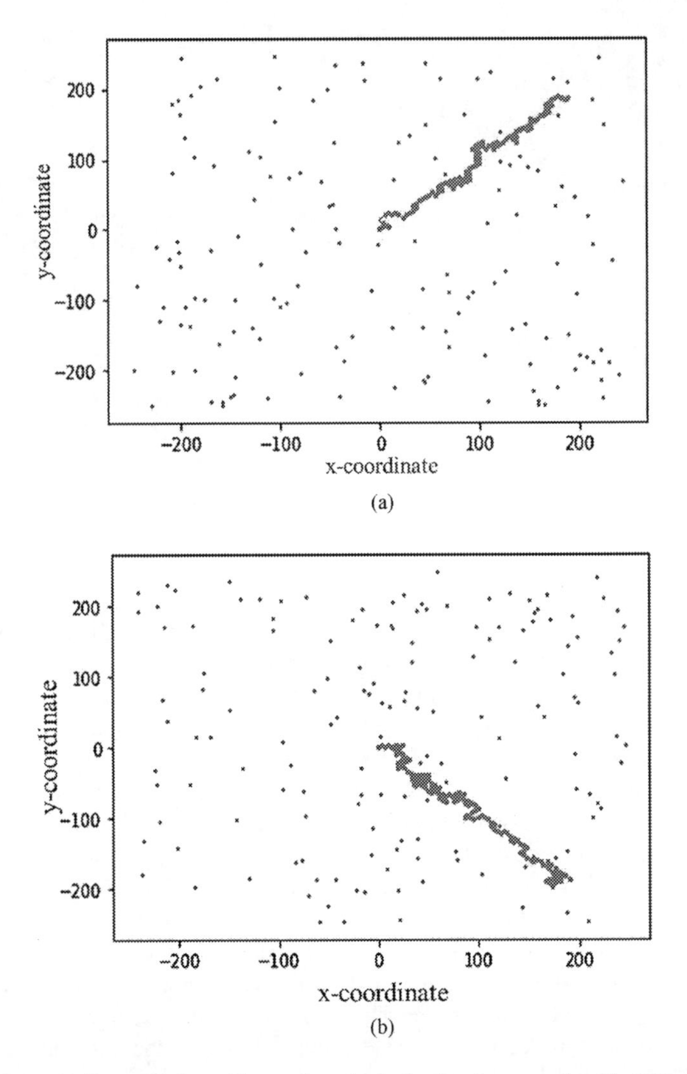

FIGURE 6.4 (a) Trace of the subject using the look-ahead approach with (200 units, 200 units). (b) Trace of the subject using the look-ahead approach with (200 units, −200 units), i.e., (1,000 ft.–1,000 ft) as the destination.

The convergence time for LA corresponding to the figures and Table 6.2 is in the range of 194.05 ± 35.30 steps. It should be noted that for each of these experiments, the subject was able to reach the destination without violating social distancing norms.

Even though this method ensures success with a higher probability, the envelope of trace isn't unique. This is a result of the stochastic choice of future steps, which is the basis of the LA method [11–15].

TABLE 6.2
Comparison of Look-Ahead, Differential Evolution, and Particle Swarm Optimizer for Self-Aware Social Distancing Paradigm

Method	Case Scenarios	x-Coordinate of destination	y-Coordinate of destination	Experimental Parameters	Average Total Benefit	Average Age # of Steps	Mean of Average Benefit Per Step	Average Total Penalty
LA	Scenario 1	200	−200	Look_ahead = 3	18279.52937	243.2	76.48964399	0
LA	Scenario 2	200	200	Look_ahead = 3	12675.59	159.1	80.08088533	0
LA	Scenario 3	200	−200	Look_ahead = 3	15018.08	187.4	80.47340007	0
LA	Scenario 4	200	200	Look_ahead = 3	14249.46	186.5	76.58325364	0
DE	Scenario 1	200	−200	Mutation = (0.5, 1), recombination = 0.7	173947.5	1687.9	103.0556622	0
DE	Scenario 2	200	200	Mutation = (0.5, 1), recombination=0.7	160791.7	1565.7	102.6964583	0
DE	Scenario 3	200	−200	Mutation = (0.5, 1), recombination=0.7	90416.4	876.5	103.1560204	0
DE	Scenario 4	200	200	Mutation = (0.5, 1), Recombination = 0.7	113992.33	1099.5	103.6764588	0
PSO	Scenario 1	200	−200	$c1$, $c2$, w = 0.3_0.3_0.9	6590.77	65.9	100.0116823	0
PSO	Scenario 1	200	−200	$c1$, $c2$, w = 0.5_0.3_0.3	7170.801	71.7	100.0113054	0
PSO	Scenario 1	200	−200	$c1$, $c2$, w = 0.5_0.3_0.9	6810.783	68.1	100.0114948	0
PSO	Scenario 1	200	−200	$c1$, $c2$, w = 0.5_0.9_0.9	7050.811	70.5	100.0115035	0
PSO	Scenario 1	200	−200	$c1$, $c2$, w = 0.7_0.3_0.9	7040.811	70.4	100.01152	0
PSO	Scenario 2	200	200	$c1$, $c2$, w = 0.3_0.3_0.9	6430.774	64.3	100.0120474	0
PSO	Scenario 2	200	200	$c1$, $c2$, w = 0.5_0.3_0.3	6660.79	66.6	100.0118596	0
PSO	Scenario 2	200	200	$c1$, $c2$, w = 0.5_0.3_0.9	6610.78	66.1	100.011802	0
PSO	Scenario 2	200	200	$c1$, $c2$, w = 0.5_0.9_0.9	6770.807	67.7	100.0119203	0

(*Continued*)

TABLE 6.2 (*Continued*)

Comparison of Look-Ahead, Differential Evolution, and Particle Swarm Optimizer for Self-Aware Social Distancing Paradigm

Method	Case Scenarios	x-Coordinate of destination	y-Coordinate of destination	Experimental Parameters	Average Total Benefit	Average Age # of Steps	Mean of Average Benefit Per Step	Average Total Penalty
PSO	Scenario 2	200	200	$c1, c2, w$=0.7_0.3_0.9	6890.814	68.9	100.0118138	0
PSO	Scenario 3	200	−200	$c1, c2, w$=0.3_0.3_0.9	6590.774	65.9	100.0117471	0
PSO	Scenario 3	200	−200	$c1, c2, w$=0.5_0.3_0.3	6750.787	67.5	100.0116607	0
PSO	Scenario 3	200	−200	$c1, c2, w$=0.5_0.3_0.9	6850.796	68.5	100.0116273	0
PSO	Scenario 3	200	−200	$c1, c2, w$=0.5_0.9_0.9	6940.804	69.4	100.0115948	0
PSO	Scenario 3	200	−200	$c1, c2, w$=0.7_0.3_0.9	7010.824	70.1	100.0117564	0
PSO	Scenario 4	200	200	$c1, c2, w$=0.3_0.3_0.9	6480.768	64.8	100.0118634	0
PSO	Scenario 4	200	200	$c1, c2, w$=0.5_0.3_0.3	6720.787	67.2	100.0117109	0
PSO	Scenario 4	200	200	$c1, c2, w$=0.5_0.3_0.9	6840.813	68.4	100.0118853	0
PSO	Scenario 4	200	200	$c1, c2, w$=0.5_0.9_0.9	6980.808	69.8	100.011578	0
PSO	Scenario 4	200	200	$c1, c2, w$=0.7_0.3_0.9	6850.804	68.5	100.0117396	0

Discussion: The variance in the solutions for each LA is because of the generation of candidate solutions using a uniform distribution (refer Section 6.4). In case of multiple neighbors qualifying within the END norms, a subset of neighbors is considered at random for the first step. For these experiments, the number of neighbors is five, according to Table 6.1. For a different run of the algorithm, the choice of neighbors may differ. Similarly, for a future step to be taken by a subject, the stochastically chosen neighbors vary for different execution runs of the algorithm. Therefore, these kinds of approaches don't ensure retracing the path of a subject for different runs of the LA algorithm. Another observation is the intra-trace repetition while deciding to proceed toward the pre-decided destination. The ability of the LA algorithm to decide the present step based on possible future steps ensures that penalties are avoided. However, the avoidance of penalties takes precedence over the preferred navigation direction. This leads to repetition of steps in the vicinity of the subject's current location. Therefore, though there are no penalties for intermediate candidate solutions (each step), the accumulated benefit is affected by the number of steps taken. This redundancy in the number of steps taken leads to reduced mean benefit. The physical significance of this redundancy is the uncertainty of a subject to proceed in the direction of the destination without hesitation. Although the LA algorithm induces immediate uncertainty in propagation, its awareness reduces deviation of the trace envelope from the shortest path. This means that increase in the number of LAs guarantees less deviation from the shortest path along the preferred direction of the chosen destination.

To mitigate the repetitive nature of the LA algorithm, as mentioned before, evolutionary methods are explored. Figure 6.5 shows the traces of two runs while applying DE. The convergence time is 800–1,600 steps, approximately. It can be observed from Figures 6.5 and 6.7 that the mean benefit is maximized by using DE. However, Figure 6.8 shows that the experiment takes substantially more time to converge as compared to LA. For implementation of the algorithm, like the LA mechanism, directional benefit is fixed. This favors solution in the direction of the destination from the present position. These experiments were executed with standard values of genetic operators, namely, mutation factor of 0.5 and 1 (for two pairs of difference vectors), and a recombination rate of 0.7.

Discussion: Experiments have been conducted by reducing the fixed directional benefit to monitor the convergence duration. It is observed that the convergence duration is inversely proportional to the directional benefit. A detailed analysis of these results shows that the delay in convergence for lower fixed benefits is because of the benefit's effect on the function value. Because the function value is directly proportional to the total benefit, a low benefit makes the difference between the optimal solution's and the derived solution's function value less drastic, hence taking more steps to reach the same destination.

Besides the convergence duration, important observations can be made from the trace followed by the subject while applying DE. The envelope of the trace for multiple runs of DE has higher similarity as compared to trace of the LA method. Another observation is the high deviation of a subject's trace from the shortest path. The similarity in trace and higher trace deviation can be ascribed to the ability of mating-based evolutionary algorithms to maximize the benefit. It can be seen in the explanation of DE that a candidate solution is produced from mutation and recombination

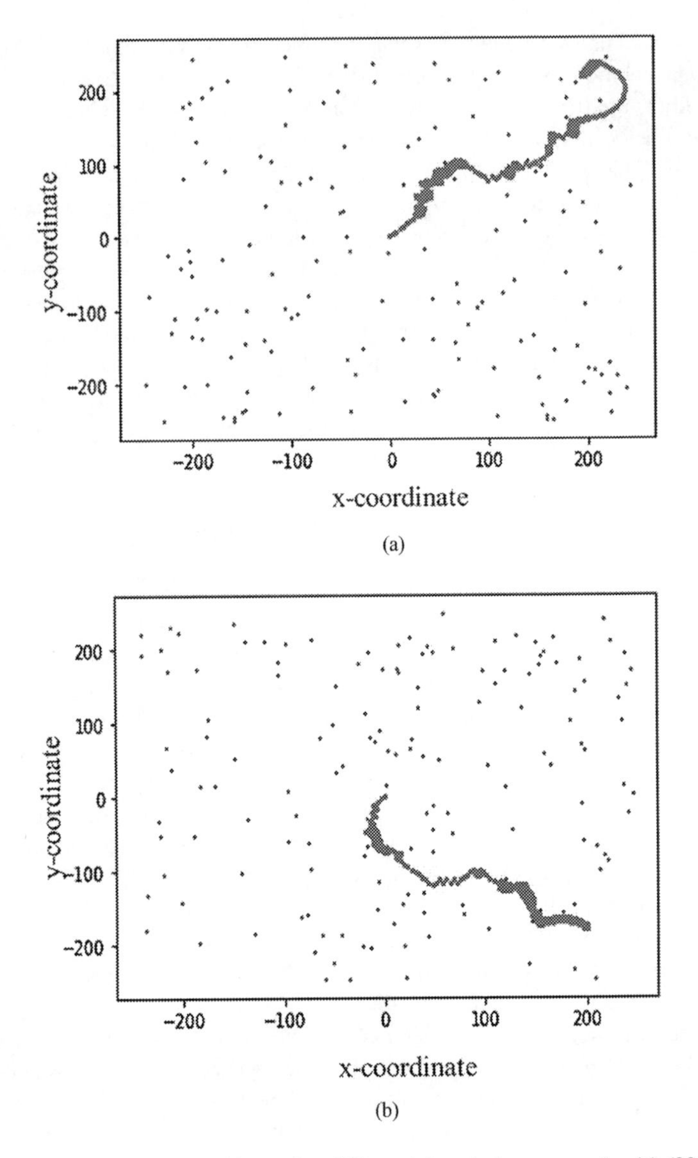

FIGURE 6.5 (a) Trace of the subject using differential evolution approach with (200, 200) as the destination. (b) Trace of the subject using differential evolution approach with (200, −200) as the destination.

of multiple candidate solutions. For this particular application, candidate solutions are limited by the geographical spread of the neighbors. Therefore, the probability of candidate solutions generated near the existing solutions increases, which leads to retracing the subject's path for different runs. However, because of the derivative-free approach of DE, it can explore a wider variety of candidate solutions without incurring significant computation costs.

Unlike LA, the deviation of the envelope of the path from the shortest path is quite high for DE. This entails least possibility for penalty but increases the convergence time substantially. This is addressed using a leader-based evolutionary method. The trace given in Figure 6.6 reflects the ability of swarm optimizer and speed of convergence. It can be seen in Table 6.2 that, in different scenarios, the number of steps taken to reach the destination is minimum while using PSO. Therefore, the accumulated benefit is greater than LA even though LA has a near straight envelope.

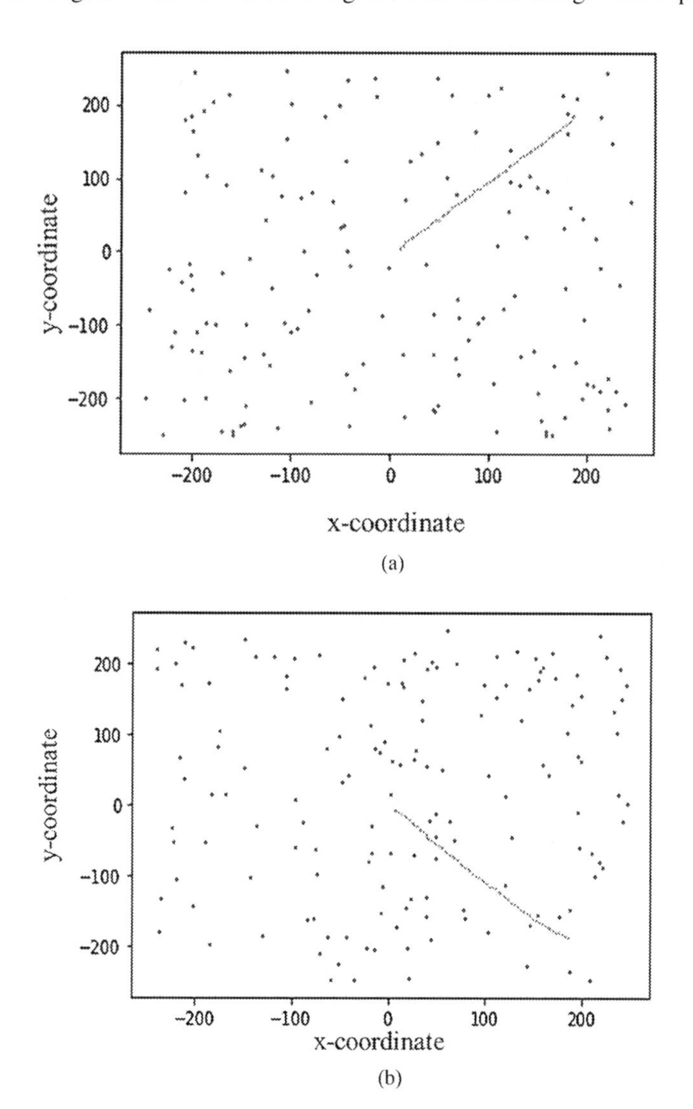

FIGURE 6.6 (a) Trace of the subject using particle swarm optimizer approach with (200, 200) as the destination. (b) Trace of the subject using particle swarm optimizer approach with (200, −200) as the destination.

Discussion: It should be noted that the decision-making is still stepwise, i.e., PSO is applied on the current state of the subject and the best solution provides the location of the next step. The experiment is designed in such a way that the number of generations is different for each step, i.e., the decision to take a step is made when there is no or least change in the best global function value. One of the reasons for faster convergence of the PSO algorithm for this application lies in the importance of inertia. A weight associated with inertia ensures that the candidate solutions are biased toward the direction of propagation of the subject. The direction of propagation of a subject is determined by the past steps. Physically, this means that the uncertainty of the subject to move forward is reduced. This reduces the repetitive nature of the trace of a subject, unlike LA and DE. Such behavior is also enforced by directional benefit, which is the same for all the algorithms.

The detailed comparative performance of LA, DE, and PSO in self-aware social distancing paradigm is given in Table 6.2, and Figures 6.7 and 6.8. The performances are recorded and averaged over several runs for each method and scenarios. The highlights from Table 6.2 are as follows. First, the subject experiences no penalty while using any of the methods employed. Second, mean of average benefits per step shows that DE maintains and maximizes the distance of the subject from its neighbors though the number of steps to reach the destination is considerably more. Third, the LA mechanism converges sooner as compared to DE but the mean of average benefit per step is low. Finally, distancing is well maintained when PSO is used while maximizing the accumulated benefit.

Figures 6.7 and 6.8 contain the comparison of LA, DE, and PSO based on average steps taken and average benefits accumulated. The steps taken using PSO, as

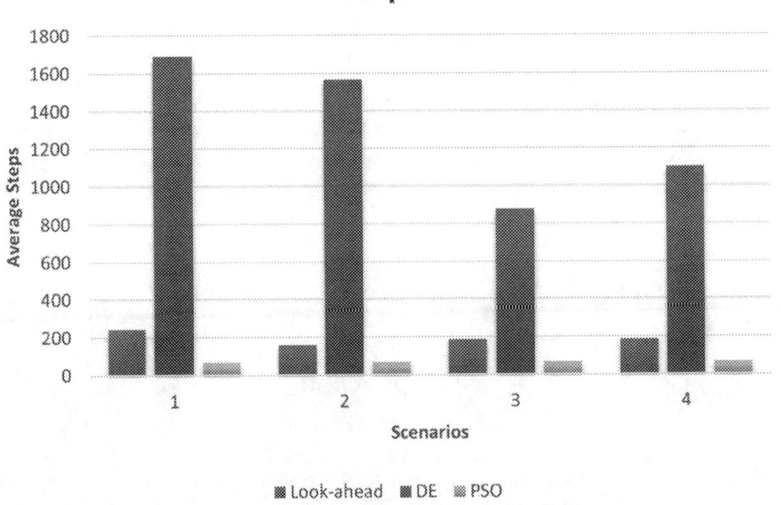

FIGURE 6.7 Average steps taken while using different methods.

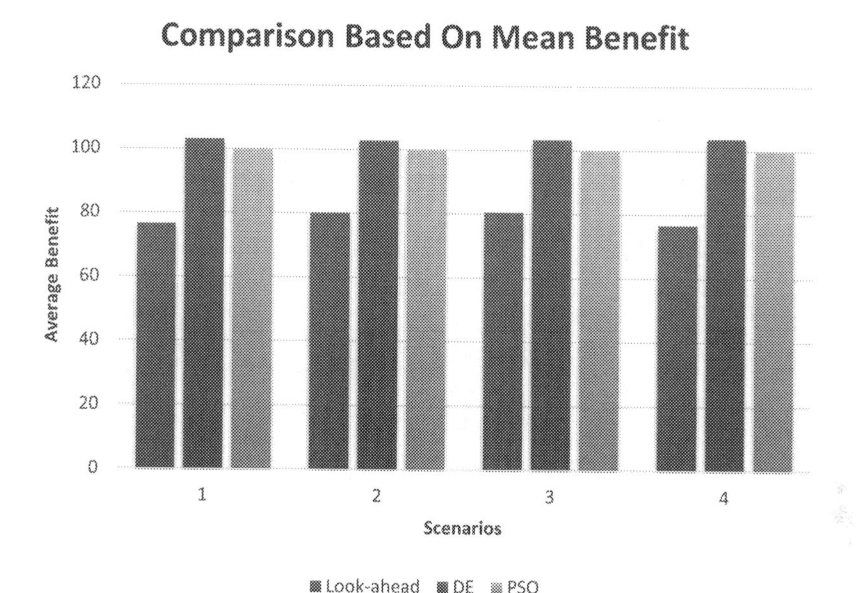

FIGURE 6.8 Mean benefit in different scenarios for look ahead (LA), differential evolution (DE), and particle swarm optimizer (PSO).

mentioned before, are the minimum with considerable average benefit. Swarm optimization's ability to consider the effect of best solutions from the previous generations makes it suitable to find the best solution locally, i.e., taking a step.

Below mentioned results from Figures 6.9 and 6.10 are used to depict the effect of changing the hyperparameters, namely, the weights associated with the local solution, global solution, and the inertia.

The standard hyperparameters setting while employing PSO for the experiments were $c1 = 0.5$, $c2 = 0.3$, and $w = 0.9$. Exploration with adverse parameters show that the least number of steps is recorded with $c1 = 0.3$, $c2 = 0.3$, and $w = 0.9$.

High weight associated with the inertia makes the next candidate solution that much closer to the destination. This is also because of the benefit for each intermediate solution based on the change in distance from the neighbors and the directional benefit effecting the best solution from every generation. Physically, the first term in velocity update equation of PSO (refer equation 6.12) represents the subject's tendency to maintain its current direction and speed, while the other two terms represent the influence of the subject's own experience and the experience of the swarm on its movement. The position update equation of PSO, as shown in equation (13), simply updates the subject's position by adding the new velocity to its current position. This operation moves the subject to a new position in the propagation space, which is closer to the intended destination.

Though it is visible that PSO doesn't maximize the diversity in path as compared to DE, the drastic reduction in the number of steps increases the mean benefit.

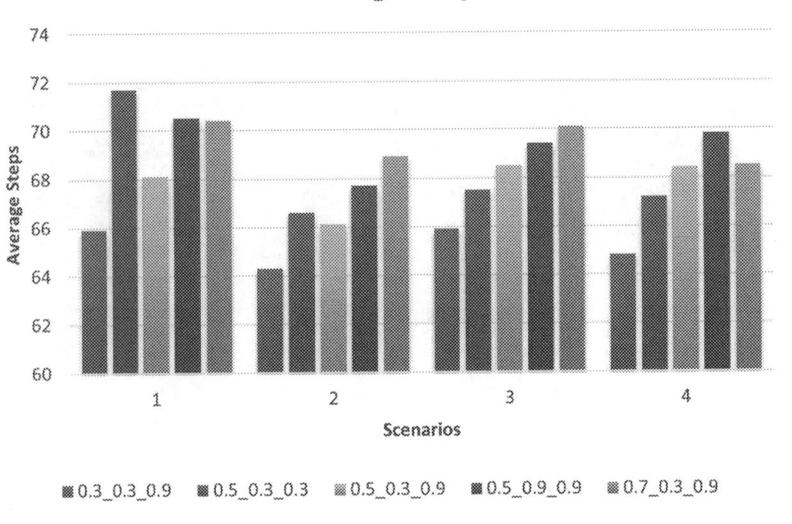

FIGURE 6.9 Average Steps taken while using particle swarm optimizer (PSO) with standard and adverse hyperparameters.

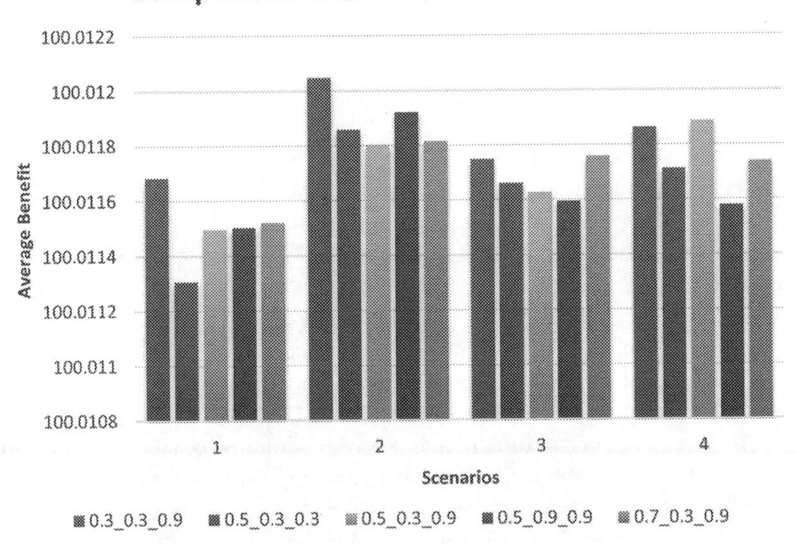

FIGURE 6.10 Mean benefits in different scenarios while using particle swarm optimizer (PSO) with standard and adverse hyperparameters.

Therefore, it can be derived that the weight associated with local best solution from the previous generation should be less so that the next best candidate solution produces the lowest function value. For the same set of adverse parameters, the highest mean benefit per step is observed, which reinforces the reduction of the weight associated with the local best solution.

6.9 CONCLUSION

In this work, we propose a self-navigation tool for the physically impaired and old people in a crowd, while following social distancing norms. A soft computing–enabled mobile IoT framework is developed for the implementation of this self-maneuvering approach. The impact of evolutionary algorithms is explored where we pose the self-aware social distancing paradigm as a variant of the knapsack problem. A LA method is proposed, which addresses the problem by changing the locality of the search space by varying the level of LA. To check the validity of mating-based evolutionary algorithms in such problems, DE is used to maximize the mean benefit, which reduces the chance of penalty. With extensive experimentation, it has been observed that DE ensures higher benefit as compared to the LA method but slower convergence. Leader-based PSO is used to find intermediate solution locally, i.e., the decision for each step. The number of generations was controlled using a minimum variance approach. It can be observed from the results that swarm optimizer speeds up the convergence while maintaining a decent mean benefit.

6.10 FUTURE PROSPECTS

Future works in this research includes the possibility of finding a feasibility equation, which uses distance from each neighbor as an item in a knapsack. Here, the distances form each neighbor can be a gene. The string of genes, i.e., the chromosome will contain all the distances from the nearest neighbors. The feasibility of an intermediate candidate will be validated if the distance from all the selected neighbors extends to one point in the search space. This can be posed as a constraint and will help us design the feasibility in the decision space and to modulate the length of the chromosome just by considering different number of neighbors.

Apart from that, to investigate the extent of DE in solving such problems, we can explore the effects of generating more than one token vector for crossover using the top solution. This may give the candidate solution more diversity while preserving the good alleles.

Furthermore, amalgamating PSO in each LA step also seems plausible. The contextual information from the LA comes with a lot of computational overhead because of the selection of the initial solution based on uniform distribution and the level of LAs. This poses a challenge for low-power and low-complexity IoT sensors with computation limitations, which adds implementation restrictions for the LA mechanism. The use of PSO can restrict the generation of candidate solution based on a distribution, or it can use the movement of the best solutions to produce candidate solutions.

Future work on this research also considers developing a prototype of IoT network for implementation of the concepts developed in this work to be deployed in real-world scenarios. Moreover, a detailed study of the network channel characteristics and its implications on the performance of the decision-making software is another possible extension of this work.

REFERENCES

[1] Xie, Yue, Aneta Neumann, and Frank Neumann. "Specific single-and multi-objective evolutionary algorithms for the chance-constrained knapsack problem." In: *Proceedings of the 2020 Genetic and Evolutionary Computation Conference*, ACM 2020. pp. 271–279.

[2] Salkin, Harvey M., and Cornelis A. De Kluyver. "The knapsack problem: A survey." *Naval Research Logistics Quarterly* 22.1 (1975): 127–144.

[3] Patvardhan, Chellapilla, Sulabh Bansal, and Anand Srivastav. "Solving the 0-1 quadratic knapsack problem with a competitive quantum inspired evolutionary algorithm." *Journal of Computational and Applied Mathematics* 285 (2015): 86–99.

[4] Chu, Paul C., and John E. Beasley. "A genetic algorithm for the multidimensional knapsack problem." *Journal of Heuristics* 4.1 (1998): 63–86.

[5] Sutton, Richard S., and Andrew G. Barto. *Reinforcement Learning: An Introduction*. MIT Press, Cambridge, MA, 2018.

[6] Wiering, Marco A., and Martijn Van Otterlo. "Reinforcement learning." *Adaptation, Learning, and Optimization* 12.3 (2012).

[7] Price, Kenneth V. "Differential Evolution." *Handbook of Optimization*. Springer, Berlin, Heidelberg, 2013., pp. 187–214.

[8] Fleetwood, Kelly. "An introduction to differential evolution." In: *Proceedings of Mathematics and Statistics of Complex Systems (MASCOS) One Day Symposium*, ACM, 26th November, Brisbane, Australia, 2004.

[9] Kennedy, James, and Russell Eberhart. "Particle swarm optimization." In: *Proceedings of ICNN'95-International Conference on Neural Networks* (Vol. 4). IEEE, New York, NY, 1995.

[10] Poli, Riccardo, James Kennedy, and Tim Blackwell. "Particle swarm optimization." *Swarm Intelligence* 1.1 (2007): 33–57.

[11] Carullo, Alessio, and Marco Parvis. "An ultrasonic sensor for distance measurement in automotive applications." *IEEE Sensors Journal* 1.2 (2001): 143.

[12] Kelemen, Michal, Ivan Virgala, Tatiana Kelemenová, Ľubica Miková, Peter Frankovský, Tomáš Lipták, and Milan Lörinc. "Distance measurement via using of ultrasonic sensor." *Journal of Automation and Control* 3.3 (2015): 71–74.

[13] Sahoo, Ajit Kumar, and Siba Kumar Udgata. "A novel ANN-based adaptive ultrasonic measurement system for accurate water level monitoring." *IEEE Transactions on Instrumentation and Measurement* 69.6 (2019): 3359–3369.

[14] Ali, Mian Mujtaba, Shyqyri Haxha, Munna M. Alam, Chike Nwibor, and Mohamed Sakel. "Design of internet of things (IoT) and android based low cost health monitoring embedded system wearable sensor for measuring spO_2, heart rate and body temperature simultaneously." *Wireless Personal Communications* 111 (2020) 2449–2463.

[15] Olivier, Pierre, AKM Fazla Mehrab, Stefan Lankes, Mohamed Lamine Karaoui, Robert Lyerly, and Binoy Ravindran. "Hexo: Offloading hpc compute-intensive workloads on low-cost, low-power embedded systems." In: *Proceedings of the 28th International Symposium on High-Performance Parallel and Distributed Computing*, ACM, pp. 85–96, 2019.

7 Combined 3D Mesh and Generative Adversarial Network–Based Improved Liver Segmentation in Computed Tomography Images

Mriganka Sarmah and Arambam Neelima
National Institute of Technology, Nagaland

7.1 INTRODUCTION

Three-dimensional (3D) reconstruction is a digital paradigm of visualizing real-world objects on digital devices such as screens and 3D holograms [1]. Considering real-world objects, internal body parts of human beings, even though constituents of nature, are not observable with bare eyes. Some of these organs, such as liver, heart, kidney, lungs, and brain, perform some of the most complex and extraordinarily important tasks for the survival of its owner. With the modern life style engulfing humans, never seen and heard of diseases have started to evolve and coexist in our environment in the form of viruses. At certain times, organ failures lead to death. Viruses have been responsible for organ failure and death in the past decade [2]. Once affected, the patient needs to be treated immediately. Doctors prescribe patients with various imaging tests to have a proper diagnosis [3]. For doctors, radiology images convey meaningful images, but for a patient these images are difficult to understand and visualize. The reason is the fuzzy nature of the generated images, tissues, and lesions. Most of the times, these images are in black and white or gray scale. In such scenarios, computers and artificial intelligence (AI) have stood and sorted these tasks for both practitioners and patients. For example, to identify inherited retinal diseases (IRD), optical tomography is used as a modality and Deep Learning (DL) techniques as the problem solver [4].

DL methods [5] are implemented to solve computationally intensive tasks, such as prediction, segmentation, classification, and regression [6]. The segmentation task is

DOI: 10.1201/9781003405368-7

a process to extract a particular region of interest (ROI) in an image and identify that as an individual identity and a separable piece of data [7]. Medical images consist of extra information, which otherwise useful does not make the computation task easier. For example, visible regions of the heart in computed tomography (CT) of the liver. This information is treated as noise and efforts are required in order to remove the visible heart from the liver image set before the DL program is trained. These pre-processing tasks are trivial and most often are done by an expert or under some expert guidance. However, semi-automatic and fully automatic methods are also gaining popularity [8]. One such fully automatic process central to the reason and motivation behind the proposed model is a cascaded neural network [9] named UFE-Net. A semi-automatic process requires manual intervention at some stage during the segmentation operation. Growing seeds into regions manually while using automatic categorization to create distinct boundaries between the liver and the tumor was the idea put out by Yang et al. [10]. Fully automatic processes are more welcome [11] but they work in set constraints only.

The most popular of all segmentation algorithms is the U-Net [12]. It is a semi-automatic network and its modified version U-Net+FP reduction+elimination (UFE-Net) [9] is an automated segmentation process. These two models are studied, and it is found that even though UFE-Net is a fully automatic process the computation cost is expensive compared to U-Net. Performance of the both the networks still remain competitive but UFE-net performs better with relative volume difference (RVD) and Hausdroff distance (HD) metric. The goal of this chapter is to reduce the computation cost of UFE-Net.

As commonly seen, test images are randomly chosen for finding the model accuracy in predicting the region boundary or segmentation mask, but unlike in real life the complete set of CT volume of a patient is segmented, of course one slice at a time. Therefore, a study on segmentation through the use of 3D mesh throws new insights and challenges where the entire volume of test masks are corrected unlike one at a time. This study poses real-time issues, and small minor false predictions (FPs) can be overcome if the nearby surface is error-free or correctly reproduced. The recent use of 3D model in segmentation was studied by Raju et al. [53]. Their model relied on graphical transformations and superposition of the mesh over the CT. The limitation was that, for each segmentation task, the volume contracted or expanded until the boundary coincided. Our proposed approach can achieve competitive results in reduced time when compared to UFE-Net [9] and Raju et al. [53].

7.2 RELATED WORK

7.2.1 Convolutional Neural Network

To understand the working of the convolutional neural network (CNN), let us consider the problem of object recognition. In Reference [13], it is found that there is a relationship between the human vision perception mechanism and the purpose of the convolution. Basically, humans identify objects based on certain features, such as shapes, size, corners, edges, surface textures, etc. Thus, once all these features can be learned by the computer in parallel, it would not only reduce the volume of data but also make AI work exactly how the human visual cortex functions. This relationship is also clearly described in Reference [14]. Hence, the goal of any AI algorithm for medical image segmentation is to simultaneously detect from the same image, its edges, sharpness changes, color variations, gradient change, blur changes, etc.

To understand further, let us gain a little insight into digital signal processing. If it is assumed that x_1 $\{t\}$ and x_2 $\{t\}$ are two continuous signals, then their convolution is shown by the operation in equation (7.1).

$$y(t) = x_1\{t\} * x_2\{t\}, \tag{7.1a}$$

$$= \int_{-\infty}^{\infty} x_1(n).x_2(t-n)dn \tag{7.1b}$$

For more on digital signal processing, refer Reference [15]. From equation (7.1b), it can be understood that for any time t, the convolution operator convolves the previous signal values with signal values that are most recent. This idea of convolving two signals if extended to image would be the same except for the time variable, and the independent factor will be the nth pixel where the previous pixels would be convolved with the most recent pixels around the nth pixel. If edges of x_1 are to be extracted, then the convolution operation would easily do it. Figure 7.1 shows three matrices. The middle matrix is a kernel to generate vertical edges only. The common choice of kernel size is 3×3, 5×5, 7×7, and so on. Lee et al. [16] proposed filter design and additive operation in row and column order to generate new images. As the convolution operand x_2 acts like a filter allowing only a certain feature to be identified, the bank of kernel filters in CNN is called feature bank or feature maps. A typical CNN architecture is shown in Figure 7.2 [17]. As the architecture is built to detect local features by convolutions, the output of one convolution layer produces an image whose dimensions differ from the input image. The convolution operation is repeated upon the output image and the net result is forwarded to another hidden layer and so on. The output image after few repeated convolutions is numerically excellent but is not perceivable by the human eye. All the different features that are computed are called depths of the layers. Each of the layers has $m \times n$ feature kernel.

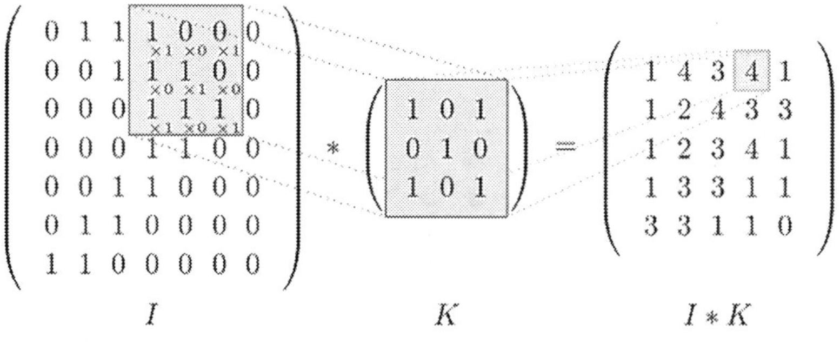

$$\underset{I}{\begin{pmatrix} 0 & 1 & 1 & 1 & 0 & 0 & 0 \\ 0 & 0 & 1 & 1 & 1 & 0 & 0 \\ 0 & 0 & 0 & 1 & 1 & 1 & 0 \\ 0 & 0 & 0 & 1 & 1 & 0 & 0 \\ 0 & 0 & 1 & 1 & 0 & 0 & 0 \\ 0 & 1 & 1 & 0 & 0 & 0 & 0 \\ 1 & 1 & 0 & 0 & 0 & 0 & 0 \end{pmatrix}} * \underset{K}{\begin{pmatrix} 1 & 0 & 1 \\ 0 & 1 & 0 \\ 1 & 0 & 1 \end{pmatrix}} = \underset{I * K}{\begin{pmatrix} 1 & 4 & 3 & 4 & 1 \\ 1 & 2 & 4 & 3 & 3 \\ 1 & 2 & 3 & 4 & 1 \\ 1 & 3 & 3 & 1 & 1 \\ 3 & 3 & 1 & 1 & 0 \end{pmatrix}}$$

FIGURE 7.1 Example of convolution operation. The colored squares in the right image are the result of convolution operation of a vertical edge filter.

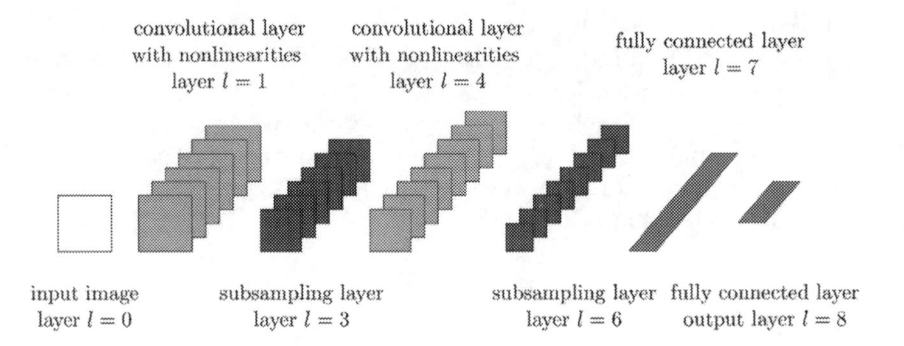

FIGURE 7.2 Convolutional neural network (CNN) architecture.

Thus, a total $m \times n \times d$ number of feature values for a depth d at layer l (shown as black color in Figure 7.2). The subsampling layer averages the pixel value by the neighboring four pixels around the pixel p further reducing the feature dimension to $m_1 \times n_1$. This continues and finally the sampled feature maps are connected to a fully connected network (FCN). The first fully connected (FC) layer (shown as orange color in Figure 7.2) is also called a flattened layer. The first FC layer is connected to another FC layer. The final layer corresponds to all the classes to which the input belongs. Only one out all the values in FC layer six will be allowed to be active and the remaining are made inactive, which will predict the particular class of the image.

7.2.2 U-Net

Olaf Ronneberger [12] presented the most famous of neural network architecture for medical image segmentation and named it as U-Net. The basic principle behind U-Net is the convolution operation and a set of filters, which increases after every layer of convolution that automatically identifies useful features. Figure 7.3 shows the U-Net architecture. The network from left to right is a continuous chain of convolution operation, Rectified Linear Unit (ReLU) operation, maxpool operation, un-pool operation, concatenate operation, FC ReLU operation, and softmax. The network operates top to bottom until the input image is reduced to the original size and then gradually expands back to a comparatively bigger size before being flattened and assigning the softmax excitation [18,19].

The network is originally designed for image dimensions 572×572. Convolution operation reduces the image to size 570×570. Because at level 0 there are two consecutive convolutions, the image further reduces to 568×568. The number of kernel or feature map used at this level is 64. There are $568 \times 568 \times 64$ neurons. Each neuron at the convolution stage is activated by the ReLU function. At level 1, the image is halved by the pooling operation. Out of the many pooling types [20], max pooling is done. Basically, one pixel's information is taken as the maximum of the neighboring four pixels. The pool window moves with a stride of two from left to right and also from top to bottom. For every four pixels, one information is generated. Thus, the number of pixels after pooling reduces to $\frac{width}{2}, \frac{height}{2}$. At this level, the image is

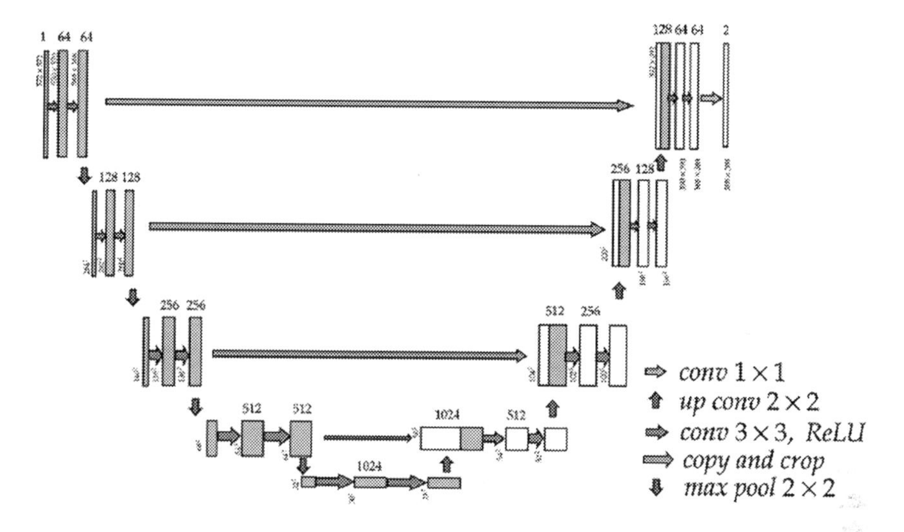

FIGURE 7.3 U-Net architecture as described in the original paper [12].

further convoluted into dimensions of 282×282 with the number of filters increased to 128. With two consecutive convolutions, the image is now 280×280×128 neural information. The convolution operation at every level is succeeded by a ReLU activation. This process continues until level 4 from where the expansion phase begins. At level 4, the image of size 28×28 is spread across 1,024 feature channels. This image is un-pooled or up sampled into larger dimension image of 56×56 [21, 22]. At level 3, half the number of feature maps are replaced by feature maps from the same level while down sampling and convoluted into 54×54×512 channel. At level 2, the image is up sampled into dimension 104×104 and 256 feature maps are concatenated from the previous feature map at level 2 while down sampling. After subsequent convolutions, the image is 100×100 dimension. At level 1, another set of 100×100 feature map from the same level is concatenated to give the image a new dimension of 200×200. Subsequent convolution operations reduce the image to a dimension of 196×196. At level 0, another set of 196×196 feature maps are concatenated to form a new image with size 392×392. Subsequent convolutions reduce the image to size 388×388. The feature channels reduce from 128 to 64 and finally to 2. The two output channels correspond to foreground and background of the generated mask, respectively.

7.2.3 More U-Net-Like Deep Networks

Wang et al. [23] developed a transformer network [24] to extract global features in addition to local features. This is useful for a semantic relationship between different segments of an image. Image transformer network is used in image completion task. Iqbal et al. [25] implemented a U-Net based multimodal segmentation scheme using an added attention gate before the skip connections. Oktay et al. [26] proposed attention-UNet for pancreas segmentation. The attention-gate works much like subdividing an image into grids. FCN is deployed to relate neighboring grids in order to focus on the pancreas automatically. The attention-gate helps to detect and remove smaller objects as false positives in the deeper layers of the network models. Sun et al. [27] developed a multi-attention based U-Net (MA-UNet), where the segmentation of similar objects from a single image is done in the attention layer using feature space mean distance, variance, and energy distribution as a loss metric. Han et al. [28] used modified convolution operation with larger kernel dimensions than normal convolutions. Also, the max pooling layer was deleted and instead an attention-like module was incorporated to filter noise and irrelevant features. Huang et al. [29] introduced a multiscale pyramidal structure attention module across different channels in a network layer. Group-based attention and channel-based attention are performed consecutively in the Group Channel Attention (GCA) module. Ullah et al. [30] performed forward and backward correspondence with the next and previous input slice before forwarding to the attention module in the Residual U-Net (Res-Unet)-like structure [31].

7.2.4 UFE-Net

Araujo et al. [32] proposed a cascaded U-Net architecture for liver segmentation. The method is of special interest to this chapter, as the core objective is to reduce the computation cost because of additional U-Nets within the architecture. The model deploys three U-Nets at various stages. First-stage U-Net extracts the whole lower region bounding box, including additional organs in its vicinity (if any). Using diagonal projection histogram [33], a diagonal across the image is projected, which hypothesizes the possible location of the liver to one side of the diagonal. This process is fully automatic and a threshold operation based on the histogram obtained can remove any unwanted connected organs with the liver. The second- and third-stage U-Nets are used to segment the liver and tumor region separately and fuse both the segments into one larger object by performing an OR operation.

7.2.5 Generative Adversarial Network

Generative adversarial network (GAN) [34] (Figure 7.4) is a class of two-stage deep neural network that tries to generate synthetic data as realistic as the trained input data. This task is done by the generator stage (Figure 7.5). The discriminator is a block within GAN that tries to identify the generated data as original or fake (Figure 7.6). A recent review on GAN [35] mentions some of the GAN structures out of which few are mentioned in Table 7.1[36–38]. A recent GAN method [39] to

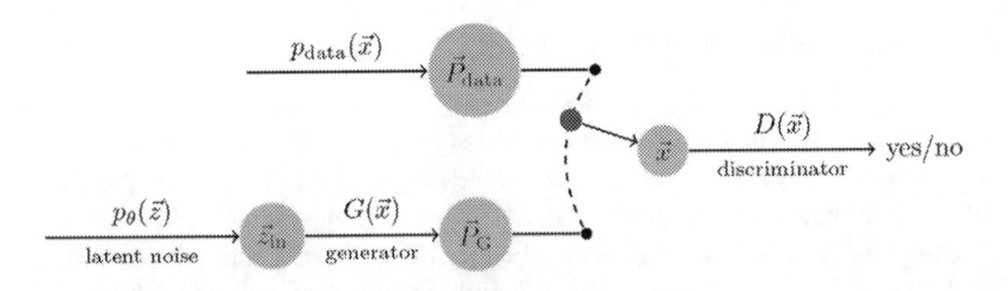

FIGURE 7.4 Generative adversarial network (GAN) architecture.

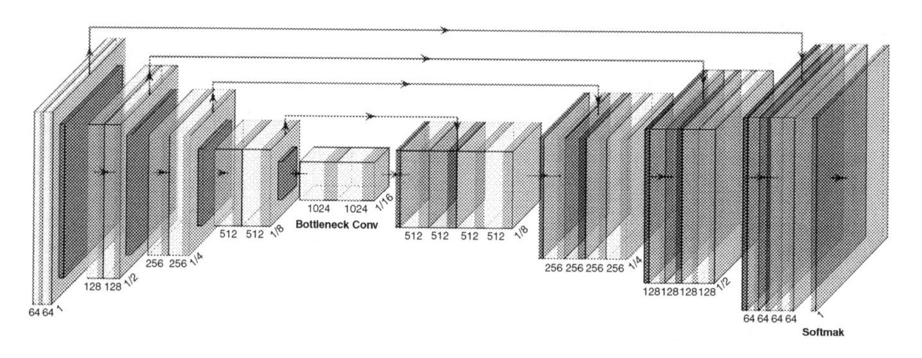

FIGURE 7.5 Generative adversarial network (GAN) generator network.

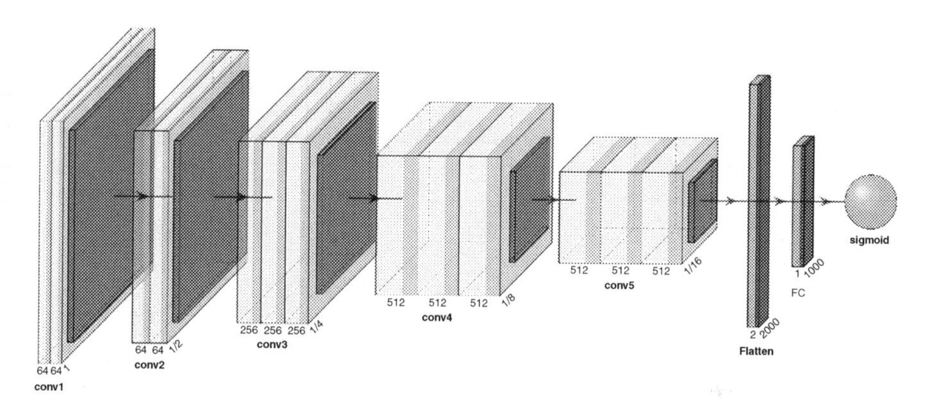

FIGURE 7.6 Generative adversarial network (GAN) discriminator network.

TABLE 7.1
Some Common GAN Variations

Type	Description
Vanilla GAN [34]	The most basic GAN where the generator is trained with corrupted input image and ground truth (GT) target image. The discriminator is trained to discriminate the generator output as fake (value=0). The discriminator is also trained with GT images as real (value=1). The generator adjusts its weights and generates real-like images to fool the discriminator.
Progressive GAN [36]	The GAN network starts with a low-resolution image of dimension 4 ×4 and gradually increases up to 1,024 ×1,024. The discriminator takes the 1,024 ×1,024 output as input and measures the discriminator loss. At successive stages, the lower resolution images are taken as input until the dimension reaches 4 ×4.
Conditional GAN [37]	Conditional GAN are trained adversarial networks with some auxiliary information. The auxiliary information is added to the training images in the form of class labels or latent code. The discriminator succeeds if it identifies the corrupted image as false.
Cycle GAN [38]	In cycle GAN, an input domain image is made indistinguishable from an output domain image and vice versa. Generally, in earlier GAN, the discrimination takes place in one direction (input to output). Hence, because of the cyclic nature, the name is cycle GAN.

segment liver using mask region-based convolution neural network (mask-RCNN) [40] was proposed. Mask-RCNN is an improvement over fast RCNN [41] and original RCNN [42]. Fast RCNN modifies VGG16 [43] pre-trained network to accept two input vectors: one being the image and another the image of ROI to segment. The max pool layer of VGG16 is dropped and replaced by the ROI pool. The output is a two-vector tensor of the object classification and also the bounding box. VGG16 classifies objects well but for images with more than one ROI, RCNN can auto-regress a rectangular box around the ROI along with classification. Mask RCNN colors different objects in an image with different colors such that not only a rectangular box is defined, but the bounding box can be of any polygonal shape. The pooling stage is also modified with the ROI align module. In this module, the larger ROI is divided into sub-ROIs of dimension 7×7. GAN trains a learning model to equalize ground truth data P_{Data} and generated data P_G as near as possible. The learning scheme is provided in equation (7.2), where Div is divergence, z is noise, and $G(z)$ is synthetic data.

$$G = \arg\min_G Div\left(P_G(x), P_{Data}(x)\right) \tag{7.2a}$$

$$V(G,D) = E_{x \to PData}\left[\log D(x)\right] + E_{x \to P(G)}\left[\log\left(1 - D\left(G(z)\right)\right)\right] \tag{7.2b}$$

$$D = \arg\max_D V(G,D) \tag{7.2c}$$

The GAN generator network produced synthetic images of liver CTs with a mask for the liver and trained the mask RCNN to segment the generated image. This method is particularly useful to train networks for any unseen damage or change in shape and size of the liver, which otherwise is not present in ground truth. Figure 7.7 shows some synthetic masks generated by vanilla GAN used in the proposed model.

7.2.5.1 GAN-Based Liver Segmentation

Demir et al. [44] proposed a vanilla GAN and a cycle GAN–based transformer neural network to segment the liver. Transformer networks are built on conventional U-Net with added attention layer. The GAN could identify a generated transformer mask as real or fake. Their results achieved a DICE score of 0.9433. He et al. [45] proposed a Deep Convolution GAN (DCGAN) combined with 3D U-Net and Deep Convolution Neural Network (DCNN). The 3D U-Net is used to identify real and fake segmentation masks, whereas DCNN is used to generate fake CT images. Their results achieved a DICE score of 0.942. Hong et al. [46] proposed labeling unlabeled MR images of the liver based on labeled CT images. Their training network comprised attention U-Net. They achieved a dice score of 0.912 ± 0.037. Liu et al. [47] proposed modeling of synthetic liver tumor images such that these can be trainable to improve lesion segmentation. For the discriminator, Wasserstein GAN [48] is used and a U-Net structure is used as the generator. A PCA-based mask generation method is used to describe various lesion shapes. The generated masks are then converted into liver lesions through a lesion synthetic network.

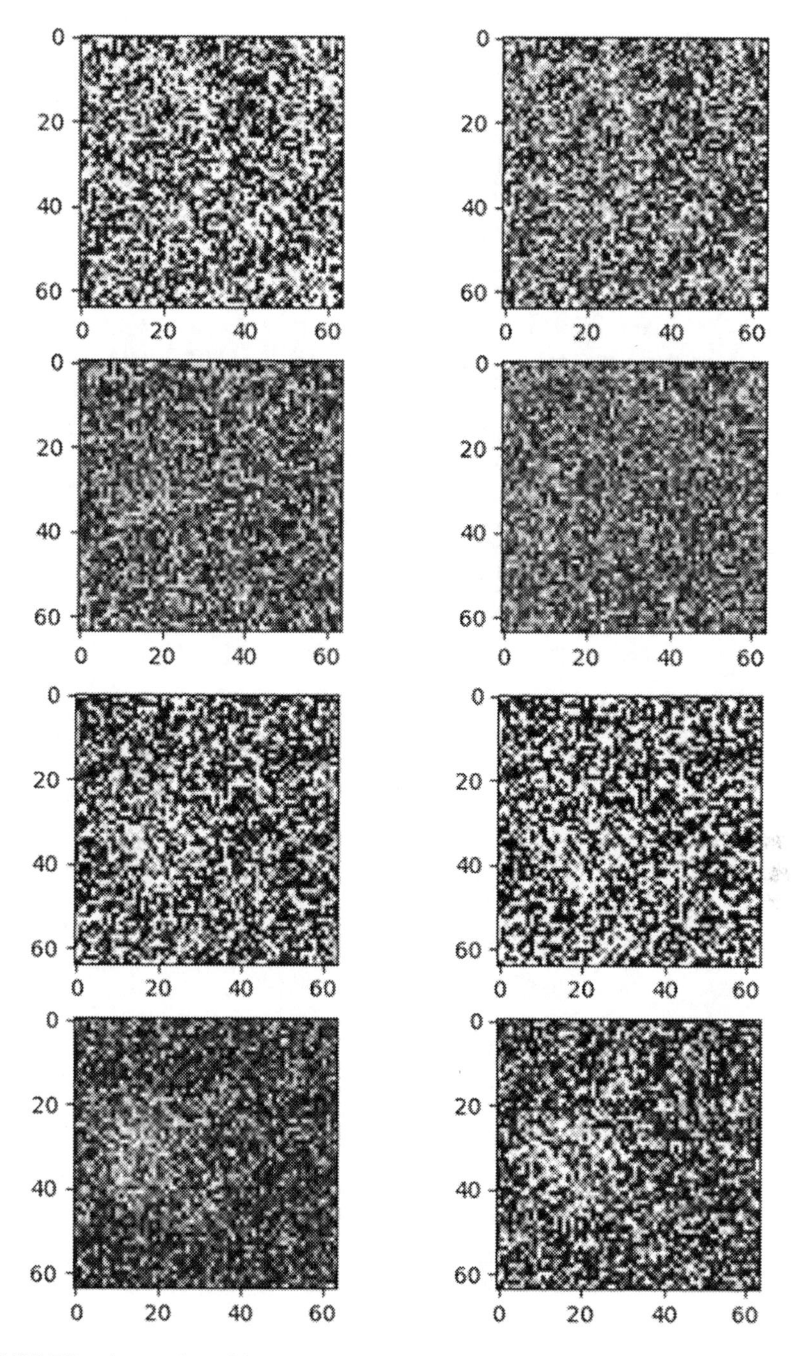

FIGURE 7.7 A snapshot of 2×2 batch of synthetic liver masks used to train the generative adversarial network (GAN) in the proposed model: (a) initial noise; (b) masks generated at iteration 500; (c) masks generated at iteration 2,000; (d) masks generated at iteration 2,500; (e) masks generated at iteration 3,000; (f) masks generated at iteration 3,800; and (g) masks generated at iteration 4,000.

(*Continued*)

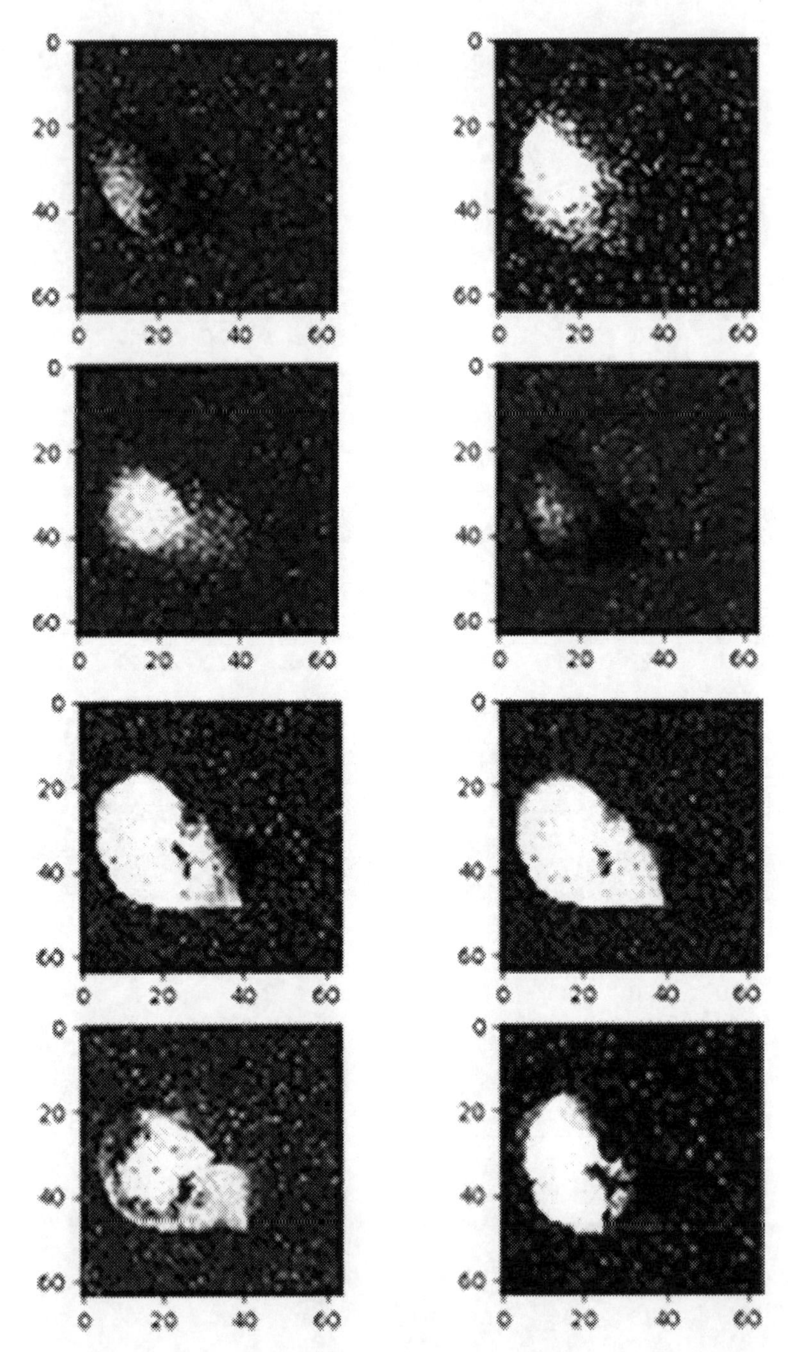

FIGURE 7.7 ***(Continued)*** A snapshot of 2×2 batch of synthetic liver masks used to train the generative adversarial network (GAN) in the proposed model: (a) initial noise; (b) masks generated at iteration 500; (c) masks generated at iteration 2,000; (d) masks generated at iteration 2,500; (e) masks generated at iteration 3,000; (f) masks generated at iteration 3,800; and (g) masks generated at iteration 4,000.

7.2.6 Digitally Reconstructed Radiograph

A digitally reconstructed radiograph (DRR) is a 2D reconstruction of medical images such that it appears real. A DRR is generated when the 3D surface of the desired ROI or organ is deemed to be projected on to a screen [49]. The 3D model is reconstructed from a mean statistical shape model (SSM) [50]. SSMs are made by aligned registration of multiple data points on a unit circle. In a recent study [51], Gaussian curvature was used to model object shapes and features. These features change the SSM shape descriptors to generate new models [52].

7.2.6.1 Model-Based Liver Segmentation

Raju et al. [53] proposed a SSM-based liver segmentation. Their main idea is super imposing a shape prior over an image with the means of translation, scaling, and rotation. They approach the problem much like the problem of pose estimation [54]. Their segmentation module introduces a Deep Implicit Statistical Shape Model (DISSM), and the dice score achieved on the cross-dataset is 0.959 ± 0.7.

7.3 PROPOSED MODEL

As a prerequisite, all the m training CT volumes contain n equal number of CTs and masks. Each volume, CT, and mask is identified by a $volumeidentifier \in \{volumeidentifier_0, volumeidentifier1, \ldots, volumeidentifier_m\}$, $CTidentifier \in \{CTidentifier_0, CTidentifier_1, \ldots, CTidentifier_n\}$, and $maskidentifier \in \{maskidentifier_0, maskidentifier_1, \ldots, maskidentifier_n\}$. The proposed model (Figure 7.8) is divided into three phases. The first phase (Stage I) is a simple

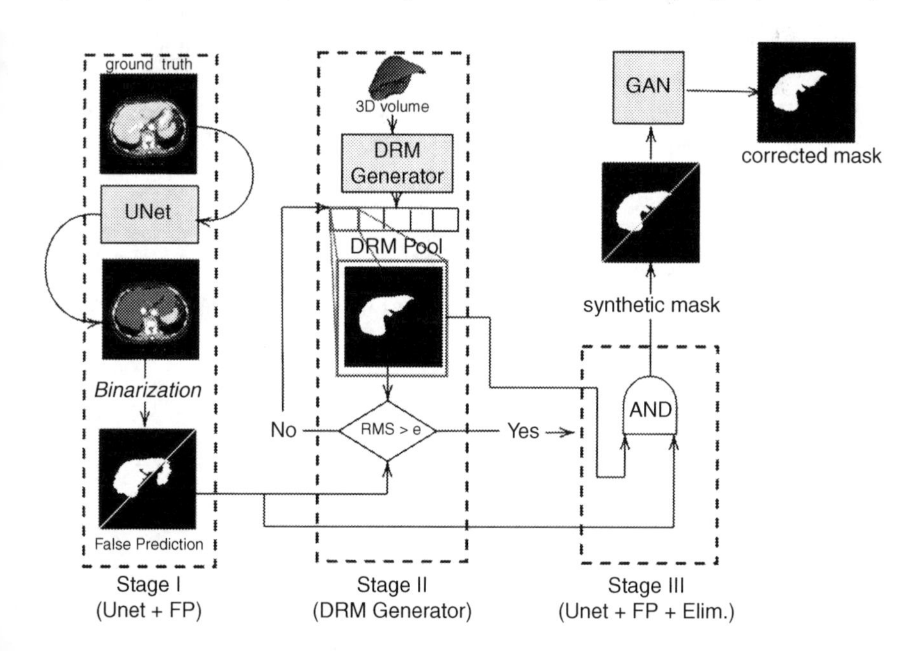

FIGURE 7.8 The proposed model.

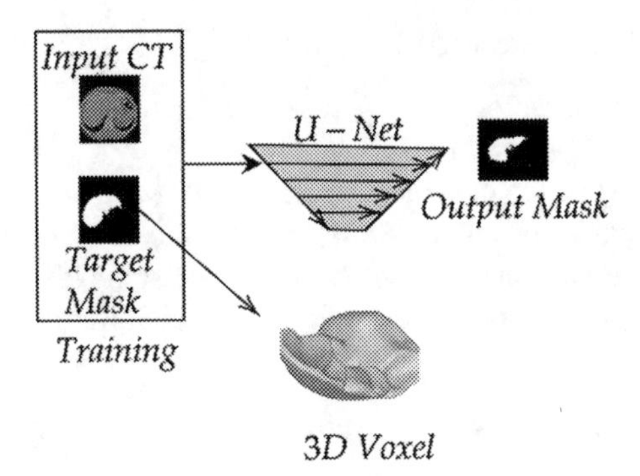

FIGURE 7.9 Voxelization of training masks into a 3D mesh ready to insert into digitally reconstructed mask (DRM) pool.

segmentation mask prediction with U-Net architecture. The U-Net is trained and validated with input liver CTs and target segmentation masks for $volumeidentifier_{TRAIN}$. The target segmentation masks of the CT volume are also used to create the 3D volume (Figure 7.9). The mask predicted at this stage is not final, as a few situations may contain unwanted noise around the liver (Figure 7.10). Such results are called false predictions (FPs).

Recently, Han et al. [55] addressed this issue with three different U-Nets trained on coronal, sagittal, and axial slices of the same CT volume. In our proposed model, the FP elimination is reserved for the next stage (Stage II). Prior to the start of Stage II, Digitally Reconstructed Mask (DRM) pool is prepared. The voxels are stored in a pool of DRM as slices of segmentation masks (Figure 7.11). The pool is informed about the $maskidentifier_{TEST}$ and its container $volumeidentifier_{TEST}$. The pool iterates over the *v* volumes within the DRM pool for the nearest root mean squared distance (RMSD) comparing each of the designated *maskidentifier*. This step is valid if $maskidentifier_{TEST} = maskidentifier_{POOL}$. If a case is matched, the pool returns a $volumeidentifier_{MATCHED}$. For each *n* slice within the selected $volumeidentifier_{MATCHED}$, the average Statistical Similarity Index Measure (SSIM) is found to measure the pool performance. Stage III performs noise elimination and FP reduction by performing logical AND operation of $volumeidentifier_{TEST}$ with $volumeidentifier_{MATCHED}$ on the entire *n* test slices [56–58].

Algorithm 1:Proposed Algorithm

```
        Begin Stage I
        DRMpool← [ ]
        volumeidentifier← m
        whilevolumeidentifier≥ 0 do
            GTCT ← [ ]
            CTslice←len(volumeidentifier)      while
            CTslice≥ 0 do
                CTidentifier← volumeidentifier[CTslice]
                GTCT.append(CTidentifier)
                CTslice− −
            end while GTmask← [ ]
            maskslice← len(volumeidentifier)
            whilemaskslice≥0do
                maskidentifier←volumeidentifier[maskslice]
                GTmask.append(maskidentifier) maskslice− −
            end while
            TrainUnet(GTCT,GTmask)
            3Dvolume←3Dmesh(GTmask)
            DRMpool.append(3Dvolume)
            volumeidentifier− −
        end while
        End Stage I
        Begin Stage II
        Require:volumeidentifier_TEST
        Require:CTidentifier_TEST
        Initiate: U − Net return maskidentifier_TEST

        volumeidentifier_DRMpool← v
        while volumeidentifier_DRMpool≥ 0 do
            if len(volumeidentifier_TEST)len(volumeidentifier_DRMpool) then
                RMSD_min← 1
                volumeidentifier_MATCHED← None
                maskidentifier_MATCHED← len(volumeidentifier_DRMpool)
                while maskidentifier_MATCHED≥ 0 do
            if maskidentifier_TEST= maskidentifier_MATCHEDthen
            RMSD_this←calcRMSD
            SSIM_this← calcSSIM
              if RMSD_this<RMSD_minthen
                RMSDmin← RMSDthis
                volumeidentifier_MATCHED← volumeidentifier_DRMpool
              end if
            end if
          maskidentifier_MATCHED− −
      end while
  end if
  volumeidentifier_DRMpool− −
end while
EndStageII
Begin Stage III
foreachmaskidentifier_MATCHEDin volumeidentifier_MATCHED
    maskidentifier_NEW← maskidentifier_MATCHEDAND maskidentifier_TEST
    GAN(maskidentifier_NEW)
endfor
End Stage III
```

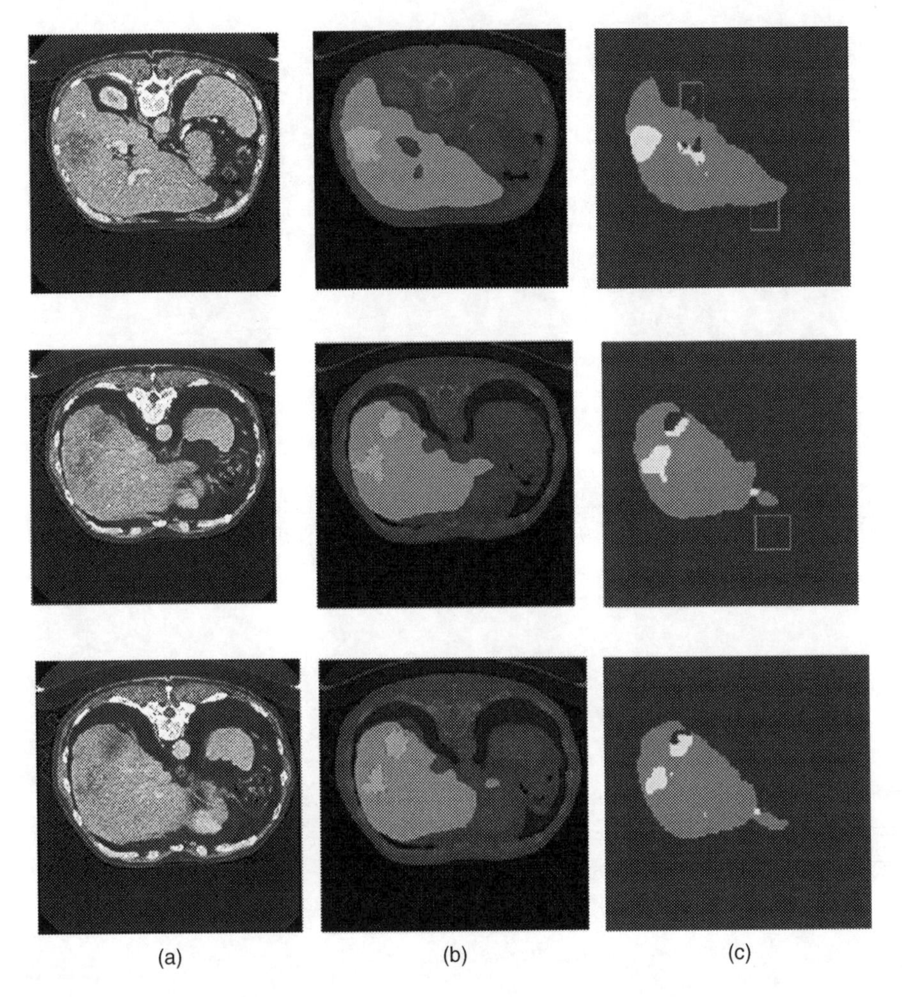

FIGURE 7.10 (a) Ground truth computed tomography (CT). (b) Ground truth annotation of liver mask. (c) Predicted mask by U-Net. Volume: 28 of LiTS dataset.

7.4 RESULTS AND DISCUSSION

Figure 7.12(a) shows the overall model segmentation accuracy. The overall accuracy of the segmentation obtained after 600 epochs is 95.34%. Between 200 and 400 epochs, the accuracy is dropped to 60% because of insufficient pool records. Figure 7.12(c) shows the GAN accuracy after 600 epochs at 78.88%. The model classifies fake liver masks as real. Consequently, the segmentation outperforms the GAN prediction. Figure 7.12(b) shows the GAN loss at training and validation. Initially, training loss is high, but it approaches the validation loss after 600 epochs. Validation and training loss remains low and shows a decreasing trend after 400 epochs. Figure 12(d) shows the generator and the discriminator losses, and both the modules compete with inverse trend such that an increase in generator loss is reciprocated with a decrease in discriminator loss. This is evident from Figure 12(d) at epoch 200, 300, 400, and 500. At the 200th epoch, the generator loss is 65.45% and discriminator loss

FIGURE 7.11 A complete digitally reconstructed radiograph (DRR) liver mask volume within digitally reconstructed mask (DRM) pool.

is 18%. At 300 epochs, the generator loses significantly with increased loss of 100% and discriminator wins with 5% loss. At 400 epochs, the generator improves with 63.99% loss and discriminator loss increases to 20%. From the 500th epoch onward, discriminator loss remains constant without much noticeable changes. However, generator loss increases till 600th epoch and then lowers to 65.43%.

7.4.1 Metrics and Measures

The metrics used to measure the similarity are RMSD, SSIM, and DICE.

7.4.1.1 RMSD

RMSD measures the Euclidean distance of the pooled surface (x_j) with the U-Net predicted mask (x_i). N is the total number of points within the volume.

$$RMSD = \sqrt{\frac{\sum_{i=1}^{N}\left(x_i - \widehat{x}_i\right)^2}{N}} \tag{7.3}$$

7.4.1.2 SSIM

SSIM is a windowed operation on a grid of $N \times N$ within two comparing images: μ is mean, σ^2 is variance, and $c_1$and c_2 are stabilizing constants.

$$SSIM(x, y) = \frac{\left(2\mu_x\mu_y + c_1\right)\left(2\sigma_{xy} + c_2\right)}{\left(\mu_x^2 + \mu_y^2 + c_1\right)\left(\sigma_x^2 + \sigma_y^2 + c_2\right)} \tag{7.4}$$

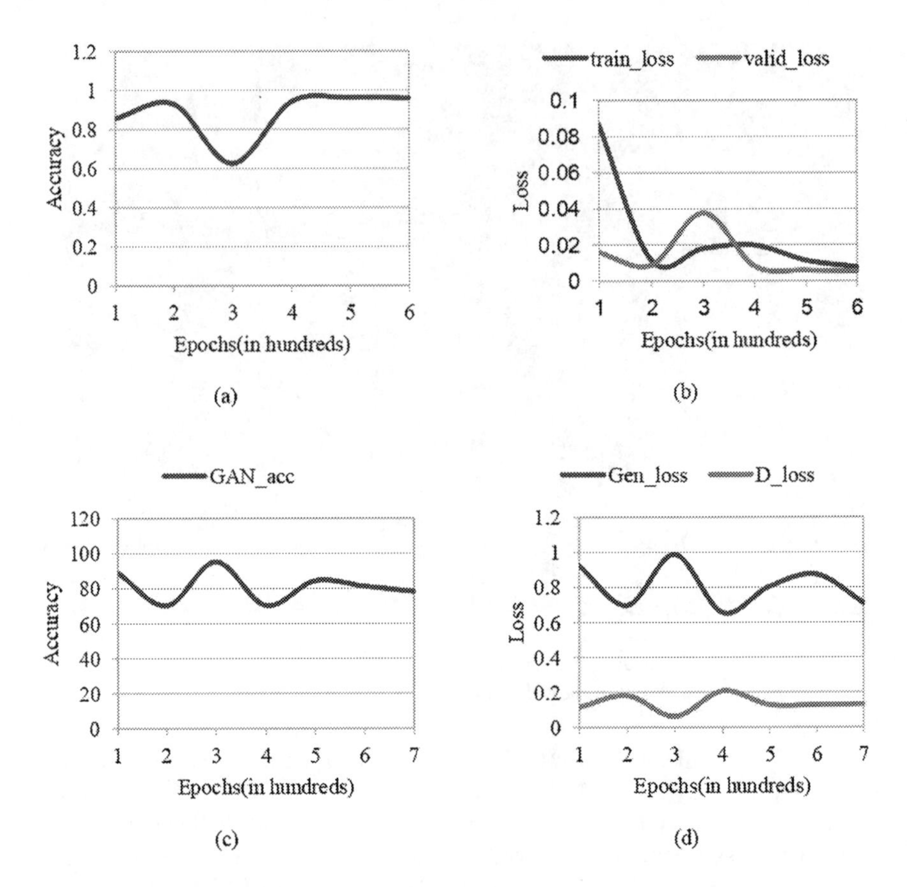

FIGURE 7.12 (a) Accuracy of the proposed model,(b) comparison of generator loss and validation loss, (c) generative adversarial network (GAN) accuracy, and (d) comparison of generator loss and discriminator loss.

7.4.1.3 DICE

Named after its founder Lee Raymond Dice, DICE score measures the number of pixels contained within the common boundary.

$$DICE = \frac{2|X \cap Y|}{|X| + |Y|} \tag{7.5}$$

7.4.2 Performance Comparison with Existing Methods

Table 7.2 shows the proposed model when tested on the LiTS dataset. The method obtained a DICE score of 0.963. This is better than some state-of-the-art segmentation methods. The proposed model comprises a set of volumetric priors to guide the segmentation process. The advantage is reduced time in segmentation.

TABLE 7.2
Performance Comparison of the Proposed Model with Existing Methods

Method	Dataset	Dice	DL Based	GAN Based	3D Based
Nanda et al. [56]	LiTS	0.955	Yes	No	No
Li et al. [57]	LiTS	0.958	Yes	No	No
Wang et al. [58]	LiTS	0.957	Yes	No	No
Araujo et al. [32]	LiTS	0.956	Yes	No	No
Demir et al. [44]	LiTS	0.9433	Yes	Yes	No
He et al. [45]	LiTS	0.942	Yes	Yes	No
Raju et al. [53]	MSD	0.959	Yes	No	Yes
Proposed	LiTS	0.963	Yes	Yes	Yes

TABLE 7.3
Performance Comparison of the Proposed Method Based on Time of Execution

Method	Dataset	Segmentation Time	Total Masks	Training Time	Image Size	Time for Segmenting One Slide	Complete Volume Segmentation Time
Araujo et. al [32]	Private	22 h	4,000	580 min	40 ×40	27 s	108 min
Raju et al. [53]	Private	22 h	4,000	260 min	40 ×40	6 s	40 min
Proposed	Private	22 h	4,000	140 min	40 ×40	4.95 s	33 min

Table 7.3 shows that for generating a hand-crafted training mask on a mini-LiTS dataset, 22 hours are required. With a reduced image size of 40 × 40, the proposed model performed the segmentation of an entire volume in 33 minutes. This is significantly less than the study by Raju et al. [53].

7.5 CONCLUSION AND FUTURE WORK

In the work described here, a proposed model with a fusion of DL, adversarial and volumetric data is described in order to perform segmentation of a full CT volume simultaneously. Such an approach is not available in the existing literature. The designed model builds a set of artificially generated segmentation pool. The model outperforms the existing state-of-the-art in whole volume segmentation task.

In future, whole volume segmentation can be studied without incorporating adversarial networks and measure model performance. The idea of whole volume segmentation using DL-based approach can also be studied.

REFERENCES

[1] Ramachandiran CR, Chong MM, Subramanian P. 3D hologram in futuristic classroom: A review. *Periodicals of Engineering and Natural Sciences (PEN)*. 2019;7(2):580–586.

[2] Cai D, Ma X, Zhou Y, et al. Multiple organ failure and death caused by staphylococcusaureus hip infection: A case report. *Open Life Sciences*. 2022;17(1):1129–1134.

[3] Kollias D, Arsenos A, Kollias S. Ai-mia: COVID-19 detection and severity analysis through medical imaging. In *European Conference on Computer Vision*. Springer Nature Switzerland, Cham, pp. 677–690.

[4] Esengönül M, Marta A, Beirão J, et al. A systematic review of artificial intelligence applications used for inherited retinal disease management. *Medicina*. 2022;58(4):504.

[5] Ulku I, Akagündüz E A survey on deep learning-based architectures for semantic segmentation on 2d images. *Applied Artificial Intelligence*. 2022; 36(1):45.

[6] Wang Q, Ma Y, Zhao K, et al. A comprehensive survey of loss functions in machine learning. *Annals of Data Science*. 2022;9(2):187–212.

[7] Liu Z, Tong L, Chen L, et al. Deep learning based brain tumor segmentation: A survey. *Complex & Intelligent Systems*. 2022; 9(1):10011026.

[8] Issa J, Olszewski R, Dyszkiewicz-Konwińska M. The effectiveness of semi-automated and fully automatic segmentation for inferior alveolar canal localization on cbct scans: A systematic review. *International Journal of Environmental Research and Public Health*. 2022;19(1):560.

[9] Araújo JDL, da Cruz LB, Diniz JOB, et al. Liver segmentation from computed tomography images using cascade deep learning. *Computers in Biology and Medicine*. 2022;140:105095.

[10] Yang Z, Zhao Yq, Liao M, et al. Semi-automatic liver tumor segmentation with adaptive region growing and graph cuts. *Biomedical Signal Processing and Control*. 2021;68:102670.

[11] Wang J, Zhang X, Lv P, et al. Automatic liver segmentation using efficientnet and attention-based residual u-net in ct. *Journal of Digital Imaging*. 2022; 35(6):1479–1493.

[12] Ronneberger O, Fischer P, Brox T. U-net: Convolutional networks for biomedical image segmentation. In: *Medical Image Computing and Computer-Assisted Intervention-MICCAI 2015: 18th International Conference*, Munich, Germany, October 5–9, 2015, *Proceedings, Part III-18*. Springer, 2015, pp. 234–241.

[13] Denker J, Gardner W, Graf H, et al. Neural network recognizer for hand-written zip code digits. *Advances in Neural Information Processing Systems*. 1988;1: 323–331.

[14] Gu J, Wang Z, Kuen J, et al. Recent advances in convolutional neural networks. *Pattern Recognition*. 2018;77:354–377.

[15] Mneney S. *An Introduction to Digital Signal Processing*. CRC Press, 2022.

[16] Lee SU. *Design of SVD/SGK Convolution Filters for Image Processing*. University of Southern California Los Angeles Image Processing Inst, 1980.

[17] LeCun Y, Boser B, Denker JS, et al. Backpropagation applied to handwritten zip code recognition. *Neural Computation*. 1989;1(4):541–551.

[18] Yegnanarayana B. *Artificial Neural Networks*. PHI Learning Pvt. Ltd, 2009.

[19] Sibi P, Jones SA, Siddarth P. Analysis of different activation functions using back propagation neural networks. *Journal of Theoretical and Applied Information Technology*. 2013;47(3):1264–1268.

[20] Tao Z, XiaoYu C, HuiLing L, et al. Pooling operations in deep learning: From "invariable" to "variable". *BioMed Research International*. 2022; 2022, 4067581, 17 pages. https://doi.org/10.1155/2022/4067581.

[21] Fattal, R. (2007). Image upsampling via imposed edge statistics. *ACM Transactions on Graphics*, 26(3), 95. doi:10.1145/1276377.1276496.

[22] Esmaeilzehi A, Ahmad MO, Swamy M. Updresnn: A deep light-weight image upsampling and deblurring residual neural network. *IEEE Transactions on Broadcasting*. 2021;67(2):538–548.
[23] Wang L, Li R, Zhang C, et al. Unetformer: A unet-like transformer for efficient semantic segmentation of remote sensing urban scene imagery. *ISPRS Journal of Photogrammetry and Remote Sensing*. 2022;190:196–214.
[24] Vaswani A, Shazeer N, Parmar N, et al. Attention is all you need. *Advances in Neural Information Processing Systems*. 2017;30.
[25] Iqbal A, Sharif M, Khan MA, et al. FF-UNet: A u-shaped deep convolutional neural network for multimodal biomedical image segmentation. *Cognitive Computation*. 2022;14(4):1287–1302.
[26] Oktay O, Schlemper J, Folgoc LL, et al. Attention u-net: Learning where to look for the pancreas. 2018. *arXiv preprint arXiv:1804.03999*
[27] Sun Y, Bi F, Gao Y, et al. A multi-attention UNet for semantic segmentation in remote sensing images. *Symmetry (Basel)*. 2022;14(5):906.
[28] Han Z, Jian M, Wang GG. ConvUNeXt: An efficient convolution neural network for medical image segmentation. *Knowledge-based Systems*. 2022;253(109512):109512.
[29] Huang Z, Zhao Y, Liu Y, et al. GCAUNet: A group cross-channel attention residual UNet for slice based brain tumor segmentation. *Biomedical Signal Processing and Control*. 2021;70(102958):102958.
[30] Ullah Z, Usman M, Jeon M, et al. Cascade multiscale residual attention CNNs with adaptive ROI for automatic brain tumor segmentation. *InfSci (NY)*. 2022;608:1541–1556.
[31] Liu N, He T, Tian Y, et al. Common-azimuth seismic data fault analysis using residual unet. *Interpretation*. 2020;8(3):SM25–SM37.
[32] Araújo JDL, da Cruz LB, Diniz JOB, et al. Liver segmentation from computed tomography images using cascade deep learning. *Computers in Biology and Medicine*. 2021;140(105095):105095.
[33] Hazra A, Choudhary P, Inunganbi S, et al. Bangla-Meitei mayek scripts handwritten character recognition using convolutional neural network. *Applied Intelligence*. 2021;51(4):2291–2311.
[34] Goodfellow I, Pouget-Abadie J, Mirza M, et al. Generative adversarial networks. *Communications of the ACM*. 2020;63(11):139–144.
[35] Iqbal A, Sharif M, Yasmin M, et al. Generative adversarial networks and its applications in the biomedical image segmentation: A comprehensive survey. *International Journal of Multimedia Information Retrieval*. 2022;11(3):333–368.
[36] Karras T, Aila T, Laine S, et al. Progressive growing of gans for improved quality, stability, and variation; 2017. https://arxiv.org/abs/1710.10196.
[37] Mirza M, Osindero S. Conditional generative adversarial nets; 2014. https://arxiv.org/abs/1411.1784.
[38] Zhu JY, Park T, Isola P, et al. Unpaired image-to-image translation using cycle-consistent adversarial networks. In: *Proceedings of the IEEE International Conference on Computer Vision*. 2017; pp. 2223–2232.
[39] Wei X, Chen X, Lai C, et al. Automatic liver segmentation in CT images with enhanced GAN and mask region-based CNN architectures. *BioMed Research International*. 2021:9956983.
[40] He K, Gkioxari G, Dollar P, et al. Mask r-cnn. In: *Proceedings of the IEEE International Conference on Computer Vision (ICCV)*. 2017; pp. 2961–2969.
[41] Ren S, He K, Girshick R, et al. Faster R-CNN: Towards real-time object detection with region proposal networks. *IEEE Transactions on Pattern Analysis and Machine Intelligence*. 2017;39(6):1137–1149.

[42] Girshick R, Donahue J, Darrell T, et al. Rich feature hierarchies for accurate object detection and semantic segmentation. In: *2014 IEEE Conference on Computer Vision and Pattern Recognition*. IEEE, New York, NY, 2014.
[43] Simonyan K, Zisserman A. Very deep convolutional networks for large-scale image recognition. 2014. arXiv preprint arXiv:1409.1556.
[44] Demir U, Zhang Z, Wang B, et al. Transformer based generative adversarial network for liver segmentation. In: *Image Analysis and Processing. ICIAP 2022 Workshops: ICIAP International Workshops*, Lecce, Italy, May 23–27, 2022, *Revised Selected Papers, Part II*. Springer, New York, 2022, pp. 340–347.
[45] He R, Xu S, Liu Y, et al. Three-dimensional liver image segmentation using generative adversarial networks based on feature restoration. *Frontiers in Medicine*. 2022;8:794969.
[46] Hong J, Yu SCH, Chen W. Unsupervised domain adaptation for cross-modality liver segmentation via joint adversarial learning and self-learning. *Applied Soft Computing*. 2022;121:108729.
[47] Liu Y, Yang F, Yang Y. A partial convolution generative adversarial network for lesion synthesis and enhanced liver tumor segmentation. *Journal of Applied Clinical Medical Physics*. 2023;e13927.
[48] Gulrajani I, Ahmed F, Arjovsky M, et al. Improved training of wassersteingans. In: *Advances in Neural Information Processing Systems*, 2017;30:11p.
[49] Gajny L, Girinon F, Bayoud W, et al. Fast quasi-automated 3D reconstruction of lower limbs from low dose biplanar radiographs using statistical shape models and contour matching. *Medical Engineering and Physics*. 2022;101(103769):103769.
[50] Heimann T, Meinzer HP. Statistical shape models for 3D medical image segmentation: A review. *Medical Image Analysis*. 2009;13(4):543–563.
[51] Liu H, Tai XC, Glowinski R. An operator-splitting method for the gaussian curvature regularization model with applications to surface smoothing and imaging. *SIAM Journal on Scientific Computing*. 2022;44(2):A935–A963.
[52] Wiputra H, Matsumoto S, Wagenseil JE, et al. Statistical shape representation of the thoracic aorta: Accounting for major branches of the aortic arch. *Computer Methods in Biomechanics and Biomedical Engineering*. 2022;1–15. doi: 10.1080/10255842.2022.2128672
[53] Raju, A., Miao, S., Jin, D., Lu, L., Huang, J., & Harrison, A. P. (2022). Deep Implicit Statistical Shape Models for 3D Medical Image Delineation. In: *Proceedings of the AAAI Conference on Artificial Intelligence*, Katia Sycara,Vasant Honavar, Matthijs spaan, 36(2), 2135-2143. https://doi.org/10.1609/aaai.v36i2.20110.
[54] Andriluka M, Pishchulin L, Gehler P, et al. 2d human pose estimation: New benchmark and state of the art analysis. In: *Proceedings of the IEEE Conference on computer Vision and Pattern Recognition*. 2014, pp. 3686–3693 doi: 10.1109/CVPR.2014.471.
[55] Han L, Chen Y, Li J, et al. Liver segmentation with 2.5 d perpendicular unets. *Computers & Electrical Engineering*. 2021;91:107118.
[56] Nanda N, Kakkar P, Nagpal S. Computer-aided segmentation of liver lesions in ct scans using cascaded convolutional neural networks and genetically optimised classifier. *Arabian Journal for Science and Engineering*. 2019;44(4):4049–4062.
[57] Li J, Ou X, Shen N, et al. Study on strategy of ct image sequence segmentation for liver and tumor based on u-net and bi-convlstm. *Expert Systems with Applications*. 2021;180:115008.
[58] Wang J, Lv P, Wang H, et al. Sar-u-net: Squeeze-and-excitation block and atrous spatial pyramid pooling based residual u-net for automatic liver segmentation in computed tomography. *Computer Methods and Programs in Biomedicine*. 2021;208:106268.

8 Applying Privacy by Design to Connected Healthcare Ecosystems

Naomi Tia Chantelle Freeman
Noroff University College

8.1 INTRODUCTION

With the proliferation of connected healthcare devices, patients' personal health information (PHI) is perceived as being increasingly at risk of being compromised (Kalloniatis et al. 2011). These devices, which include wearables, health apps, and other Internet of things (IoT) devices, generate a vast amount of data that can be used to monitor patients' health, track their behavior, and even diagnose medical conditions. Although these devices offer significant benefits to patients, such as improved diagnosis and treatment, they also present significant privacy risks for sensitive data when the handover between these connected healthcare system nodes is not handled with best practice, particularly when soft computing techniques are added into the system.

Operationalizing privacy in connected healthcare is an essential step toward mitigating these risks (Cavoukian 2012). Privacy is a fundamental right that guarantees individuals the right to control how their personal data are collected, used, and disclosed. Although this may have been a high ideal or grandiose statement in the past, the introduction of ISO 31700, which has a key focus on embedding and operationalizing privacy in consumer applications (International Organization for Standardization 2023), moves the ideal of privacy as a right into the arena of policy on an international level. In healthcare, privacy is critical to maintaining the trust between patients and healthcare providers (Caulfield and Donnelly 2013). Patients expect that their own personal data and information will be kept both confidential and secure, and healthcare providers have a legal and ethical obligation to protect this information. The details may vary between jurisdictions, but 71% of countries worldwide already have data-protection laws of some kind in place according to the UN (2001), and a further 9% of countries had data-protection laws in draft as of 2001.

In this chapter, I will explore the concept of operationalizing privacy in connected healthcare. Specifically, I will examine the different strategies that can be used to protect patients' privacy in the context of connected healthcare, using a case model of a pediatric healthcare institution with multiple organizational contexts and data demands within it. I will also explore the challenges that must be overcome to implement these strategies effectively, and the limitations of a generalized approach.

DOI: 10.1201/9781003405368-8

This chapter is structured as follows: To begin, I will offer a brief overview of connected healthcare and the privacy risks associated with it. Following this overview, I will explore the concept of operationalizing privacy in healthcare and discuss the different strategies that can be used to achieve this. Next I will examine the challenges associated with implementing these strategies. Finally, I will summarize the findings and provide recommendations for healthcare providers and healthcare setting IT policymakers to protect patients' privacy in connected healthcare.

8.2 BACKGROUND/LITERATURE REVIEW

8.2.1 Privacy as a Barrier to Data Sharing and Innovation

Everyday talks about implementing privacy and security measures to center the user (or, in this case, the patient) often lead to someone remarking that all of the policy measures are slowing down development, innovation, and the ability to improve current services. In contrast to this, an online industry article (Pohl 2018) says

> Data security and privacy are one of the main barriers to data sharing in healthcare. However, I consider that the biggest reason for the lack of interoperability and open APIs is related to business interests of different stakeholders who are trying to hold the data as much as they can since data is money and power. For example, interoperability has been solved in banking years ago, where data are equally sensitive.

Pohl introduces a key idea that the business interests and goals have to be aligned with patient or data outcomes in order for programs to transform digital records into connected health systems.

8.3 INTEROPERABILITY: CLOUD VS. CONNECTED HEALTHCARE SYSTEMS

Interoperability refers to the ability of different systems or devices to exchange information and use that information effectively. In the context of cloud systems, interoperability refers to the ability of different cloud-based systems or applications to communicate and share data seamlessly. In connected healthcare systems, interoperability refers to the ability of different healthcare systems and devices to share patients' health data efficiently and securely. An important distinction that this chapter offers, in contrast to the Eze et al.'s (2018) paper, is that connected healthcare is not cloud systems. Connected healthcare offers a much wider and broader set of connections and nodes beyond the specifics of cloud computing infrastructure.

8.4 CLOUD SYSTEMS INTEROPERABILITY

Cloud systems are typically made up of multiple applications and services that are hosted in the cloud and accessed through the Internet. Interoperability in cloud systems is critical to ensure that these applications can communicate and work together seamlessly.

This interoperability is achieved through the use of standard protocols and interfaces that enable different applications to exchange data and interact with each other.

In cloud systems, interoperability also extends to the ability to move data and applications between different cloud service providers. This is important because organizations may need to move their data and applications to different cloud service providers for a variety of reasons, including cost, reliability, and compliance with regulatory requirements. Eze et al. (2018) offered an excellent case to support this idea of cloud systems, highlighting the connection of external partners and stakeholders to the internal medical data through a series of consents.

8.5 CONNECTED HEALTHCARE SYSTEMS INTEROPERABILITY

In contrast, interoperability in connected healthcare systems is focused on data sharing. Interoperability is essential to ensure that patients' health data can be shared securely and efficiently between different healthcare providers and systems, as well as a variety of devices and access points (Caulfield and Donnelly 2013). Sharing these data is critical to providing effective healthcare services, such as coordinating care between different providers and ensuring that patients' health data are up-to-date and accurate.

Interoperability in connected healthcare systems is often challenging because of the complexity of healthcare data and the large number of different healthcare systems and devices that need to exchange information. This moves beyond institutions and into areas such as connected ambulance services that can now share patients' vitals in real time (Alkhatib and Buchanan 2023). This complexity is further compounded by the need to ensure that patient data are shared securely and in compliance with national privacy regulations such as HIPAA in the United States, and international regulations and recommendations such as GDPR and ISO standards.

To achieve interoperability in connected healthcare systems, standard protocols and interfaces have been developed, such as the Health Level Seven (HL7) and Fast Healthcare Interoperability Resources (FHIR) standards (FHIR n.d.). These standards provide a common language for different healthcare systems and devices to exchange data and communicate with each other. These are not the core focus of this chapter. However, they have introduced key concepts in achieving good standardization in this area.

8.6 APPLYING PRIVACY BY DESIGN: A TRIANGLE OF LEGAL, TECHNICAL, AND ORGANIZATIONAL APPLICATION

The current best practice in applying privacy to connected healthcare settings involves a comprehensive approach that includes technical, organizational, and legal measures. These measures aim to protect patients' PHI from unauthorized access, use, and disclosure. In this section, I will describe some of the best practices that have emerged in recent years. Many papers read in the literature review were found to have a focus on pre-pandemic activities, but the foundations and principles in the papers are important to learn from and iterate on.

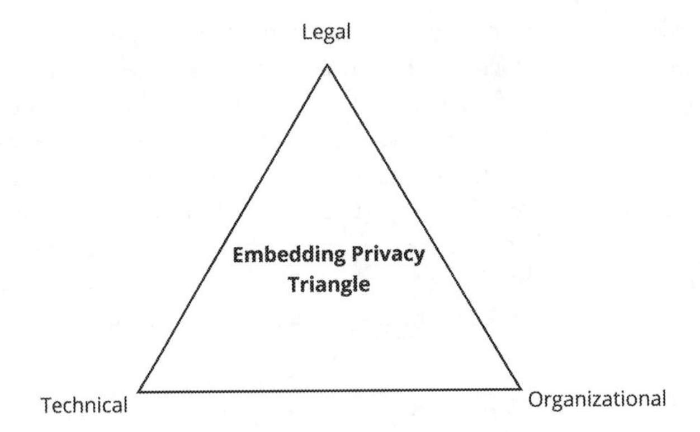

FIGURE 8.1 Embedded privacy is a balance of organizational, legal, and technical implementation.

8.7 TECHNICAL APPROACHES AND METHODOLOGY: BEYOND TELEHEALTH

Technical measures are one of three critical measures for protecting PHI in connected healthcare settings. Standard technical measures can include the use of encryption, access controls, and data minimization, among others. Encryption is the process of converting data into a code that can only be read by authorized parties (Mukherjee et al. 2021). Access controls, such as passwords and biometric authentication, ensure that only authorized individuals can access PHI (Hagland 2022). Data minimization involves collecting only the minimum amount of data necessary to provide healthcare services (Caulfield and Donnelly 2013).

One approach that has been taken, as outlined by Eze et al. (2018), is to use SQL to attach metatags to pieces of data to implement the necessary policies associated with the data. The core feature of the tagging system developed by the University of Ottawa researchers is to ensure that patients' consent is attached to every piece of data that is moved. However, the recommendation comes with some challenges as well, because not only the patient consent is considered. In their paper, they outlined that the primary tagging comes from agreements between healthcare organizations and external partners, wherein a second party only receives as much information as they themselves take (Eze et al. 2018). As an example, if Organization B does not typically take information about a patient's middle name, then even if Organization A has the information about the middle name, the tagging method declares that that piece of information should not be shared with Organization B, even before a question is posed to the patient about consent for that piece of data. Although this approach does offer some interesting considerations, it is not technically sound and loses a great deal of data along the pipeline, making it difficult for researchers to aggregate and make use of enough data to conduct good research and also making it difficult for patients themselves to access their own complete records, depending on the place in the connected health chain they find themselves in at any given time.

The paper was delivered in 2018, and since that time the connected healthcare system has continued to evolve, as have the demands and needs in the system post-pandemic. With far fewer doctors available (Hagland 2022), the demands on the digital systems increase to take on innovative practices to reduce time spent by patients, chronic and otherwise, in the doctor's office, as well as to reduce the recurrence of symptoms or diseases that are not chronic.

8.8 REGULATORY ENVIRONMENT

Legal measures are also critical to protecting PHI in connected healthcare settings. These measures include data-protection laws and regulations, such as the General Data Protection Regulation (GDPR) and the Health Insurance Portability and Accountability Act (HIPAA). These laws require healthcare providers to protect PHI and provide patients with certain rights, such as the right to access their data and the right to request that their data be deleted.

In creating a GDPR compliant platform, Tsohou et al. (2020) offered a method for surveying different types of stakeholders and extracting and defining their particular functional and regulatory needs for the different kinds of data they are managing. This is from a European perspective and so does not include HIPAA regulations as a focus. Another approach, in a Canadian healthcare context, focused more on privacy by design regulations (Eze et al. 2018), while the study by Kuziemsky et al. (2018) focused more on the business outcomes rather than the regulatory environment in which it occurs. Finally, Alkhatib and Buchanan (2023) referred to the regulatory bodies themselves: NHS UK, Australian Capital Territory (ACT), American Health Information Management Association (AHIMA), and the Canadian Institute for Health Information (CIHI) as having documents and guidelines for IoT and data management that should be referred to.

GDPR	HIPAA	PIPEDA	FIPPA
• Implementation: 2018 • Compulsory: data collectors and processors of EU & UK citizen and resident data • Applies to all personal data collection • of EU & UK citizens or people residing in the EU or UK • only applies to living, not deceased persons	• Implementation: 1996 • Compulsory: B2B service providers to US hospitals + other covered entities • Applies to sensitive data collection • Protection offered for 50 years after death	• Implementation: 2001-2022 • Compulsory: key provider industries in the private sector, Canada • Applies to health information, banking, telecommunications and other industries where personal data can be collected or stored and then used commercially • only applies to living, not deceased persons but also to named living persons in the data	• Implementation: 1992-2022 • Compulsory: public sector offices but not federal, crown or private offices (Canada) • Applies to data collected by public offices in Canada • usually a request must make the data be released within 30 days

FIGURE 8.2 An overview to compare some of the key data-protection regulations.

GDPR	HIPAA	PIPEDA	FIPPA
• Demographic information: racial and ethnic information, political opinions, religious or ideological convictions or trade union membership, information about a person's sex life or sexual orientation • Medical information (including genetic data and genetic processing or biometric data for the purpose of uniquely identifying a person, and health data) • Personal data includes identification of a person through data combinations	• Names & initials • Addresses • Dates related to identity or service provided • Contact information (including phone, fax and email addresses) • Social security numbers • Medical record information including account numbers, personal & beneficiary health insurance information • Certification & license numbers • Vehicle identification info • IP address • URLs • Device identifiers and serial numbers • Biometric identifiers (finger, retinal and voice prints) • Full-face photos and all comparable images • Any other unique identifying characteristics or numbers	• Demographic information: name, age, social insurance number, nationality, ethnicity, martial status • Contact information (including phone, fax and email addresses) • Financial information: income, banking, credit and loan records, any merchant and consumer disputes • Medical information (including medical history, DNA identifiers, blood type and any records and personal data) • Personal history information (educational, employment, disciplinary actions, evaluations, intentions, opinions or comments)	[SECTIONS 36 and 37] • Under FIPPA, a public body may collect personal information only if: • collection of the information is authorized by or under a statute or regulation of the province or Canada; • the information relates directly to and is necessary for an existing program or activity of the public body; or • the information is collected for law enforcement purposes or crime prevention.

FIGURE 8.3 Each set of regulations protects different kinds of data in different ways, as summarized in this figure.

Therefore, the standard for managing legal in this area is not yet solidified. The recommendation put forth here in this chapter is to begin by splitting the areas of data into their different kinds of functionality and then seek out and apply the correct regulatory frameworks. This should occur before software building begins but after the data architecture and design and data exploration has been completed.

Other regulations to consider are incoming ISO 31700, UN updates on the right to be forgotten and the right to privacy, as well as GDPR updates on AI blackboxing and explainability.

8.9 EMBEDDING PRIVACY

Organizational measures are also essential to protecting PHI in connected healthcare settings. These measures include staff training, risk assessments, and privacy impact assessments (PIAs). Staff training ensures that healthcare providers understand their responsibilities regarding PHI and how to protect it (Information Commissioner's Office 2021). Risk assessments identify potential risks to PHI and develop strategies to mitigate these risks (Cavoukian 2012). PIAs assess the privacy risks associated with new technologies and processes before they are implemented (Eze et al. 2018).

In "A Soft Computing Approach for Privacy Requirements Engineering: The PriS Framework," researchers outlined that PriS considers privacy points as organizational goals (Kalloniatis et al. 2011). This chapter aims to provide a contrasting perspective, that embedding privacy enters the organization as a discussion at legal and strategic levels, and is then implemented at the technical level of an organization. Therefore, privacy is fundamentally a technical pursuit that can be enhanced or hindered by organizational culture, goals, ambitions, values, and readiness for change management principles.

Healthcare IT leadership and personnel must ensure that they are implementing these measures to protect patients' PHI and maintain their trust in the healthcare system. In addition, healthcare providers must be proactive in keeping up-to-date with new privacy regulations and technologies to continue to provide secure healthcare services. Privacy and data management in connected health systems is a joint effort between healthcare providers who see patients every day and the technical personnel setting up the systems in the background.

In the study by Eze et al. (2018), data-sharing agreements formed a core part of the method and approach. The data-sharing agreements were hierarchically below the organizational consent level. Both of these were implemented by adding metatags to the data. Therefore, the technical implementation becomes the actual embedding of privacy into the system. In the approach presented in this chapter, I want to propose flipping the model, to align with privacy by design principles, to center the patient rather than the organization or external partners.

8.10 COMMON CHALLENGES AND COMPLAINTS IN EMBEDDING PRIVACY IN CONNECTED HEALTHCARE

Connected healthcare systems are designed to facilitate the sharing of patient data between different healthcare providers and institutions. Although this can improve patient outcomes and streamline healthcare delivery, it also raises significant challenges in terms of maintaining data and privacy across the big data value chain in healthcare settings. Some of the current challenges in this regard are discussed below.

Data Collection: One of the challenges of maintaining data and privacy in connected healthcare is the sheer volume of data being collected. Healthcare providers are collecting increasing amounts of data from a wide range of sources, including electronic health records (EHRs), wearable devices, and health apps. Ensuring that these data are collected in a secure and privacy-preserving manner is a significant challenge.

Data Storage: Once data are collected, they need to be stored in a secure and privacy-preserving manner. This is particularly challenging in the context of connected healthcare, as data may be stored in multiple locations, including on-premises servers, cloud-based storage, and mobile devices. Each of these storage locations presents unique security and privacy risks.

Data Sharing: Connected healthcare systems are designed to facilitate the sharing of patient data between healthcare providers and institutions. However, this raises significant privacy concerns, particularly if the data are not properly de-identified or if there are insufficient access controls in place to ensure that only authorized personnel can access the data.

Data Analysis: Once data are collected and shared, it may be used for a wide range of purposes, including clinical research, quality improvement initiatives, and population health management. However, this also raises privacy concerns, particularly if the data are used in a way that could potentially identify individual patients.

Data Retention: Connected healthcare systems may retain patient data for extended periods of time. Ensuring that these data are properly secured and protected throughout its lifecycle is a significant challenge, particularly as data-retention policies may vary depending on the jurisdiction.

Regulatory Compliance: Healthcare providers are subject to a wide range of regulatory requirements, including HIPAA in the United States and GDPR in the European Union. Ensuring that connected healthcare systems are compliant with these requirements is a significant challenge, particularly as the regulatory landscape is constantly evolving.

In summary, maintaining data and privacy across the big data value chain in healthcare settings presents significant challenges for healthcare providers. Addressing these challenges will require a multifaceted approach that incorporates technical, organizational, and regulatory measures to ensure that patient data are collected, stored, and shared in a secure and privacy-preserving manner.

8.11 BIG DATA VALUE CHAIN: CONTEXTUALIZING CONNECTED HEALTHCARE

A big data value chain is a sequence of processes and activities that are involved in the creation, management, and analysis of large volumes of data. The value chain typically includes data collection, storage, processing, analysis, and visualization, as described above. The goal of the value chain is to extract value from the data and use it to inform decision-making and drive business outcomes.

In the context of connected healthcare, the big data value chain plays a critical role in the management and analysis of patient data. The value chain begins with data collection, which can include data from EHRs, wearables, medical devices, and other sources. These data are typically stored in a variety of locations, including on-premises servers, cloud-based storages, and mobile devices.

Once the data have been collected and stored, it can be processed and analyzed using a variety of techniques, including machine learning, data mining, and natural language processing. These techniques can be used to identify patterns and insights in the data that can inform clinical decision-making, improve patient outcomes, and drive operational efficiencies.

The results of data analysis are typically visualized using dashboards, reports, and other tools that provide stakeholders with actionable insights into the performance of the healthcare system. These insights can be used to inform strategic planning, quality improvement initiatives, and other business decisions.

The big data value chain in connected healthcare settings is subject to a number of challenges, including privacy concerns, data security risks, and regulatory compliance requirements. Ensuring that patient data are collected, stored, and analyzed in a secure and privacy-preserving manner is critical to maintaining the trust of patients and stakeholders in the healthcare system.

In summary, the big data value chain is a critical component of connected healthcare systems. Understanding the various stages of the value chain and the challenges associated with each stage is important for healthcare providers and researchers seeking to leverage big data to improve patient outcomes and drive operational efficiencies in healthcare settings.

8.11.1 Operationalizing Privacy Compliance

I present our approach to operationalizing connected healthcare systems between internal quasi-organizations and institutions within one larger hospital organization. The goal of the prototype/project was to remodel data storage during the process of digitization of administrative records at a children's rehabilitation hospital in Canada. The focus of the project was to align healthcare adjacent data in the Volunteer Resources Department to the Freedom of Information and Protection of Privacy Act (FIPPA) privacy and data security standards in storage of public health records in the province of Ontario, because a significant number of volunteers were also medical patients in the same building. Therefore, they had medical records in the protected database of this hospital and volunteer records in the nonprotected database, accessed by the same internal network and access management system, with different permissions for different staff roles.

This children's rehabilitation hospital offers inpatient and outpatient services, for both continuing rehabilitation of chronic patients and community services. The hospital additionally features a foundation and a research hub, as well as a school, summer camps and activities, and community access to the swimming pool, gardens, and camps.

The case and project followed a human-centered design (HCD) approach, where incorporating the user's perspective into software development is considered of paramount importance in order to achieve a functional and usable system (Maguire 2001). This approach aligns well with the privacy by design principles that were core to our build, wherein the user is centered in the design of privacy and data protection. These both align with the organization's goal to empower diverse populations to access care and service, and are aligned with general communication guidelines for the organization to create plain language, transparent and easy-to-understand guidelines for healthcare needs for children. Overall, these approaches best support the patient participatory care delivery model offered in the hospital, giving a diverse range of end users the ability to share data across multiple touchpoints to move to accessing healthcare data in any context in a connected healthcare system.

8.12 CASE/CHALLENGE

Often data would be in wholly different silos. In this case, with health administration, data present themselves in a new way. For example, a child patient with a redacted disease, symptom, or presentation has the same name, address, and birth date of the child in the separate volunteer database who happens to be the Chair of the Child Cirrhosis Committee, which we can link to the hospital name and address or to the child's own name and address.

In addition, while medical consent was well established, had strict guidelines, and well-developed processes, in the administrative arm of the hospital, it was a simple checkbox to say that we can store your form. At that time, the forms were not being processed as if they had sensitive data contained in them. During the digitization process, we realized this challenge and addressed it.

8.13 METHOD

To develop user adaptive systems, it is an established practice to collect user data through questionnaire and interview-based approaches. To ensure accurate data collection, focus groups were organized to validate the elicited requirements and data-collection instruments. As Alkhatib and Buchanan (2023) noted, "The advent of innovative wearable medical devices in remote patient monitoring is changing the production and delivery of patient generated health data (PGHD)." Although this project was undertaken at a significant time before the study by Alkhatib and Buchanan, the pediatric rehabilitation health space had innovations introduced sooner, with systems change seen long before full scale or commercialization was reached. We identified key stakeholders and prepared questionnaires for each user category to capture their needs in various aspects such as legal, functional, security, privacy, and acceptance, but also related to groups within the departments, such as family services, orthopedics, volunteer services, and program leaders, for the various activities and camps. Using these data, we translated user needs into clusters of regulatory groups, because the survey of needs had begun with the legal and privacy demands from FIPPA. However, as we added in function-specific needs that went beyond electronic healthcare systems and into unique device types, particular outpatient partner services, and data needs and into research, we discovered a more function-first approach is best.

First, define the cluster of functions—for example, a child submitting schoolwork is a very different activity than a child attending a community pool event. Then, with the clusters of functions, identify the key regulatory frameworks required to cover that type of activity with those kinds of people—whether that be their role in the organization or their age. Finally, create databases for the storage of particular kinds of data that are produced by and used in each type of activity, and ensure that the relevant database has only read or write access depending on the role of the person trying to access the data.

Organizationally, there is one source of truth database. The model diagram in this chapter focuses specifically on the discussed areas: child (in various roles), research data needs, and administrative data needs. A more complete diagram would also have finance and operations information going into this source of truth that would be wholly non-connected to most of the databases that can be seen in the model presented here. Most databases should only have write access to the source of truth, and not read or edit access. Rather, databases can read from the core source of truth, then apply processes to aggregate, minimize, deidentify, or otherwise clean the data, before writing them to a new database for the related task, such as research or marketing. That secondary database becomes the primary place for updates and edits to data that need to be working data for the area. Building in this way allows

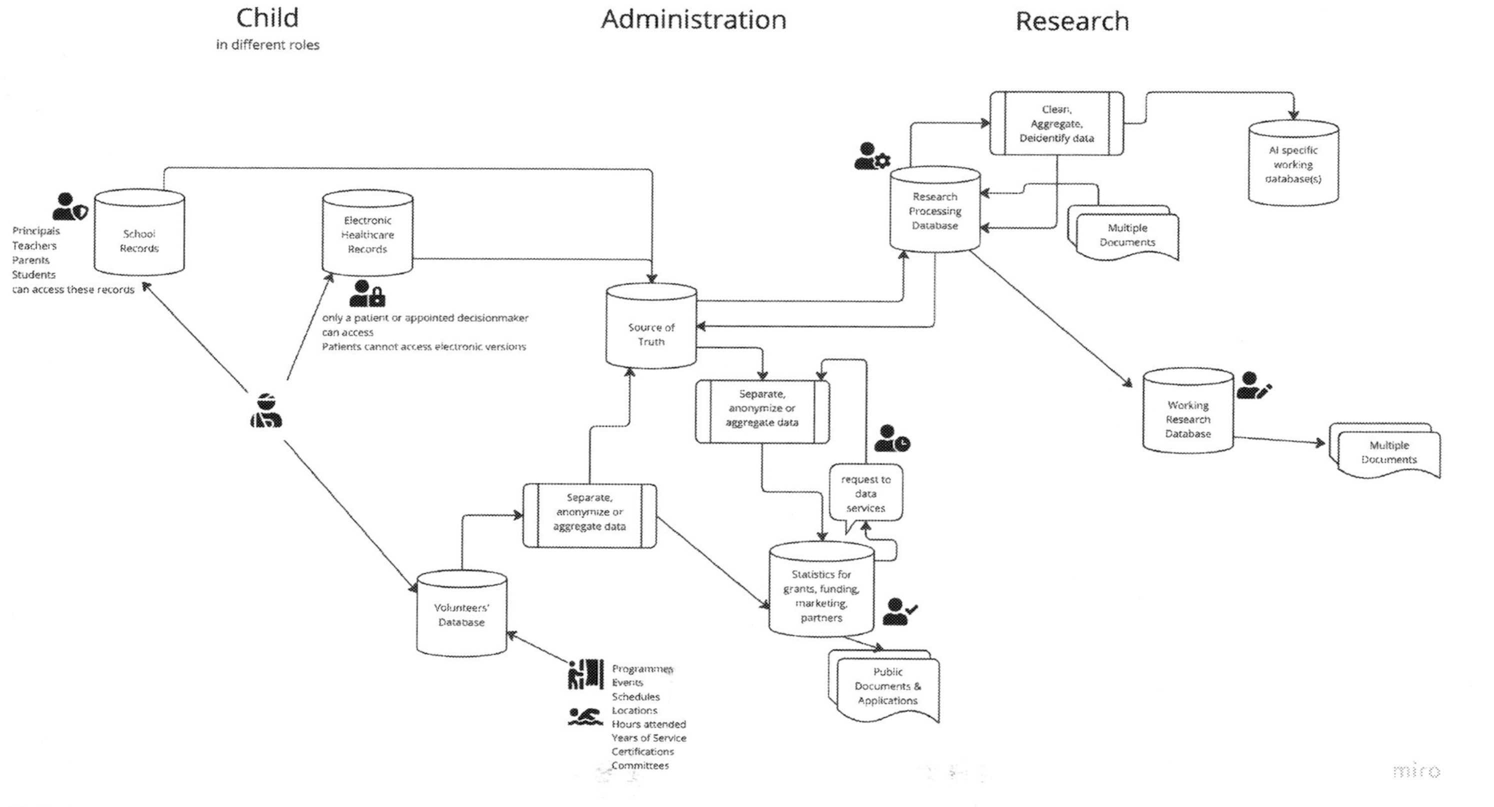

FIGURE 8.4 A simplified outline of the model of databases to separate concerns and reduce impact of breaches.

sharing of data between databases that have already had data manipulated for reuse, and reduces the number of persons who can access any particular kind of data at any time. A researcher role, for example, may be able to access two research databases, but not anything in electronic healthcare records. Instead, the EHRs have been stripped, anonymized, aggregated, and stored somewhere separate from the original information. A researcher would also have no need to go into volunteer services, and the reverse is of course true as well. Administrators in the school or volunteer services require no access to EHRs or research areas.

Importantly, although not explicitly diagrammed, any secondary internal database can make calls to retrieve data from another secondary database, to ensure there is not one single point of failure. One drawback of this approach is that some duplication can occur. Therefore, the system will need a way to manage this write process to correct any duplication. An automatic version control can be one way to manage this. Additionally, external partners have only one access point through the public-facing records storage, which is managed by staff in the organization.

8.14 PRIVACY COMPLIANCE MODEL

8.14.1 FIPPA

According to the Government of Manitoba (2022),

> FIPPA applies to all records in the custody or under the control of a public body. A 'record' is information in any form and includes information that is written, photographed, recorded or stored in any manner on any storage medium or by any means including graphic, electronic or mechanical means. It does not include electronic software or any mechanism that produces records.

However, many Acts can override the FIPPA model in different contexts, excluding key information, such as teaching materials in public schools, court records of judges' personal notes, anything related to mental health or to family services (Government of Ontario 2014). Therefore, in a context where we are working with medical patients who are pediatric, with a variety of other connected institutions in the same location, all of which are governed by different layers of the laws, it becomes necessary to create a model for understanding what data can be shared where and when, without the complexity of returning to the hierarchy of policy documents each time a decision is needed. This ecosystem is very much a connected health ecosystem, with a much closer brand and geography layout than I have encountered in many other papers and approaches. Of course, there are external stakeholders, external to the hospital ecosystem, but each department becomes an external stakeholder to other departments, in terms of data storage and access.

Although FIPPA may not apply to your project's regulatory context, countries such as Japan and Brazil also have unique regulatory data demands much like Canada does. FIPPA also offers an important lesson in understanding what is meant by a record, and this definition may be different in your regulatory context, locally or across your connected health system providers. It is important to find the most lengthy and involved definition of a record when you begin designing your privacy and data security system, and build for that.

8.15 CLEAR CONSENT

In addition to that point, rather than treating the whole system as a system that has one type of consent, our approach was to separate information by type of regulation or legal framework that governs it: education (at the provincial level) and healthcare records (non-federal). Then, we looked at the research arm, which has evolving needs, as big data and AI research have become more important in understanding healthcare contexts. These permissions are governed by ethics, best practice, and international regulations. International regulations include formal and already agreed upon regulations such as GDPR as well as upcoming regulations and best practice and recommendations from organizations like the UN, which has begun introducing key guidelines for data management, including the principle of "the right to be forgotten". Finally, we looked at the data that was left, which was primarily for administrative needs. Here, it was clear that there was a lot of bloat around the fields collected for data, and many departments were storing more than one copy of the data to maintain the purity of the data or to maintain a manipulation they themselves had created of the data so that they fit the purpose and needs of the work to be done in that area. Learning more about this administrative data brought us to the realization that one source of truth was needed across the entire organization. Departments could then have their own departmental databases for working with data. The data can then be cleaned and aggregated and kept as working data within the departments without impacting the original data. These departmental databases have read-only access from the source of truth.

Consider each handover of data as a new gate for consent. Instead of offering monolithic privacy coverage, rely on privacy by design principles and create short, clear, plain language statements about the type of activity that is going to happen on the other side of that gate, how it benefits the patient or end user, and how long the organization will keep that kind of data for.

8.16 SEPARATING DATA

Important to this separation by department and by regulations and governing practices was the ability to separate data and send them to different streams for different uses within the organization. The example I have illustrated focuses on what kind of information is received by a volunteer department in a pediatric hospital. In the volunteer department, there is not only data about individuals who may be in patient care but also information about all of the times and dates of every program, from speech therapy to swimming and other rehabilitation programs are received. This information does not need to be stored and so is stripped from the data when it is being sent on to other data storage systems within the larger organization. Some information, such as the number of hours that a volunteer has collected, can be important for school curriculum requirements, parole and community service requirements, social services and unemployment benefits requirements, and of course, for issuing the volunteer certificates outside of any legal requirements any volunteer may have to show their work. However, prior to the transformation, all of the dates and times of each volunteer engagement, as well as the location of the engagement and the start and end time, were included. In the new system, we processed data by calculating the hours, and then keeping the total hours and deleting the rest of the data.

This separation was key to maintaining GDPR and privacy by design regulations. Prior to the updates and new data processing and storage, it became possible to identify a pediatric patient's medical appointment information as well as correctly identify their medical diagnosis in many cases using the healthcare adjacent volunteer management database system in this connected healthcare context.

The key takeaway for your connected health system is as follows: as the person enters the next gate, what information that is already attached to them needs to be used to achieve the goal on the other side of the gate. Any data that are not necessary for the goal or outcome of the next activity should be processed, discarded, or aggregated at this point.

8.17 STORING DATA IN DIFFERENT LOCATIONS

Along with separating the data, ensuring that data are stored in different locations with good encryption and decoupling of sensitive data is key to maintaining a strong privacy environment for your patients and end users.

8.18 DISCUSSION

Although our project was not specific to core healthcare data in the connected health system, it presented us a model to minimize data and offer a prototype model for transformation of the collection and processing of the healthcare data for pediatrics, both in the hospital and in the province at large. In the pediatric context, all patients with chronic conditions are likely to be moved to another healthcare setting and into the connected healthcare system by the time they are 18 years of age, and so it is best to build with that transfer of records and potential exposure to commercial applications to empower patients to continue their care beyond pediatric years. In the context of our hospital, we also had a research tower that was often inviting participants in for studies, both from our rehabilitation hospital and from the main children's hospital downtown. Therefore, it was common enough knowledge and practice between the research arm and the clinical arm of the building to know how to obfuscate and aggregate data and records for patient protection. Of course, since that time, this has developed into a complete and published data-management strategy that allows external partners to understand data storage, sharing, and processing as well as rights and responsibilities the same way. The new approach here was that administrative staff were also trying to protect the data and information, so that risks for breaches of sensitive data were greatly reduced.

8.19 IMPLEMENTATION PRINCIPLES FOR EMBEDDING PRIVACY BY DESIGN

In "Privacy and the Internet of Things (IoT) Monitoring Solutions for Older Adults," Alkhatib and Buchanan (2023) note in their recommendations section that,

> While each reviewed study proposed a solution to protect users' privacy, no clear method on how the developers of these solutions reached their understanding of privacy has been identified [...] This raises questions about how developers reach their understanding of privacy and thus propose solutions for it".

Therefore, in this chapter, I propose key implementations that developers can note today to implement their privacy principles, based on implementations in a children's hospital context in 2015.

1. Minimize data collection.
2. Based on our core legal and regulatory frameworks, privacy by design and GDPR, it is important to limit the amount of user data to reduce any possible impact on patients should a breach occur.
3. In the pediatric hospital I worked in, this piece of work became important in the volunteer resources department, where we had many pieces of information about our volunteers stored: first name, last name, age, date of birth, volunteer roles and committees, personal address, awards received, school name, qualifications, and certificates. After much discussion with senior coordinators (who are not technical) about why these pieces of information were stored, we came to understand that we did not need as much information as was there. There were heated discussions about fields such as age, because coordinators used this information to sort volunteers into role types. For example, a 16-year-old can only perform some tasks in the pool as a volunteer if they have certain certifications, but a 17-year-old can work in a variety of areas and not the pool if they do not have certifications.
4. Together we transformed the system, using age as an input for volunteers to check their roles in a front-end focused form. They could save and send or print for a coordinator to review in the application process, and then destroy and not store afterward.
5. These principles become important in connected healthcare systems, because information isn't just being sent between internal hospital departments, where the patient becomes a volunteer and then moves back to the patient role. Instead, the information can go to commercial apps, to broader EHR storage, or even to another provider. Minimizing the data stored in the first place greatly reduces the risk of impact or serious concern for sensitive information being breached.
6. The key question is: what role does this piece of data play in achieving the goal of this task?
7. If you can't answer that question, it is likely not a necessary piece of data. Even with an answer you feel is justified, stop and consider: is there another way to achieve this goal that uses less data or less sensitive data?
8. Empower people (parents or patients).
9. This recommendation is more from the perspective of leadership, but developers can help describe in plain terms what data is being stored, why it is being stored, how that benefits the patient or caregiver, and how long the data will be kept.
10. In a connected healthcare system, it is important for each provider to have available the list of other providers in the system, as well as note which devices are transmitting data live or if that are collected inside the primary care center and then stored in local or cloud storage.

11. An easy way to implement this is with a color-coded sticker system. Decide, for example, whether transmitting devices get a blue dot and non-transmitting devices get a yellow dot. A cardiogram, for example, does not transmit information live but will be handed back in to the healthcare provider and that data will then be extracted and stored.
12. For systems beyond the control of the provider providing the data information, it is important to have information available about the diverse nature of devices not issued by the organization itself, and to reiterate that patients can always choose what data to share, as well as reiterate the benefits to the patient for sharing the data.
13. Create one source of data truth that cannot be changed.
14. All data are delivered to and retrieved from this one repository of truth. The data are moved from the source of truth to a new storage area for each different kind of project or use. For example, a repository of aggregate data for research, which only requires certain parts of the original data. Another repository of data might be for patient-specific use and therefore maintains the identifiers necessary to operationalize the data in a variety of apps and uses.
15. This main source of truth has high security and very restricted data observability.
16. Encrypt data.
17. Data encryption alone is not a privacy measure. However, data that are being transported between different systems and storages, as well as into the connected health ecosystem, requires encryption so that it is not breached in transport.

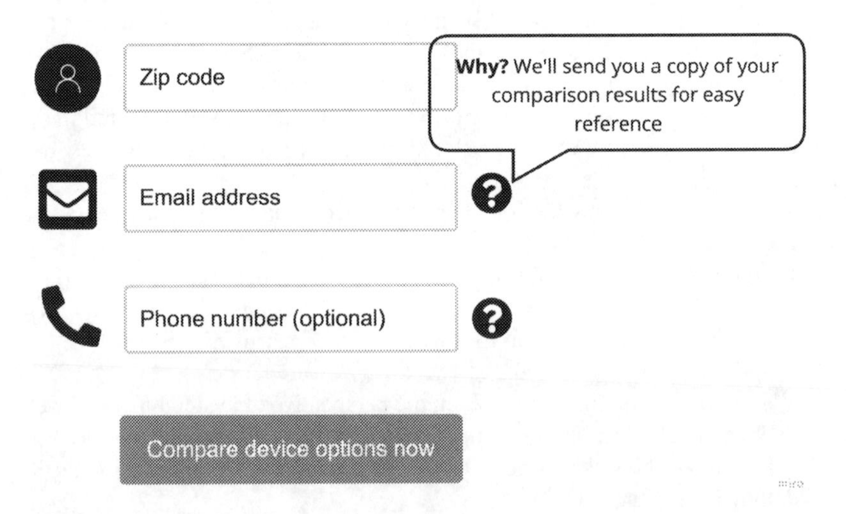

FIGURE 8.5 Small changes in technical implementation like where you put information on a webpage for end users can have a huge impact on the success of a privacy and data-protection strategy and project.

18. Just-in-time notices.
19. This is a highly useful and effective tool for connected health contexts. Rather than anticipate every node of the ecosystem that your patient will touch before they begin their journey, you can offer clear notices throughout the journey as they encounter new nodes and have a small pop-up box (in accordance with accessibility standards) that explains why certain pieces of information are being taken.
20. Implementing just-in-time notifications meets principle six of the privacy by design of general guidelines, which are focused on plain language, transparency, and visibility.
21. Leave boxes empty by default—no pre-filling form fields.
22. This recommendation comes specifically from the time when the web platform was core to the patient digital experience. However, it is still helpful to remember across device types to not default to consent. Remember that patients may in fact be, for example, unconscious. As technical, policy, and research folks, we can sometimes be forgetful about the context these data are being received in or from. Of course, an unconscious person is not filling out a web form on a desktop. An unconscious person may, however, have their data entered into a tablet, which sends information into an EHR system and into a patient app that the patient has not been made aware of yet, and offers commercialized data opportunities, as well as AI learning for hospital research that creates feedback.
23. In addition, although there are many nuances that we cannot explore in this section, when managing pediatric patients, it is necessary from a legal perspective that the correct person has given consent. Depending on the jurisdiction, this can vary widely, as well as the age that consent can be given for different types of information and different kinds of applications.
24. The key takeaway here is: did we set up our system to default to consent? Check the connections between your applications and access to your APIs and devices.
25. You can create two settings for the data if you need to—one is emergency data that are core data to accomplishing the goal. The other is, once a patient or caregiver is able to give consent, more data are added to the profile.
26. Simplify consent.
27. In the connected health ecosystem, and especially with the introduction of soft computing techniques, many unknowns can find their way into your data ecosystem. It is the role of each provider to inform in a transparent way, with clear and simple language, what is being done with the data, how long it will be kept, and who can access it beyond the patient.
28. In these new ecosystems, where data come in and out of the primary healthcare setting or hospital, it is impossible to know everything about every piece of data. Cisco offers a model for this though.
29. When faced with the question: can I do this? The simplest answer is: did you ask? If you discover you need a piece of information quickly and it is not

explicitly covered, you can print out a small one paragraph form that says: we are taking x type of data to use in some other context than the one you originally agreed to. This will help you to _____ or help us to _____. Industry standard says we will keep this information for 7 years before deleting it. Do you consent to us taking, processing, and storing these data for that purpose?

30. Of course, across systems, it is impossible to manually write notes for each instance. A few edge cases, however, can be covered this way. Be sure to be bringing edge cases back to your technical team for consideration and review so that they can work to add these cases to more regular coverage, if necessary.

8.19.1 Some Remarks on New Research Directions

At the time this case/project was undertaken, much of the data entry and organization was manual. A new research direction to improve the process of data cleaning and matching can be techniques in fuzzy matching for decision-makers.

Fuzzy matching is a technique used in data analysis and text mining to find similarities between two or more strings of data that are not exact matches. It is useful when dealing with data that contains variations in spelling, typos, or other errors that can occur naturally or because of data entry issues. Fuzzy matching algorithms assign a similarity score to each pair of strings based on how closely they match. The scores can then be used to group similar strings together, identify duplicates, or categorize data. This technique can help decision-makers in various ways, such as the following.

Improved data quality: By identifying and grouping similar strings, fuzzy matching can help to identify and correct data entry errors, ensuring that decision-makers have access to accurate and reliable data.

Efficient data analysis: Fuzzy matching can reduce the time and effort required for manual data cleaning and matching, allowing decision-makers to focus on analyzing the data and making informed decisions.

Better patient service delivery: Fuzzy matching can be used to match patient data across different systems and platforms, allowing organizations to provide more personalized and efficient service.

Important to note, beyond improving data analysis and quality with fuzzy matching, key research in using blockchain to reorganize data so that they belong to the patient entirely can be an interesting approach. Systems will continue to evolve beyond connected health into fully context-aware systems. Here, the principles would still apply to build each node its own set of requirements. However, the connections between nodes in future will likely become less disparate and distinct than they are today. As more data builds up over time and more connections are made, the nodes will look less like individual data points and more like a landscape of closely related data.

The most immediate next step for research and consideration would be in differential privacy, as well as in homomorphic encryption methods.

BIBLIOGRAPHY

Alkhatib, S., and J. Buchanan. 2023. "Privacy and the Internet of Things (IoT) Monitoring Solutions for Older Adults." In: *Connecting the System to Enhance the Practitioner and Consumer Experience in Healthcare*, edited by E. Cummings, A. Ryan, and L. K. Schaper, vol. 252, pp. 8–14, IOS Press. doi:10.3233/978-1-61499-890-7-8.

Bock, S. 2022. "The Ultimate Guide to Connected Healthcare." *Trapollo*, July 13, 2022. https://www.trapollo.com/articles/connected-health/guide-connected-healthcare-solutions/.

Caulfield, B. M., and S. C. Donnelly. 2013. "What Is Connected Health and Why Will It Change Your Practice?" *QJM: An International Journal of Medicine* 106 (8): 703–707. doi:10.1093/qjmed/hct114.

Cavoukian, A. 2012. "Operationalizing Privacy by Design: A Guide to Implementing Strong Privacy Practices." https://www.schwaab.ch/wp-content/uploads/2013/09/operationalizing-pbd-guide.pdf.

Eze, B., C. Kuziemsky, and L. Peyton. 2018. "Operationalizing Privacy Compliance for Cloud-Hosted Sharing of Healthcare Data." *IEEE Xplore*, May 1, 2018. https://ieeexplore.ieee.org/document/8452636.

Fhir, Hl7. n.d. "What Is HL7 FHIR?" https://www.healthit.gov/sites/default/files/page/2021-04/What%20Is%20FHIR%20Fact%20Sheet.pdf.

Government of Manitoba. 2022. "Understanding FIPPA|FIPPA|Province of Manitoba." https://www.gov.mb.ca/fippa/understanding_fippa.html#11.

Government of Ontario. 2014. "Law Document English View." https://www.ontario.ca/laws/statute/90f31.

Hagland, M. 2022. "StackPath." https://www.hcinnovationgroup.com/policy-value-based-care/news/21289935/survey-consumers-engaged-in-connected-health-but-concerned-over-privacy.

Information Commissioner's Office. 2021. "Health Data." https://ico.org.uk/for-organisations/guide-to-data-protection/guide-to-the-general-data-protection-regulation-gdpr/right-of-access/health-data/.

International Organization for Standardization. 2023. "ISO 31700-1:2023." https://www.iso.org/standard/84977.html.

Kalloniatis, C., P. Belsis, and S. Gritzalis. 2011. "A Soft Computing Approach for Privacy Requirements Engineering: The PriS Framework." *Applied Soft Computing* 11 (7): 4341–48. doi:10.1016/j.asoc.2010.10.012.

Kuziemsky, C., S. Gogia, M. Househ, C. Petersen, and A. Basu. 2018. "Balancing Health Information Exchange and Privacy Governance from a Patient-Centred Connected Health and Telehealth Perspective." *Yearbook of Medical Informatics* 27 (1): 48–54. doi:10.1055/s-0038-1641195.

Maguire, M. 2001. "Methods to Support Human-Centred Design." *International Journal of Human-Computer Studies* 55 (4): 587–634. doi:10.1006/ijhc.2001.0503.

Mukherjee, P., L. Barik, C. Pradhan, S. Shekhar Patra, and R. K. Barik. 2021. "HQChain." *International Journal of E-Health and Medical Communications* 12 (6): 1–20. doi:10.4018/ijehmc.20211101.oa3.

Pohl, M. 2018. "HIPAA, GDPR and Connected Health—Interview with Jovan Stevovic, CEO of Chino.io." https://research2guidance.com/hipaa-gdpr-and-connected-health-interview-with-jovan-stevovic-ceo-of-chino-io/.

Research 2 Guidance. 2018. *The Largest Research Program on MHealth App Publishing*. mHealth Developer Economics.

Tsohou, A., E. Magkos, H. Mouratidis, G. Chrysoloras, L. Piras, M. Pavlidis, J. Debussche, M. Rotoloni, and B. Gallego-Nicasio Crespo. 2020. "Privacy, Security, Legal and Technology Acceptance Elicited and Consolidated Requirements for a GDPR Compliance Platform." *Information & Computer Security*. doi:10.1108/ics-01-2020-0002.

UK Public General Acts. 2018. "Data Protection Act 2018." https://www.legislation.gov.uk/ukpga/2018/12/contents/enacted.

United Nations Conference on Trade and Development. 2021. "Data Protection and Privacy Legislation Worldwide|UNCTAD." https://unctad.org/page/data-protection-and-privacy-legislation-worldwide.

Yang, Z., T. Zhang, H. Garg, and K. Venkatachalam. 2022. "A Multi-Criteria Framework for Addressing Digitalization Solutions of Medical System under Interval-Valued T-Spherical Fuzzy Information." *Applied Soft Computing* 130 (130): 109635. doi:10.1016/j.asoc.2022.109635.

9 Next-Generation Platforms for Device Monitoring, Management, and Monetization for Healthcare

Aparna Agarwal, Khoula Al Harthy, and Robin Zarine
Middle East College

9.1 INTRODUCTION

9.1.1 Overview of the Need for Next-Generation Platforms for Device Monitoring, Management, and Monetization

The emerging trends in technologies, such as Internet of things (IoT), big data management, cloud computing, artificial intelligence, machine learning, and blockchain, are without doubt impacting current practices across multiple disciplines, and the healthcare sector is no exception. These emerging technologies provide great potential for enhancing operational efficiency and effectiveness within the healthcare sector. For instance, effective monitoring of patient medical status to ensure accurate diagnosis and treatment is crucial within the healthcare sector, and the emergence of IoT technologies is gradually normalizing that. Monitoring of sensitive patient medical status using IoT is now simplified through the development of lower cost high-quality sensors and data transmission. The volume of medical data being collected through the medical monitoring devices needs high capacity and high-speed transmission medium, whereas the real-time analysis of those data for crucial diagnosis and treatment needs extremely high processing power. The big data management supplemented by the power of cloud computing offers just that, and it can also provide comprehensive data analytics to guide profitable management decision-making. Moreover, the high level of confidentiality expected within the healthcare sector cannot be overstated. The development of the blockchain technology can offer an appropriate level of data security and integrity, which should enhance patient confidence toward the healthcare sector.

DOI: 10.1201/9781003405368-9

The incorporation of artificial intelligence and machine learning within the healthcare system can strengthen the diagnostic and treatment ability of medical practitioners. Artificial intelligence will allow for collecting medical data to be easily compared to learnings from previously collected medical data, while machine learning will allow for new learnings if the collected medical data are different. These can be very impactful with regard to the efficiency and effectiveness of medical diagnosis and treatment. Integrating the potentials of the abovementioned technologies can well become the next-generation healthcare platform, which can substantially enhance the quality of healthcare, provide significant saving, and create windows of business opportunities.

In this chapter, the ongoing development of these technologies is explored with a focus on their potentials toward enhancing the healthcare sector in general. Then, the identified potentials are further discussed to show their links to specific practical implementations within the healthcare sector, be it for management purposes to promote data-driven decision-making, or device monitoring to improve diagnostic and treatment accuracy, or to create monetization opportunities by integrating those technologies into the next-generation smart health platform.

9.1.2 The Requirement and Challenges of the Three Ms (Management, Monitoring, and Monetization) in Healthcare

Effective *device management* in healthcare involves proper device maintenance, calibration, and servicing to prevent device malfunctions and failures. It also involves tracking and monitoring device usage to ensure that they are being used correctly and safely.

The following are some reasons why device management is essential in healthcare:

- **Patient safety:** Proper device management ensures that medical devices are functioning correctly, reducing the risk of harm to patients.
- **Regulatory compliance:** Maintaining devices in compliance with regulations ensures that healthcare providers are meeting legal requirements and avoiding potential fines or legal actions (Miotto et al., 2018).
- **Cost-effectiveness:** Proper device management can help healthcare organizations avoid unnecessary expenses by identifying issues early and extending the life of devices.
- **Data management:** Accurate device management can help healthcare providers effectively track and manage patient data, ensuring that they are collected and stored appropriately.
- **Operational efficiency:** Streamlining device management processes can help healthcare providers improve workflow efficiency, reducing time spent on administrative tasks and improving patient care (Gao et al., 2019; Ma et al., 2020).

The use of medical devices has become increasingly prevalent in modern healthcare, from simple devices like thermometers to more complex devices such as pacemakers and ventilators. These devices play a vital role in patient care and can help improve outcomes and quality of life for patients. With the advancements in the number of devices in the healthcare sector, the demand for device monitoring has also increased. With this approach, we need to monitor the performance and the health of the devices, including the battery life, the storage capacity, the network connection, etc. It will also entail the detection of the issues and fixing them, such as device failures or security breaches. This can help in spotting the problems early and avert the potential damage that could have been done to the patients.

One example of the need for device monitoring is the usage of insulin pumps for diabetic patients. These pumps provide insulin into the patient's body and as such need to be closely monitored to guarantee a good operational skill. If the pump fails because of any reason, it will inject too much or too little insulin into the body, resulting in serious health consequences like coma, diabetic ketoacidosis, or may even lead to death. Such issues can be rapidly discovered and treated with device monitoring thus preventing potential catastrophic problems.

Another example to be considered here is the use of ventilators in critical care units. Improper functioning of ventilators can lead to a lot of complications, such as pneumonia, lung damage, and even death. Device monitoring can help us detect these issues before and take corrective actions to avoid any complications.

In addition, device monitoring can also help the healthcare providers make more informed decisions about patient care. The tracking of data pertaining to the medical devices can provide valuable insights into the patient health and identify the trends and patterns that can help in taking informed medical decisions. For example, a patient's heart rate data collected from a monitoring device can provide valuable information about their cardiovascular health and help healthcare providers make informed decisions about medication or other interventions (Kohli et al., 2021).

Device monetization in healthcare refers to the process of generating revenue from medical devices, either through the sale of the devices themselves or by providing additional services related to their use. Although the primary focus of healthcare is patient care and treatment, healthcare providers are also under increasing pressure to manage costs and generate revenue. Device monetization can help healthcare providers achieve these goals while improving patient care and outcomes (Kohli et al., 2021).

The following are some reasons why device monetization is required in healthcare.

- **Cost recovery:** Medical devices are expensive to develop and manufacture, and healthcare providers must recover these costs to remain financially viable.
- **Improved patient care:** Improved patient care by providing access to new and innovative medical devices that might not otherwise be available. Monetizing medical devices can also fund research and development with regard to new devices that can improve patient outcomes and quality of life.
- **Revenue generation:** By monetizing medical devices, healthcare providers can create new revenue streams that can be used to fund research, clinical trials, and other areas of healthcare.
- **Value-based care:** Device monetization can also support value-based care models that focus on patient outcomes rather than the volume of services provided. By monetizing medical devices, healthcare providers can invest in devices and services that improve patient outcomes and quality of life, rather than simply providing more services.
- **Competition:** Monetizing medical devices can provide healthcare providers with a competitive advantage by offering new and innovative devices and services that are not available elsewhere.

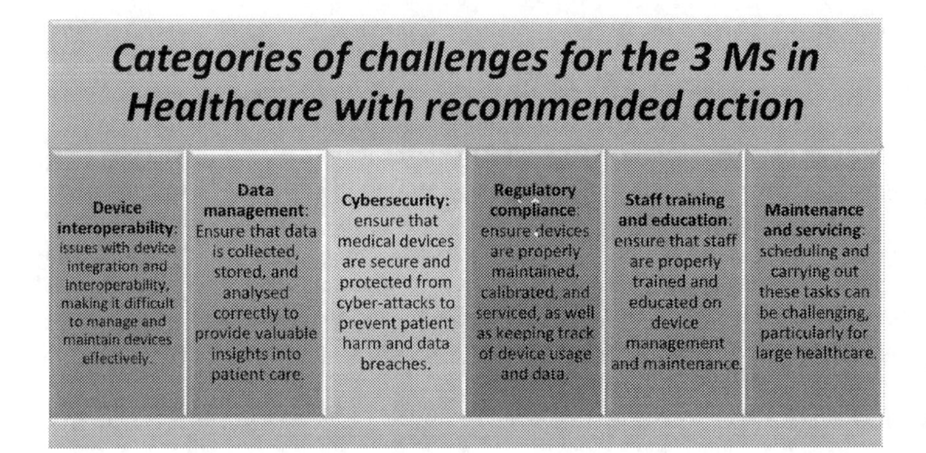

FIGURE 9.1 Categories of challenges for the three Ms in healthcare with recommended action.

9.1.3 Challenges of Three Ms in Healthcare

The following are some of the key challenges of device management in healthcare (Rau et al., 2022):

- Device interoperability
- Data management
- Cybersecurity
- Regulatory compliance
- Staff training and education
- Maintenance and servicing

The role of AI, IoT, and blockchain in shaping the future of device monitoring, management, and monetization is described in the following sections (Figure 9.1).

9.1.4 AI for Device Monitoring and Management

9.1.4.1 Introduction to AI and its Applications in Device Management and Monitoring

The advancements in machine learning, which is a subset of AI combined with the abundance of data and computer power for processing in recent times, have given a great boost to the development of AI and its applicability across various sectors including healthcare. Deep learning, which is a subfield of machine learning, uses neural networks to accomplish complicated tasks such as image and speech recognition (Suganyadevi et al., 2022). Reinforcement learning, which is another type of machine learning, concentrates on training agents to make decisions in whatever environment necessary. Natural language processing (NLP), on the other hand, which is a subfield of AI, allows for human–computer interactions using natural language. Similarly, generative models that are AI models can acquire special capabilities by using appropriate training data to learn patterns and to produce new data such as images and text. Moreover, AI-enabled robots can now perform a wide range of tasks such as inspection or assisted medical surgeries (Klodmann et al., 2021).

These advancements in AI technology have a great potential in healthcare, be it image and object recognition for security systems and facial recognition software, to predictive analytics for pattern analysis and future outcomes predictions, to virtual assistants used in personal devices for reminder settings, to robotics used in instruments for precise task performances, to fraud detection used for fraud detection and prevention, and to NLP used for sentiment analysis, language translation, and text classification (Russo et al., 2022). These advancements and applications provide distinct features that allow for the management of a variety of extremely complicated tasks with minimal supervision compared to traditional technology, making an easy argument for their fitting to the healthcare industry, which is filled with highly complicated tasks. For instance, the need for precise analysis of medical images for accurate diagnosis of patients' illness that would normally be challenging for medical experts to diagnose using conventional technology is supported by the AI image and object recognition capability; the need for continuously close-monitoring of patients' crucial medical conditions and flagging the slightest change in conditions that could otherwise jeopardize patients' recovery process can be supported by the AI-predictive analytics capability; or the need to extract hidden patterns from large volume of medical data for a more effective handling of unexpected pandemic like COVID-19 (Wang et al., 2021).

9.1.4.2 Use Cases of AI in Device Management and Monitoring

The use of AI applications in the healthcare sector helps to enhance patient outcomes, lower costs, and streamline operations. In this section, AI applications used in key areas of the healthcare sector are presented.

Medical image analysis is crucial to clinical diagnosis as it allows for timely and accurate diagnosis. As it depends heavily on AI-powered technology for its processing, it is no surprise to see a continuous increase in AI-powered analytical application use within the medical imaging analysis field (Wang et al., 2021). The huge volume of existing medical data can easily provide sufficient training datasets to train the AI algorithms and plenty of test datasets to test their performance, so that they can be confidently used to analyze medical images, such as X-rays, CT scans, MRIs, and mammograms. As a result, more accurate detection of patterns, including abnormalities and timely diagnoses of diseases, such as cardiovascular disease, cancer, and neurological disorders, can be guaranteed. AI technology continues to improve in that area especially in the use of deep learning techniques to assist medical practitioners in identifying, classifying, and quantifying patterns in medical images (Suganyadevi et al., 2022).

Moreover, AI-powered software can assist doctors in planning for surgical procedures by providing 3D images of patients' examination that helps in identifying possible surgical complications (Liu et al., 2021). This significantly improves the effectiveness and efficiency of medical images analysis compared to when it is done by a human radiologist alone. Radiologists are now being relieved from the tedious tasks that can be handled by the AI technology for medical images analysis thus allowing them to further focus on other vital areas where they are indeed required.

The same is vital to patient treatments and so AI application is also used within the field of Precision Medicine where AI technology can be used to analyze huge quantities of patient data to personalize treatment plans and to predict treatment outcomes (Awwalu et al., 2015). For instance, AI algorithms can be used to analyze patient health record, identify symptoms, and examine results to produce individualized treatment recommendations. AI can further improve the treatment plans by identifying the most effective treatments, prescription, and length of treatment centered on patient traits and illness progression (Russo et al., 2022). Moreover, AI can be used during the treatment of patients to ensure they adhere to their treatment plans by monitoring their adherence, reminding them to take their medicine, and offering assistance and education.

In addition, AI application is used for clinical decision support, where AI algorithms can be included into electronic health records (EHRs) to provide physicians with real-time decision-making support. These AI applications are useful for supporting and enhancing the decision-making process of physicians, especially in communities where physicians must care for an overwhelming numbers of patients. This AI application can also be used to provide decision support to healthcare providers, be it managing drug stock or developing clinical guidelines (Hasan et al., 2023).

Close monitoring of health data of inpatients, particularly within the intensive care unit, is of utmost importance for patients' recovery. Though several technologies are being used for the monitoring, many of them still need significant human interventions, and so the effectiveness of this patient monitoring approach depends heavily on staff workload and time management (Katiyar & Farhana, 2022). AI technology has simplified this process, where it can do the monitoring of health data, as well as provide real-time insights and recommendations for individuals and healthcare providers. Healthcare goes beyond the hospital walls. After patients are released from healthcare facilities, healthcare providers are still responsible for their recovery process and so treatment continues till that point. To facilitate such treatment and recovery process, healthcare providers are now turning to remote patient monitoring. Remote patient monitoring allows healthcare providers to monitor their patients' vital signs remotely by using AI-powered sensors supported by IoT technology (Shaik et al., 2023). This practice is becoming increasingly popular among healthcare providers of patients with chronic illness, elderly patients in-home care, and even in-patients. The AI-powered sensors can be used to monitor patients' medical conditions in real time by tracking health metrics, such as blood pressure, heart rate, and oxygen dispersion, and send them to the healthcare providers for appropriate treatment response (Majumder et al., 2017). This does not only help with improving patients' results but also reduce hospital readmissions.

Close monitoring of patients' health data using AI technology also helps with the early detection of diseases, such as diabetes, cancer, and Alzheimer. In such situations, AI algorithms can be taught to recognize patterns in medical data that reveal the early stages of diseases, and so earlier detection and treatment can take place leading to possible saving of lives (Kumar et al., 2022). Similarly, close monitoring is also extended to mental health. In this case, AI algorithms can be utilized to examine patterns in speech, behavior, and other data to detect symptoms of mental health issues (Agbavor & Liang, 2022). Again, this can assist healthcare providers recognize patients who may be at risk and offer early treatment.

Furthermore, close monitoring relates to disease surveillance, where AI technology is used to assist in predicting the likelihood of disease outbreaks. This is viewed by many as the foundation of public health in terms of tracking and preventing disease outbreaks. AI-supported predictive analytics can be utilized to analyze huge quantities of data from a variety of sources, including social media, to predict the likelihood of disease outbreaks. This can help healthcare providers to prevent, mitigate, or prepare for outbreaks to provide quicker responses and to identify training opportunities for public health experts (Aiello et al., 2020).

Wearable health-monitoring devices are used to monitor patients' commitment to their prescribed medication. AI-powered wearable monitoring devices can help patients in keeping track and following their prescribed medication schedules, which can lead to improved health results. They can also help patients in tracking and examining their own health metrics, such as activity levels, heart rate, and even sleeping patterns. The health data gathered from these devices can then be used by AI algorithms for data analysis to offer personalized health recommendations, such as better sleeping practice, water intake reminders, and tailored workout plans. AI monitoring in healthcare has the potential to transform the approach to healthcare delivery, to health progress, and to individual well-being (El-Rashidy et al., 2021).

Similar monitoring is also important for out-patients. The use of virtual health assistants can assist in ensuring that out-patients are closely following their treatment plans. The AI-powered virtual health assistants can assist patients with checking symptoms and providing basic health information. Virtual health assistants can interact with patients through voice or text, answer patients' questions, and provide patients with information regarding their health conditions, treatments, and medication (Majumder et al., 2017). They can also assist patients in scheduling of appointments and provide them with reminders for taking medication. For instance, a diabetic patient can request information regarding their glucose level from the AI-powered virtual health assistant, and using the patient's data from a wearable device or smartphone, the virtual health assistant can provide an exact reading. In addition, the virtual health assistant can offer dietary suggestions and recommend physical activities, over and above reminders for taking medication. The provision of immediate and correct information is very helpful to out-patient health management and can mitigate severe health problems. The AI-powered virtual health assistants are helping with enhancing patient engagement, improving patient satisfaction, and simplifying healthcare for healthcare providers (El Rashidy et al., 2021).

The importance of research and development within the health sector cannot be overstated. There cannot be a clearer example than the race to develop drugs for vaccination to combat COVID-19 during the year 2020. AI technology can be used to accelerate the drug discovery process by analyzing huge volumes of data to discover potential new treatments (Hasan et al., 2023). By analyzing the huge amounts of existing genetic data including medical images, the AI algorithms can assist in identifying potential drug targets and predicting drug efficacy. This can also allow for the detection of patterns and trends that would normally be challenging for humans to detect thus leading to new understandings and findings. Moreover, AI algorithms can assist in designing new medicines by predicting their effectiveness, harmfulness, and any other relevant properties centered on molecular composition (Karger & Kureljusic, 2022).

The same can be applied to clinical trial design, where AI algorithms can assist in designing and improving clinical trials by predicting most likely effective treatments based on patient data analysis (Harrer et al., 2019). Furthermore, AI algorithms can be used for drug repurposing by assisting in identifying current medicines that could be helpful for new symptoms, again through the analysis of the huge existing amount of clinical data. In addition, AI algorithms can be used to identify biomarkers that can facilitate disease diagnosis and observe treatment reaction (Selvaraj et al., 2021). The use of high-quality medical data can ensure successful application of AI technology within the medical research field, and as such can transform the drug discovery process by allowing for enhanced effectiveness, precision, and momentum (Bhattamisra et al., 2023). These AI technologies are being rolled out across pharmaceutical companies to enhance their drug development process.

Healthcare managers nowadays can also leverage the powered-AI technologies while taking crucial decisions. The AI technologies discussed in the research and development section above play vital roles in healthcare management decision-making, as to what research and development investment healthcare facilities should be included to optimize returns. AI technologies for healthcare management are also happening in the field of predictive maintenance for medical devices. In this field, AI technology can be used to predict when a medical device is likely to fail so that necessary timely maintenance can be done to reduce the likelihood of equipment downtime (Qaid et al., 2022). Mitigating fraudulent activities is also important in managing healthcare facilities and AI technology use in that area has shown great potential (Gupta et al., 2019). Such AI technology can be used to detect fraudulent healthcare practices such as false insurance claims or overbilling. AI algorithms can be applied to scrutinize claims data, detect patterns, and mark potentially fraudulent claims. Fraudulent claims can also be prevented by using AI predictive analytics capabilities to predict fraud by identifying patterns of activity that are likely to result in fraudulent behavior (Yange et al., 2020). Moreover, using AI algorithms for behavioral analysis can assist in analyzing patient and provider behavior to detect abnormalities and flag potentially fraudulent activity. Monitoring healthcare transactions in real time using AI algorithms can also help with identifying and flagging suspicious activities as they happen. Similarly, AI algorithms can be used for network analysis to detect any improper relationships between patients and providers that may lead to potentially fraudulent activities and flag them (Oskarsdottir et al., 2020). These AI technology uses within the healthcare management departments provide managers with a greater monitoring capability of the operations; enhanced decision-making abilities, better operational control and efficiency; significant cost-saving possibilities; and guide toward policy development (Liu et al., 2016).

9.1.4.3 Monetization of AI Technologies within the Healthcare Sector

Healthcare providers need to invest significant amounts of funds to incorporate all these AI technologies discussed above within the healthcare services they provide, and so they need to find ways to generate the appropriate returns from them (Kulkov, 2021). Healthcare providers can monetize their AI-supported health services in several ways. For personalized medicine, healthcare providers can make available monitoring devices or virtual assistants that patients can use at home for a fee. The fee will have to be at a markup price so that the healthcare providers can cover for the cost incurred from the AI technology providers (Majumder et al., 2017). For medical image analysis using AI technology, be it for timely and more accurate diagnosis or planning for safer surgical procedures, patients will be expected to pay a higher fee than through the traditional process.

Healthcare providers can even offer the same service to other healthcare providers who do not have AI technology at their disposal. For precision medicine, where AI technology assists in analyzing patient data for personalized treatment plans with predicted outcomes, patients will be willing to pay additional fees for such service especially those who have been going back and forth to doctors will minimal improvements (Awwalu et al., 2015). Patients who want a second medical opinion or other practicing physicians with limited facilities can also be potential customers. In research and development, having AI technology in assisting with designing new drugs, with identifying and validating new drug targets, with repurposing drugs, and with optimizing clinical trial can result in significant returns for healthcare providers (Karger & Kureljusic, 2022). Telemedicine is another area where patients will be willing to pay so that they are matched to the appropriate healthcare specialist for their health conditions. This service can also be offered to insurance providers for a fee (Oskarsdottir et al., 2020). Real-time clinical decision support using AI technology also brings value to healthcare services. Accuracy and time saving in healthcare means indirect cost savings and enhanced reputation, which is crucial for any healthcare provider. Correctly identifying high-risk patients and predicting their future health outcomes is also of interest to those types of patients and they will be willing to pay additional fees for such services. This service can be of interest to other healthcare providers to improve their own patients' outcomes and reduce their treatment costs. As it may be noted, there are significant ways to monetize the AI technologies being used within the healthcare sector so that healthcare providers can ensure appropriate returns on their investments.

9.1.4.4 Limitations of AI for Device Management and Monitoring

AI, like any other technology, has its limitations that need to be considered when implementing it for device management and monitoring in healthcare. This chapter will discuss some of the key limitations of AI in healthcare device management and monitoring and provide examples of how these limitations can impact patient care (Raghupathi et al., 2022).

- **Limited data quality and quantity:** One of the key limitations of AI for device management and monitoring is the quality and quantity of data available for analysis. AI algorithms rely on large datasets to identify patterns and trends, but the quality of data collected from medical devices can vary greatly. For example, data from some devices may be incomplete or inaccurate, making it difficult for AI algorithms to identify patterns accurately. In addition, some devices may not collect data frequently enough, limiting the amount of data available for analysis.
- **Lack of regulatory standards:** The lack of regulatory standards for AI in healthcare can also limit its effectiveness in device management and monitoring. Without clear guidelines and standards, it can be difficult to ensure that AI algorithms are accurate, reliable, and safe for patients. This lack of

regulation can also hinder the adoption of AI in healthcare, as healthcare providers may be hesitant to use technology that is not adequately regulated.
- **Dependence on human oversight:** Although AI can automate many aspects of device management and monitoring, it still relies on human oversight to interpret and act on the data collected. Healthcare providers need to have the skills and knowledge necessary to understand the data provided by AI algorithms and make informed decisions about patient care. Without adequate training and education, the benefits of AI for device management and monitoring may not be fully realized.
- **Bias and fairness:** AI algorithms can be biased if they are trained on datasets that are not representative of the entire patient population. This bias can lead to incorrect diagnoses and treatment recommendations, particularly for minority or underrepresented groups. In addition, AI algorithms can perpetuate existing biases in healthcare, such as racial disparities in diagnosis and treatment. Healthcare providers need to be aware of these biases and take steps to mitigate their impact on patient care.

9.1.4.5 Examples of the Impact of AI Limitations on Patient Care

The limitations of AI in healthcare device management and monitoring can have significant implications for patient care. For example:

- Incomplete or inaccurate data can lead to incorrect diagnoses or treatment recommendations, potentially causing harm to patients.
- Lack of regulation can lead to the adoption of AI algorithms that are not adequately tested or validated, posing a risk to patient safety.
- Dependence on human oversight can lead to errors or delays in interpreting and acting on the data provided by AI algorithms.
- Bias in AI algorithms can perpetuate existing disparities in healthcare, leading to unequal access to care and poorer health outcomes for some patient populations.

Although AI has enormous potential in healthcare device management and monitoring, it is important to recognize its limitations and take steps to mitigate its impact on patient care. The medical data that are used to train the AI algorithms should be accurate and this should be ensured by the healthcare providers. The medical data must be representative of the patient population, which means the sample size should be sufficient. The algorithms should be regulated and tested. The healthcare staff should be adequately trained and educated to ensure that they interpret and act on the data provided by the AI algorithms to benefit patient care.

9.1.4.6 Case Studies of Successful AI-Based Device Management and Monitoring Platforms

- **Viz.ai:** This platform is used to analyze the medical images as data. It helps in the diagnosis and treatment of stroke patients. The platform can detect the potential cases of stroke early and can alert the neurologists.

They can then monitor the data remotely and make the treatment decisions (Rennert et al., 2019).

- **Verily:** This is a wearable device that uses the AI to monitor the health of the patients with Parkinson's disease. The parent company is Google. The device tracks the tremors, gait, and other symptoms. This provides valuable details about the patient's health to the healthcare providers and aids in making informed medical decisions for treatment (Hausdorff et al., 2019).
- **GE Healthcare:** GE Healthcare has developed an AI-powered device management platform called Edison. The platform uses machine learning algorithms to predict when medical equipment may fail or need maintenance, allowing healthcare providers to proactively address these issues and prevent downtime (Bahl et al., 2020).
- **Vodafone:** Vodafone has partnered with medical device manufacturer "A-Medicare" to develop an AI-powered remote monitoring system for patients with chronic conditions. The system uses sensors and machine learning algorithms to monitor patients' vital signs and detect changes in their health status, allowing healthcare providers to intervene early and prevent hospitalizations (Memon et al., 2019).
- **Philips:** Philips has developed an AI-powered platform called "IntelliSpace Discovery," which is designed to help researchers analyze large amounts of medical data and identify new insights and trends. The platform can be used to support research in a variety of areas, including drug discovery, genomics, and clinical trials (Yadav et al., 2020).

9.1.5 IoT for Device Monitoring and Management

9.1.5.1 Introduction to IoT Applications in Device Management and Monitoring

The amalgamation of artificial intelligence and machine learning technologies into IoT systems has empowered the devices to learn from the medical data and make predictions. This leads to the advances in predictive maintenance, anomaly detection, and other complex IoT applications. This has also led to the development of edge computing, which allows for data processing to be done closer to the source of data, cutting the latency and improving response times.

IoT has progressed from a concept of linking everyday objects to the Internet to a complex ecosystem of connected devices, cloud computing, big data, and artificial intelligence. The future of IoT in healthcare shows potential, with developments in 5G networks, blockchain, and quantum computing expected to further alter the IoT landscape in healthcare.

IoT-based device monitoring is critical for healthcare to ensure that their IoT devices are operating optimally. There are several examples of device monitoring solutions, including cloud-based, edge-based, and predictive maintenance solutions. However, businesses may face challenges, such as data management, interoperability, security, and cost, when implementing an IoT-based device monitoring solution.

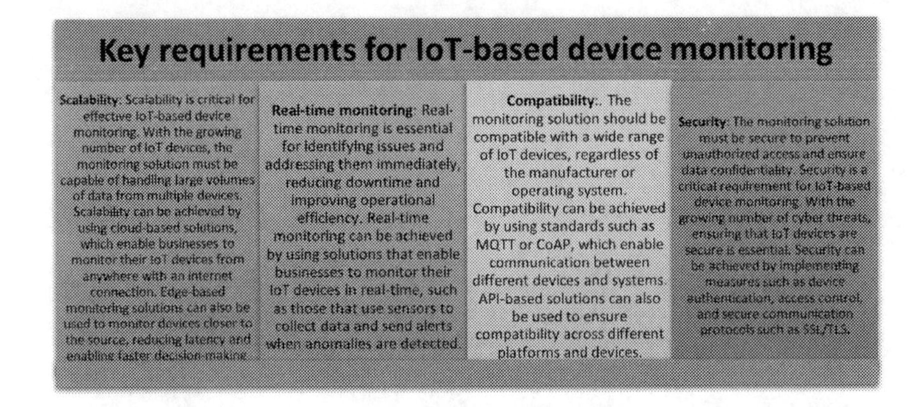

FIGURE 9.2 Key requirements of Internet of Things (IoT)-based device monitoring.

9.1.5.2 Key Requirements for IoT-Based Device Monitoring

To achieve effective device monitoring in IoT, there are certain key requirements that must be met. These include (Figure 9.2):

1. Scalability
2. Real-time monitoring
3. Compatibility
4. Security

9.1.5.3 Use Cases of IoT in Device Management and Monitoring

There are several examples of device monitoring solutions that can be used in IoT, including:

1. **Cloud-based monitoring**: Cloud-based solutions enable businesses to monitor their IoT devices from anywhere with an Internet connection, making it easier to manage and monitor large fleets of devices.
2. **Edge-based monitoring**: Edge-based monitoring solutions allow businesses to monitor devices closer to the source, reducing latency and enabling faster decision-making.
3. **Predictive maintenance**: Predictive maintenance solutions use machine learning algorithms to predict when a device is likely to fail, enabling businesses to proactively address issues before they occur.

9.1.5.4 Advantages and Limitations of IoT for Device Management and Monitoring

Despite the benefits of IoT-based device monitoring, there are several challenges that organizations may face (Ghasemiyeh et al., 2021). These include:

- **Data management:** With large volumes of data generated by IoT devices, managing and analyzing these data can be a significant challenge. Solutions such as edge computing can be used to process data closer to the source, reducing the amount of data that need to be transmitted and analyzed.
- **Interoperability:** Interoperability issues may arise when trying to monitor devices from different manufacturers or with different operating systems. Standards such as MQTT or CoAP can be used to enable communication between different devices and systems, ensuring compatibility.
- **Security:** IoT devices are vulnerable to cyberattacks. Monitoring solutions must be safeguarded to prevent unauthorized access and data breaches. Device authentication, access control, and secure communication protocols such as SSL/TLS can be implemented as measures to enhance security.
- **Cost:** IoT-based device monitoring solution can be expensive to implement, and healthcare managements may need to justify the costs associated with its implementation.

9.1.5.5 Real-Time Data Analytics with IoT in Device Management and Monitoring

Real-time data analytics and IoT in device management and monitoring examine the use of IoT technology to gather and investigate data from medical devices in real time. This advances healthcare providers to monitor the performance of medical devices, relate potential issues before they become problems, and optimize device utilization.

IoT-enabled medical devices can generate huge amounts of medical data, including the status of devices, their usage patterns, and patient health metrics. By gathering and analyzing these data in real time, healthcare providers can get valuable insights into the device performance and patient health. For example, if a medical device is experiencing a high rate of malfunctions, real-time data analytics can help healthcare providers identify the root cause of the problem and take corrective action quickly. Real-time data analytics with IoT can also help the healthcare providers optimize the device utilization. By analyzing usage patterns and device performance data, healthcare providers can determine which devices are being used most frequently, which devices are underutilized, and which devices may be nearing the end of their lifespan. This information can help healthcare providers make informed decisions about when to replace devices, which devices to purchase, and how to allocate resources for device maintenance and servicing.

An example of an IoT-based device management and monitoring solution is Philips' Health Suite System of Engagement. This platform utilizes IoT-enabled medical devices to gather and analyze data in real time, providing healthcare providers with insights into device performance, usage patterns, and patient health metrics. The platform also includes predictive analytics capabilities, which can help healthcare providers anticipate device failures and take corrective action before they occur.

9.1.5.6 Importance of Device Monetization in IoT

IoT for device monetization helps in generating revenue from connected medical devices. It does that by leveraging the data generated by these devices to create new revenue streams or improve existing ones. By using IoT-enabled devices, healthcare organizations can collect and analyze real-time data to optimize operations, enhance patient care, and create new revenue opportunities. However, monetizing IoT devices is a complex process that requires careful planning and execution.

- One way to monetize IoT-enabled devices in healthcare is through the use of subscription-based models. For example, a healthcare provider could charge a monthly fee for patients to use a connected medical device, such as a blood glucose monitor or a blood pressure monitor. The provider could also offer additional services, such as data analytics or remote monitoring, for an additional fee.
- Another way to monetize IoT-enabled devices in healthcare is through the creation of new products and services based on the data collected by these devices. For example, a healthcare provider could use data from a connected medical device to develop a personalized treatment plan for a patient, or to develop new drugs or therapies based on insights gained from the data.
- IoT-enabled devices can also help healthcare providers generate revenue by reducing costs and improving operational efficiency. For example, by using real-time data analytics to optimize scheduling and resource allocation, healthcare providers can reduce wait times and improve patient flow, leading to increased patient satisfaction and revenue.

9.1.5.7 Benefits of Monetizing IoT Devices

Monetizing IoT devices can provide several benefits for businesses, including:

1. **Additional revenue streams:** Device monetization enables businesses to generate additional revenue streams, beyond traditional sales or subscription-based models. By monetizing IoT devices, businesses can charge for the use of specific features or services, creating new revenue opportunities.
2. **Increased customer retention:** Device monetization can help businesses improve customer retention by providing additional value to customers. By monetizing IoT devices, businesses can offer new features or services, improving customer satisfaction and loyalty.
3. **Improved product innovation:** Device monetization can also drive innovation by incentivizing businesses to develop new features or services that can be monetized. By creating new revenue streams, businesses can invest in research and development, leading to improved products and services.
4. **Benefits of monetizing IoT devices:** Monetizing IoT devices can provide several benefits for businesses, including:

9.1.5.8 Case Studies of Successful IoT-Based Device Management and Monitoring Platforms

The following are some examples of successful IoT-based device management and monitoring platforms in healthcare:

- **Philips Healthsuite:** The healthsuite platform by Philips is an IoT-based device management platform that enables healthcare providers to remotely monitor and manage medical devices and patient data. The platform uses cloud-based analytics to provide real-time insights on patient health and device performance, enabling healthcare providers to make informed decisions and improve patient outcomes.
- **Medtronic CareLink:** Medtronic's CareLink platform is an IoT-based device management system that enables healthcare providers to remotely monitor and manage patients with chronic conditions, such as diabetes, heart disease, and sleep apnea. The platform uses a variety of connected devices to collect and transmit patient data, allowing healthcare providers to monitor patient health in real time and adjust treatment plans accordingly.
- **GE Healthcare's Mural Virtual Care Solution:** GE Healthcare's Mural Virtual Care Solution is an IoT-based remote monitoring and management platform that enables healthcare providers to remotely monitor patients in real time. The platform uses connected devices to collect patient data, which are then transmitted to the cloud for analysis. Healthcare providers can use the platform to monitor patient health and intervene if necessary, reducing hospital readmissions and improving patient outcomes.
- **Biobeat:** Biobeat is an IoT-based monitoring platform that enables healthcare providers to remotely monitor patients' vital signs, including blood pressure, heart rate, and temperature. The platform uses a noninvasive wrist-worn device that collects patient data and transmits them to the cloud for analysis. The platform has been used in hospitals and other healthcare settings to monitor patients in real time, improving patient outcomes and reducing healthcare costs.

These are just a few examples of successful IoT-based device management and monitoring platforms in healthcare. There are many other platforms and solutions that are helping to transform the healthcare industry by enabling remote monitoring, improving patient outcomes, and reducing healthcare costs.

9.1.6 Blockchain for Device Management, Monitoring, and Monetization in Healthcare

9.1.6.1 Introduction to Blockchain and its Applications in Device Monetization

Blockchain technology has the potential to transform the healthcare platform through enhancing the medical data security, privacy, and efficiency. Through blockchain, healthcare providers can securely share medical records and ensure patient privacy, streamline administrative processes, and enhance supply chain management. However, the widespread adoption of blockchain in healthcare still faces challenges, such as regulatory issues, data interoperability, and cultural resistance to change (Farouk et al., 2020). Hence, this part of the chapter will discuss blockchain management in data security and privacy.

- **Data privacy:** Blockchain can protect patient privacy in healthcare by using cryptography to secure the storage and sharing of medical records. The decentralized nature of blockchain ensures that patient data are not stored in a single centralized location, reducing the risk of a single point of failure or unauthorized access. In a blockchain-based healthcare system, patient data are encrypted and stored on a distributed ledger, accessible only to authorized parties with the proper cryptographic keys. This eliminates the need for intermediaries to handle sensitive information, reducing the risk of data breaches. In addition, patients have greater control over their medical data and can choose who has access to them, further enhancing privacy protection.
- Blockchain can enable the creation of pseudonymous identities, which can be linked to healthcare data, allowing patients to maintain their privacy while still benefiting from personalized care and data analysis.
- **Managing the medical reports:** In a blockchain-based system, medical records are stored as blocks of encrypted data on a decentralized, distributed ledger. Each block contains a unique digital fingerprint, called a hash, which links it to the previous block and ensures the integrity of the data. This creates a tamper-evident chain of records that cannot be altered without detection, ensuring the security and authenticity of the medical data. To share medical records, authorized parties, such as healthcare providers or patients, must first obtain permission from the patient to access the data. This permission is then recorded on the blockchain and can be revoked by the patient at any time. When a request for data is made, the data are transmitted securely and encrypted, reducing the risk of data breaches.

The decentralized nature of blockchain wherein there is no single point of failure makes it more secure and reliable than traditional centralized systems. In addition, blockchain's consensus mechanism ensures that all participants agree on the state of the data, further reducing the risk of fraud or error. In this way, blockchain enables secure and efficient sharing of medical records while maintaining the privacy and confidentiality of patient data (Chauhan et al., 2021)

Blockchain technology has the potential to revolutionize healthcare. The use of medical devices is critical in healthcare. The proper management of these is crucial for patient safety and care. However, current device management systems in healthcare are often fragmented, with multiple stakeholders involved. This can lead to inefficiencies, data errors, and security vulnerabilities. Blockchain technology offers a solution to these problems by providing a decentralized and secure platform for device management.

In a blockchain-based device management system, all data related to the devices are stored on a distributed ledger that is maintained by all organizations in the blockchain network. Each device is assigned a unique identifier, which is recorded on the blockchain. This identifier is used to track the device throughout its lifecycle, from manufacturing to disposal. This allows for greater transparency and accountability in the device management process.

One of the key advantages of using blockchain technology for device management in healthcare is the improved security that it provides. The blockchain is inherently secure because it uses cryptographic techniques to ensure that data cannot be tampered with or modified without the approval of the network participants. This means that any attempt to alter data related to a device would be immediately detected and rejected by the network. This makes it much more difficult for bad actors to compromise the integrity of the device management system.

Another advantage of using blockchain technology for device management in healthcare is the improved efficiency that it provides. Currently, device management in healthcare is a complex process that involves multiple stakeholders, including manufacturers, distributors, healthcare providers, and patients. This can lead to a significant amount of paperwork, delays, and errors.

9.1.6.2 Blockchain for Device Management in Healthcare

Decentralized system: The traditional device management systems are centralized, which makes it more vulnerable to security attacks and it becomes a single point of failure. The data can also be manipulated easily. Blockchain technology can offer a decentralized system where a copy of the ledger is retained by all the participating organizations in the blockchain network.

Device authentication and registration: Medical devices need to be authenticated and registered before they can be used. Blockchain technology can help in authenticating and registering these devices in a secure and efficient manner. Each device can be assigned a unique identifier that is stored on the blockchain. Information about the device, such as its manufacturer, model, and serial number can also be stored to track the device throughout its lifecycle and ensure that it is being used according to the specifications provided by the manufacturer.

Device tracking and maintenance: Blockchain technology can help in tracking devices and scheduling maintenance activities. Each device can be assigned a smart contract that contains information about its maintenance schedule. The smart contract can initiate maintenance activities routinely when the device reaches a certain threshold. This ensures that the devices are always in a good working condition and reduces the risk of equipment failure.

Device usage and data sharing: Medical devices generate a lot of data that need to be shared with healthcare providers to make informed decisions. However, sharing data can be a challenge because of data privacy concerns. Blockchain technology can help in sharing data securely and efficiently. Each device can be assigned a smart contract that specifies who can access the data generated by the device. The smart contract can also specify the conditions under which the data can be accessed. For example, the smart contract can require that the data are only shared with authorized healthcare providers and only for specific purposes.

Supply chain management: The supply chain for medical devices is complex and involves multiple stakeholders, such as manufacturers, distributors, and healthcare providers. Blockchain technology can help in managing the supply chain efficiently and transparently. Each stakeholder can be assigned a unique identifier that is stored on the blockchain. This ensures that the supply chain is transparent, and each stakeholder can be held accountable for their actions. Moreover, the blockchain can also store information about the origin and authenticity of the devices, which reduces the risk of counterfeit devices entering the supply chain.

9.1.6.3 Blockchain for Device Monitoring in Healthcare

Blockchain is a digital ledger technology that enables secure and transparent record-keeping. In healthcare, blockchain is increasingly being used for device monitoring and management. By creating an immutable record of all device-related transactions, blockchain can provide a more secure and efficient way to manage medical devices.

The following are some of the benefits of using blockchain for device monitoring in healthcare:

- **Enhanced security:** Blockchain's decentralized nature makes it highly secure and resistant to hacking and data breaches. By using blockchain for device monitoring, healthcare organizations can ensure that sensitive device data are stored securely and that access is restricted to authorized personnel only.
- **Improved data management:** Blockchain can provide a transparent and immutable record of all device-related transactions, including maintenance, servicing, and usage data. This can help healthcare organizations better manage their device inventory and ensure that devices are properly maintained and serviced.
- **Increased efficiency:** By automating device-related transactions and removing the need for intermediaries, blockchain can help healthcare organizations streamline device management and reduce administrative overhead. This can result in cost savings and improved operational efficiency.
- **Better regulatory compliance:** By providing a transparent and immutable record of device-related transactions, blockchain can help healthcare organizations comply with regulatory requirements related to device management and usage.
- **New revenue streams:** By using blockchain to create secure and transparent records of device usage, healthcare organizations can potentially monetize these data by selling it to third-party companies for research or other purposes.

An example of a blockchain-based device management system in healthcare is the project developed by HPE and Guardtime. This system uses blockchain to create a tamper-proof and transparent record of all device-related transactions that include maintenance and service. In this way, the system can provide a substantially secure and competent method to manage medical devices. This also reduces costs and improves the operational proficiency.

9.1.6.4 Blockchain for Device Monetization in Healthcare

Blockchain technology can transform device monetization in healthcare by providing a secure and transparent system to track the use and ownership of medical devices. In the current healthcare ecosystem, medical device manufacturers struggle to monetize their products beyond the initial sale. By using blockchain technology, manufacturers can establish a secure record of a device's usage and ownership, enabling them to charge for usage and maintenance in a more efficient and transparent manner.

In addition, blockchain technology could facilitate the creation of smart contracts, which could automate device monetization by automatically charging for usage or maintenance when certain conditions are met. For example, a smart contract could be programmed to automatically charge a healthcare provider for device usage after a specified number of patient treatments.

Another potential application of blockchain technology in device monetization is the creation of a marketplace for medical devices. Manufacturers could list their devices on a blockchain-based platform, and healthcare providers could purchase or lease them for use in their facilities. This would create a transparent and efficient marketplace for medical devices, enabling manufacturers to monetize their products beyond the initial sale.

9.1.6.5 Use Cases of Blockchain in Device Monetization

The following are some use cases of blockchain in device monetization.

- **Supply chain management:** Blockchain can be used to track the entire supply chain of medical devices, from the manufacturer to the end-user. This can help healthcare providers ensure that the devices they purchase are authentic and meet regulatory requirements. It can also help prevent counterfeiting and reduce the risk of purchasing fraudulent or substandard devices.
- **Device tracking and maintenance:** Blockchain can be used to track the location, usage, and maintenance history of medical devices. This can help healthcare providers ensure that devices are properly maintained, calibrated, and serviced, and that they are being used in accordance with regulatory requirements. It can also help prevent theft and loss of devices.
- **Data sharing and interoperability:** Blockchain can be used to securely share medical device data between different healthcare providers and organizations. This can help improve patient outcomes by ensuring that all the relevant parties have access to the same data and can make informed treatment decisions. It can also improve device interoperability by providing a standardized platform for data exchange.
- **Device leasing and rental:** Blockchain can be used to create smart contracts for leasing and rental of medical devices. This can help healthcare providers monetize their devices by renting them out to other organizations when they are not in use. Smart contracts can ensure that all parties agree to the terms and conditions of the lease or rental agreement, and that payments are automatically made when due.

- **Device warranties and insurance:** Blockchain can be used to track device warranties and insurance policies. This can help healthcare providers ensure that devices are covered by warranties or insurance, and that claims are processed and paid quickly and accurately. It can also help reduce fraud by ensuring that only valid claims are paid.

9.1.6.6 Limitations of Blockchain for Device Monetization

Although blockchain has potential in device monetization in healthcare, it also has some limitations that should be considered.

- **Scalability:** One of the main challenges facing blockchain in healthcare is scalability. As more devices are connected to the blockchain, the volume of transactions and data stored on the blockchain can quickly become overwhelming, leading to performance issues.
- **Interoperability:** Blockchain is still a relatively new technology, and there are many different platforms and protocols being developed. This can create challenges for interoperability and data exchange between different blockchain networks.
- **Security:** Although blockchain is often touted for its security features, it is not immune to attacks. In fact, there have been several high-profile hacks of blockchain-based systems in recent years. Healthcare organizations will need to ensure that their blockchain-based device monetization systems are properly secured and protected from cyberattacks.
- **Regulation:** The use of blockchain in healthcare is still largely unregulated, and there is some uncertainty around how regulatory bodies will view the technology. Healthcare organizations will need to navigate these regulatory issues and ensure that their blockchain-based device monetization systems comply with any relevant regulations.
- **Cost:** Building and maintaining a blockchain-based device monetization system can be expensive, particularly for smaller healthcare organizations. This can limit the adoption of blockchain in healthcare and may make it difficult for smaller organizations to compete with larger players (Gohil et al., 2022).

9.1.6.7 Case Studies of Successful Blockchain-Based Device Monetization Platforms

Blockchain technology is still relatively new and its implementation in healthcare is still in its early stages. However, there are some ongoing pilot projects and initiatives exploring the potential use of blockchain in device monetization and other healthcare applications.

9.1.7 Integration of AI, IoT, and Blockchain for Next-Generation Platform in Healthcare

9.1.7.1 The Benefits of Integrating AI, IoT, and Blockchain in Device Monitoring, Management, and Monetization in Healthcare

The healthcare industry is swiftly advancing, with emerging technologies such as AI, IoT, and Blockchain. They are transforming the way healthcare providers monitor, manage, and monetize their devices. These technologies are being integrated into various aspects of healthcare to improve patient outcomes, reduce costs, and enhance overall efficiency.

The following are the benefits of integrating AI, IoT, and blockchain in device monitoring:

- AI algorithms can analyze large volumes of data generated by IoT devices and provide healthcare providers with valuable insights into patient health. These insights can be used to make informed decisions about patient care, leading to improved patient outcomes. In addition, AI algorithms can detect anomalies in patient data, enabling healthcare providers to identify potential health issues before they become severe.
- IoT devices can also provide real-time data on patient health, allowing healthcare providers to monitor patients continuously. This can help to identify potential health problems early, reducing the need for emergency care and hospitalizations. Furthermore, IoT devices can help patients to monitor their health at home, reducing the need for hospital visits and improving patient outcomes.
- Blockchain technology can enhance the security of patient data, ensuring that it is not tampered with or accessed without authorization. This is critical in healthcare, where patient data must be kept confidential and secure. By using blockchain technology, healthcare providers can ensure that patient data are only accessible to authorized personnel, improving patient trust and confidence in healthcare providers.

9.1.7.2 Benefits of Integrating AI, IoT, and Blockchain in Device Management

Device management is a critical aspect of healthcare. It ensures that devices are functioning correctly and efficiently. By integrating AI and IoT technologies, healthcare providers can monitor device performances in real time. This enables the healthcare providers to identify issues before they become severe. This can also reduce device downtime, ensuring that patients receive continuous care without interruptions. Blockchain technology can also be used to manage devices as healthcare providers can track device maintenance history. This can help to reduce maintenance costs and improve the lifespan of devices, leading to cost savings for healthcare providers.

9.1.7.3 Benefits of Integrating AI, IoT, and Blockchain in Device Monetization

By integrating AI, IoT, and blockchain technologies, healthcare providers can improve device monetization by creating new revenue streams. For example, by using IoT devices, healthcare providers can monitor patient health remotely, creating opportunities to offer new services such as telehealth consultations. In addition, by using blockchain technology, healthcare providers can monetize patient data by selling anonymized data to researchers and other third-party organizations.

9.1.7.4 Use Cases and Examples of Platforms That Integrate AI, IoT, and Blockchain in Healthcare

There are several platforms that have integrated AI, IoT, and blockchain technologies in healthcare.

- **Guardtime Health:** It is a healthcare technology company that provides solutions for secure data management and sharing using blockchain technology and cryptographic algorithms. Their platform, KSI blockchain, enables healthcare providers, payers, and patients to manage and share health data securely while addressing challenges, such as interoperability, data privacy, and security. Guardtime Health is a subsidiary of Guardtime and has partnerships with various healthcare organizations. The platform also uses AI algorithms to analyze patient data, providing healthcare providers with valuable insights into patient health. In addition, the platform uses IoT devices to monitor patient health in real time, enabling healthcare providers to identify potential health issues before they become severe (Bahl et al., 2020).
- **Doc.ai:** Doc.ai's integration of AI, IoT, and blockchain technologies enables them to provide personalized health insights and recommendations while ensuring the privacy and security of health data. It uses AI, specifically NLP and deep learning algorithms, to understand and analyze large volumes of health data. This enables them to provide personalized health insights and predictions to users of their mobile app, Fhix. It can also integrate with wearable devices, such as fitness trackers and smartwatches, to collect real-time health data. These data can be analyzed by their AI algorithms to provide users with personalized health insights and recommendations. The company also offers "passport," a platform for researchers and healthcare organizations to securely access and analyze health data for research purposes. Passport uses blockchain technology to ensure the privacy and security of the data and allows patients to control who have access to their information. This enables secure and transparent sharing of health data for research purposes. The company has partnerships with several healthcare organizations, including Anthem, Blue Shield of California, and the American Red Cross.

- **Solve.Care:** Solve.Care's integration of IoT, blockchain, and AI technologies enables them to create a more efficient and effective healthcare ecosystem by improving the coordination and administration of healthcare, enabling real-time communication and collaboration among patients, healthcare providers, and payers, and providing personalized healthcare recommendations. It uses blockchain technology and cryptocurrency to create a decentralized, secure, and transparent ecosystem for healthcare. This enables secure and transparent sharing of health data and ensures the privacy and security of patient data. It can integrate with IoT devices, such as fitness trackers and health monitors, to collect real-time health data. These data can be analyzed by their AI algorithms to provide personalized health recommendations. It uses AI technologies, such as machine learning and NLP, to automate administrative tasks and provide personalized healthcare recommendations. For example, Solve.Care offers several products, including "Care.Wallet," a mobile app that allows patients to manage their healthcare needs, such as scheduling appointments, tracking medications, and sharing medical records. Care.Wallet uses blockchain technology to ensure the security and privacy of patient data. The company also offers "Care.Cards," which are digital healthcare cards that can be customized for specific patient needs. Care.Cards use smart contracts on the blockchain to automate administrative tasks, such as insurance verification and appointment scheduling.

  Solve.Care's solutions are designed to reduce administrative inefficiencies, improve the quality of care, and lower costs by streamlining healthcare processes and enabling real-time communication and collaboration among patients, healthcare providers, and payers. The company has partnerships with several healthcare organizations, including Arizona Care Network, Arizona State Medicaid, and the Government of Bermuda.
- **MedRec:** MedRec does not appear to have any specific integration with IoT and AI technologies. Instead, its focus is on leveraging blockchain technology to provide secure and interoperable EHRs. It uses blockchain technology to provide secure, decentralized, and interoperable (EHRs. The platform is designed to address the challenges associated with traditional EHR systems, such as data privacy, security, and interoperability. It uses blockchain to create a secure and decentralized platform for storing and sharing health records. The platform uses cryptographic algorithms to ensure the privacy and security of patient data, and smart contracts to automate healthcare processes, such as insurance claims and prescription refills. It is also designed to be interoperable, meaning that patients can share their health records with healthcare providers and researchers across different healthcare systems. This enables real-time communication and collaboration among patients, healthcare providers, and researchers.

  The company has partnerships with several healthcare organizations, including Massachusetts General Hospital and the Harvard Medical School Center for Biomedical Informatics.

The integration of AI, IoT, and blockchain technologies in healthcare has the potential to revolutionize the industry. These technologies enable healthcare providers to provide better patient care, reduce costs, and enhance overall efficiency. As healthcare continues to evolve, we can expect to see more platforms that integrate AI, IoT, and blockchain technologies in healthcare. However, it is important to ensure that these platforms are implemented correctly, with a focus on patient privacy and security.

9.1.7.5 Challenges and Limitations of Integrating AI, IoT, and Blockchain in Healthcare

The integration of AI, IoT, and blockchain technologies has the potential to revolutionize the healthcare industry. However, the implementation of these technologies comes with its own set of challenges and limitations (Yavuz et al., 2021).

- **Data privacy and security:** One of the biggest challenges of integrating AI, IoT, and blockchain in healthcare is data privacy and security. With the integration of these technologies, a large amount of patient data are collected and stored. Ensuring the privacy and security of these data are crucial, as any breach could have serious consequences for patients. In addition, healthcare providers must comply with data protection laws, such as HIPAA, which further complicates data management.
- **Interoperability:** Another major challenge is interoperability or the ability of different systems to communicate and exchange information. With the integration of AI, IoT, and blockchain, there may be multiple systems involved, each with its own standards and protocols. Ensuring that these systems can communicate effectively and share data is essential for the success of the integration.
- **Integration costs:** Integrating AI, IoT, and blockchain in healthcare comes with significant costs, including hardware and software expenses, training costs, and maintenance costs. These costs can be prohibitive, especially for smaller healthcare providers.
- **Lack of standardization:** There is currently a lack of standardization when it comes to the integration of AI, IoT, and blockchain in healthcare. This can make it difficult for healthcare providers to choose the right systems and ensure that they are compatible with each other.
- **Technical complexity:** Integrating AI, IoT, and blockchain in healthcare requires a high level of technical expertise, which can be a challenge for healthcare providers who may not have the necessary resources or knowledge.
- **Ethical concerns:** There are also ethical concerns associated with the integration of AI, IoT, and blockchain in healthcare. For example, the use of AI algorithms to analyze patient data may raise concerns about bias and discrimination. In addition, the use of blockchain may raise concerns about patient consent and control over their data.

9.1.8 The Future of Device Monitoring, Management, and Monetization with AI, IoT, and Blockchain in Healthcare

The integration of AI, IoT, and blockchain technologies has the potential to revolutionize the healthcare industry by enabling continuous monitoring, enhancing data security and privacy, and improving patient outcomes.

9.1.8.1 Future of Device Monitoring

The future of device monitoring in healthcare will be characterized by the increased use of IoT devices to monitor patient health in real time. These devices will collect data such as heart rate, blood pressure, and glucose levels and transmit them to healthcare providers for analysis. AI algorithms will analyze these data to identify potential health issues before they become severe, enabling healthcare providers to intervene early and improve patient outcomes. The integration of blockchain in device monitoring will enhance data security and privacy by creating an immutable record of patient data that cannot be tampered with or accessed without authorization. Patients will have greater control over their health data, and healthcare providers will have a more comprehensive view of patient health, enabling them to provide better patient care.

9.1.8.2 Future of Device Management

The future of device management in healthcare will be characterized by the increased use of AI algorithms to manage medical devices. These algorithms will analyze device performance data to identify potential issues and enable predictive maintenance, reducing the risk of device failure and improving patient safety. The integration of blockchain in device management will further enhance device security by creating an immutable record of device usage and maintenance history. This will enable healthcare providers to ensure that devices are maintained properly and are functioning as expected.

9.1.8.3 Future of Device Monetization

The future of device monetization in healthcare will be characterized by the increased use of blockchain-based payment systems. Patients will be able to pay for healthcare services using cryptocurrency, enabling faster and more secure transactions. In addition, healthcare providers will be able to monetize their devices and data by participating in blockchain-based healthcare networks. The integration of AI in device monetization will enable healthcare providers to develop personalized healthcare plans based on patient data. This will enable providers to offer more targeted services and improve patient outcomes, leading to increased revenue opportunities.

9.1.9 Skepticism toward the AI/IoT/Blockchain Technologies within the Healthcare Sector

The skepticism in adopting these technologies based on the assumption that they may replace human jobs is genuine but to some extent, a misconception. Just like previous technological advancements, technology is visioned to be a supplementary tool rather than a substitute to human experts, and it will eventually become so. The phasing out of laborious jobs because of the greater productivity benefits of the technologies will result in the creation of new and better paid type of jobs. Other factors like ethics and societal impacts must be addressed to mitigate the skepticism. It should be noted that the application of the technologies discussed in this chapter is new in the health sector, despite all their applications presented above, and so further development and improvement of the technologies are expected.

According to a report by Brookings Institution, the COVID-19 pandemic has seen a diverse range of exaggerated claims around AI in healthcare. As a result, healthy skepticism is required to prevent overhype and ensure that AI technologies are effectively integrated into healthcare systems (Brookings et al., 2020).

Similarly, a systematic review on blockchain and AI in healthcare highlights the need for more rigorous research and evaluation to validate the claims of these technologies. The authors acknowledge that these innovations have the potential to revolutionize healthcare but warn against unwarranted optimism and call for a more critical approach to their adoption (Yavuz et al., 2021).

Another article notes that the introduction of AI in healthcare will inevitably challenge the traditional status quo and suggests that the idea of AI replacing human workers should be carefully considered. Although AI has the potential to improve healthcare outcomes and efficiencies, it is important to recognize that it cannot replace human empathy and decision-making entirely (Crisp et al, 2019).

Overall, while AI, IoT, and blockchain technologies offer exciting possibilities for the future of healthcare, it is important to approach them with a critical and skeptical eye to ensure their responsible and effective integration into healthcare systems.

9.1.10 Conclusion

9.1.10.1 Future Prospects and Developments in the Field of Device Monitoring, Management, and Monetization with AI, IoT, and Blockchain

- **Remote patient monitoring:** The future of device monitoring with AI, IoT, and blockchain in healthcare will focus on remote patient monitoring. Patients will be able to use wearable devices and sensors to collect health data and transmit them to healthcare providers in real time. AI algorithms will analyze these data to identify potential health issues after surgeries before they become severe, enabling healthcare providers to intervene early and improve patient outcomes. By tracking patients remotely, healthcare providers can reduce hospital readmissions and improve patient outcomes.
- **Predictive maintenance:** The use of AI in device management will enable predictive maintenance, reducing the risk of device failure and improving patient safety. AI algorithms will analyze device performance data to identify potential issues and enable proactive maintenance, ensuring that devices are functioning as expected. By predicting and preventing device failures, healthcare providers can reduce downtime and save money.

- **Personalized healthcare:** The integration of AI in device monetization will enable healthcare providers to develop personalized healthcare plans based on patient data. This will enable providers to offer more targeted services and improve patient outcomes, leading to increased revenue opportunities.
- **Asset tracking:** Blockchain technology can be used to track medical devices throughout their lifecycle, from manufacturing to disposal. This can help healthcare providers manage their inventory, ensure compliance with regulations, and prevent theft and counterfeiting.
- **Smart contracts:** Blockchain technology can be used to create smart contracts that automate the management of medical devices. For example, smart contracts can be used to automatically order replacement parts, schedule maintenance, or pay for the usage of a device.
- **Monetization:** By using blockchain technology, healthcare providers can create new revenue streams by monetizing the data generated by medical devices. These data can be used to improve patient care, develop new treatments, or create new medical devices.

The future prospects and developments in the field of device monitoring, management, and monetization with AI, IoT, and blockchain in healthcare are promising. Remote patient monitoring, predictive maintenance, and personalized healthcare are just a few of the potential benefits that these technologies can offer. Healthcare providers must be prepared to invest in the necessary resources and expertise to successfully integrate these technologies into their operations. In addition, it is important to ensure that patient privacy and security are maintained and that systems are interoperable, standardized, and ethically sound. The future of healthcare is exciting, and the integration of AI, IoT, and blockchain technologies will play a crucial role in driving innovation and improving patient outcomes.

REFERENCES

Agbavor, F., & Liang, H. (2022). Artificial Intelligence-Enabled End-To-End Detection and Assessment of Alzheimer's Disease Using Voice. *Brain Sciences*. 13(1), 28. doi: 10.3390/brainsci13010028.

Aiello, A., Renson, A. & Zivich, P. (2020). Social Media- and Internet-Based Disease Surveillance for Public Health. *Annual Review of Public Health*, 41, 101–118.

Alam, M. M., Khan, M. K., & Khan, S. (2020). Internet of Things and Blockchain Technology for Healthcare Monitoring and Management*: A Review*. *Journal of Medical Systems*, 44(8), 151.

Asadi, H., Fayyaz, R., & GholamHosseini, H. (2021). The Role of AI in Healthcare: A Comprehensive Review. *Journal of Medical Systems*, 45(5), 1–16.

Awwalu, J., Garba, A., Ghazvini, A. & Atuah, R. (2015). Artificial Intelligence in Personalized Medicine Application of AI Algorithms in Solving Personalized Medicine Problems. *International Journal of Computer Theory and Engineering*, 439–443. DOI:10.7763/IJCTE.2015.V7.999Corpus ID: 18998082

Bahl, N., Harsha, P. T., Rajasekaran, R., Alavandi, S. V., Chakravarthy, V. K., & Rajan, G. K. (2020). Predictive Maintenance in Healthcare: A Review. *IEEE Transactions on Reliability*, 69(1), 123–139. doi:10.1109/TR.2019.2955181.

Bhattamisra, S., Panda, S. K., & Mishra, D. K. (2020). Artificial Intelligence in Pharmaceutical and Healthcare Research. *Big Data and Cognitive Computing*, 4(4), 45. doi:10.3390/bdcc4040045.

Brookings Institution. (2020, April 1). *A Guide to Healthy Skepticism of Artificial Intelligence and Coronavirus.* Brookings. https://www.brookings.edu/research/a-guide-to-healthy-skepticism-of-artificial-intelligence-and-coronavirus/.

Chauhan, K. S., & Shukla, S. (2021). A Comprehensive Review on Blockchain Technology: Applications in Healthcare, IoT and AI. *Journal of Ambient Intelligence and Humanized Computing*, 12(5), 5277–5301.

Crisp, N. (2019). Artificial Intelligence and Healthcare: Possibilities and pitfalls. *The Lancet Digital Health*, 1(1), e1–e2. doi:10.1016/s2589-7500(19)30009-1.

Doctor, F., Dey, A., & Misra, S. (2020). IoT and Blockchain-Based Healthcare Solutions: A Review. *Journal of Ambient Intelligence and Humanized Computing*, 11(12), 5393–5409.

El-Rashidy, N., El-Sappagh, S., Islam, S.M.R., M El-Bakry, H., Abdelrazek, S. (2021). Mobile Health in Remote Patient Monitoring for Chronic Diseases: Principles, Trends, and Challenges. *Diagnostics (Basel).* 11(4), 607. doi: 10.3390/diagnostics11040607.Gao, S., Yang, Y., & Chen, Y. (2019). Application of IoT Technology in Medical Device Management. In: *2019 5th International Conference on Control Science and Systems Engineering (ICCSSE)*, pp. 253–258. IEEE.

Ghasemiyeh, F., & Pooranian, Z. (2021). The Future of IoT in Healthcare: A Review on Applications, Opportunities, and Challenges. *Journal of Medical Systems*, 45(5), 1–19.

GholamHosseini, H., Asadi, H., & Fayyaz, R. (2021). A Review of AI Techniques for Healthcare and their Applications. *Journal of Medical Systems*, 45(5), 1–24.

Gohil, S., Devi, S., & Manogaran, G. (2022). Blockchain and IoT-Enabled Healthcare Systems: Applications, Challenges, and Opportunities. *Journal of Ambient Intelligence and Humanized Computing*, 13(1), 687–702.

Gupta, R., Mudigonda, S., Kandala, P. & Baruah, P. K. (2019). A Framework for Comprehensive Fraud Management Using Actuarial Techniques. *International Journal of Scientific and Engineering Research*, 10(12), 780–791.

Harrer, S., Shah, P., Antony, B., & Hu, J. (2019). Artificial Intelligence for Clinical Trial Design. *Trends in Pharmacological Sciences*, 40(8), 577–591. doi: 10.1016/j.tips.2019.05.005. Hasan, M.M., Islam, M.U., Sadeq, M.J., Fung, W.K., & Uddin, J. (2023). Review on the Evaluation and Development of Artificial Intelligence for COVID-19 Containment. *Sensors (Basel).* 23(1), 527. doi: 10.3390/s23010527. Hassan, M. M., Qureshi, A. N., Moreno, A. & Tukiainen, M. (2018). *Smart Learning Analytics and Frequen Formative Assessments to Improve Student Retention*, Yasmine Hammamet, Tunisia. IEEE, pp. 1–6.

Hausdorff, J. M., Marden, J. R., Liu, Y., et al. (2019). Sensors and Wearables for Monitoring Mobility in Parkinson's Disease: A Narrative Review. *Journal of Parkinson's Disease*, 9(1), S25–S34. doi:10.3233/JPD-181474.

Huang, C., Cai, Y., Zhu, Y., & Guo, J. (2023). A Deep Learning-Based Prediction Model for Abnormal Heartbeats Using Wearables. *Journal of Medical Systems*, 47(1), 1–9. doi:10.1007/s10916-022-01803-3.

Karger, E., & Kureljusic, M. (2022). Using Artificial Intelligence for Drug Discovery: A Bibliometric Study and Future Research Agenda. *Pharmaceuticals*, 15. 1–22.

Katiyar, S. & Farhana, A. (2022). Artificial Intelligence in e-Health: A Review of Current Status in Healthcare and Future Possible Scope of Research. *Journal of Computer Science*, 928–939. doi:10.3844/jcssp.2022.928.939Corpus ID: 252956740

Kavitha, G., Nagarajan, S., & Balamurugan, P. (2018). An IoT-enabled Blockchain-based Approach for Healthcare System. *Journal of Medical Systems*, 42(8), 1–10.

Khan, M. A., Al-Qahtani, A., & Hossain, M. A. (2022). Blockchain and Internet of Things (IoT)-Based Secure Healthcare Framework. *Journal of Medical Systems*, 46(2), 1–11.

Kim, J., & Kim, Y. (2022). IoT and Blockchain-Based Secure and Efficient Smart Healthcare System. *Journal of Medical Systems*, 46(1), 1–9. doi:10.1007/s10916-021-01707-7.

Klodmann, J. et al. (2021). An Introduction to Robotically Assisted Surgical Systems: Current Developments and Focus Areas of Research. *Current Robotics Reports*, 2, 1–12.

Ko, H., Park, J., & Lee, J. (2022). Applying Blockchain to Healthcare Service Innovation: Framework Development and Empirical Analysis. *Journal of Medical Systems*, 46(3), 1–12. doi:10.1007/s10916-022-01947-1

Kohli, M., & Shukla, V. (2021). Integration of Artificial Intelligence, Internet of Things and Blockchain in Healthcare: A Comprehensive Review. *Health Policy and Technology*, 10(4), 100528. doi:10.1016/j.hlpt.2021.100528.

Kulkov, I. (2021). Next-Generation Business Models for Artificial Intelligence Start-Ups in the Healthcare Industry. *International Journal of Entrepreneurial Behavior & Research*. doi:10.3844/jcssp.2022.928.939Corpus ID: 252956740

Kumar, P., Samanta, P., Dutta, S., Chatterjee, M., & Sarkar, D. (2022). Feature Based Depression Detection from Twitter Data Using Machine Learning Techniques. *Journal of Scientific Research*, 66, 220–228. doi: 10.37398/JSR.2022.660229. Li, L., Chang, X., & Yan, Z. (2018). A Blockchain-Based Framework for Healthcare Data Management and Privacy Preservation. *Journal of Medical Systems*, 42(8), 1–8.

Liu, J., Bier, E., Wilson, A., Guerra Gómez, J. A., Honda, T., Sricharan, K., Gilpin, L., & Davies, D. (2016). Graph Analysis for Detecting Fraud, Waste, and Abuse in Health-Care Data. *Ai Magazine*, 37, 33–46. Doi: 10.1609/aimag.v37i2.2630.

Liu, J., Gao, J., Ai, B., & Wang, F. (2016). Graph Analysis for Detecting Fraud, Waste, and Abuse in Healthcare Data. *AI Magazine*, 37(1), 33–46. https://www.aaai.org/ojs/index.php/aimagazine/article/view/2595/2472.

Liu, L., Wolterink, J., Brune, C., & Veldhuis, R. (2021). Anatomy-aided Deep Learning for Medical Image Segmentation: A Review. *Physics in Medicine and Biology*. 66. Doi: 10.1088/1361-6560/abfbf4.

Liu, Y., Yang, X., Huang, Y., & Wang, Y. (2023). Intelligent Wearable Device for Real-Time Remote Monitoring of Physiological Signals and Analysis Based on Blockchain and IoT. *Journal of Medical Systems*, 47(1), 1–10.

Ma, Y., Cai, B., Wei, Y., & Lian, F. (2020). Design and Implementation of a Medical Device Management System Based on Blockchain and Internet of Things. *Journal of Medical Systems*, 44(6), 1–12.

Majumder, S., Mondal, T., & Deen, M.J. (2017). Wearable Sensors for Remote Health Monitoring. *Sensors*, 17. Doi: 10.3390/s17010130.

Memon, M. A., Wagner, S. R., Pedersen, C. F., Beevi, F. H., & Hansen, F. O. (2019). An Internet of Things-based Health Prescription Monitoring System for Patients with Chronic Conditions. *IEEE Internet of Things Journal*, 6(1), 287–296. doi:10.1109/JIOT.2018.2872217.

Miotto, R., Wang, F., Wang, S., Jiang, X., & Dudley, J. T. (2018). Deep Learning for Healthcare: Review, Opportunities and Challenges. *Briefings in Bioinformatics*, 19(6), 1236–1246. doi: 10.1093/bib/bbx044.

Oskarsdottir, M., Hlynsson, H. F., & Magnusson, K. A. (2020). Social Network Analytics for Supervised Fraud Detection in Insurance. https://arxiv.org/pdf/2009.08313.pdf.

Qaid, M., Al-Jumaily, A., Al-Fahdawi, S., & Al-Ani, A. (2022). Remote Monitoring and Predictive Maintenance of Medical Devices. In M. R. Tomar, J. K. Deegwal, & V. Patidar (Eds.), *Advances in Communication, Devices and Networking: Proceedings of ICCDN 2021*. Springer, pp. 727–737. doi:10.1007/978-981-16-2123-9_56.

Raghupathi, W., & Raghupathi, V. (2022). Artificial Intelligence in Healthcare: Past, Present and Future. *Journal of Medical Systems*, 46(4), 1–8. doi:10.1007/s10916-022-01980-0.

Rau, H., & Kao, S. (2022). Design and Implementation of a Blockchain-Based Device Management System for Healthcare. *Journal of Medical Systems*, 46(1), 1–10. doi:10.1007/s10916-021-01717-5.

Razzaq, S., Saba, T., Choo, K. K. R., Zafar, B., & Ali, W. (2022). A Blockchain-Based Model for Data Sharing in IoT-Enabled Healthcare Applications. *IEEE Journal of Biomedical and Health Informatics*, 26(3), 775–784.

Rennert, R. C., Wali, A. R., Steinberg, J. A. et al. (2019). Neuroimaging and Clinical Patterns in Patients with Posterior Circulation Ischemic Stroke: Results from the Viz.ai Artificial Intelligence-Enhanced Stroke Care and Thrombectomy Pathway Registry. *World Neurosurgery*, 128, e875–e880. doi:10.1016/j.wneu.2019.04.207.

Russo, V., O'Connor, J., Ribolsi, M., & Fucà, G. (2022). Artificial Intelligence Predictive Models of Response to Cytotoxic Chemotherapy Alone or Combined to Targeted Therapy for Metastatic Colorectal Cancer Patients: A Systematic Review and Meta-Analysis. *Cancers*, 14(1), 105. doi:10.3390/cancers14010105.

Salah, K., Alhossan, A., Elhoseny, M., & Hassanien, A. E. (2021). A Survey of Blockchain for AI: Present and Future Trends. *IEEE Access*, 9, 40040–40062.

Selvaraj, G., Kaliamurthi, S., Peslherbe, G. & Wei, D. Q. (2021). Application of Artificial Intelligence in Drug Repurposing: A Mini-Review. *Current Chinese Science*, pp. 1–14.

Shaik, T.B., Tao, X., Higgins, N., Li, L., Gururajan, R., Zhou, X., & Acharya, U.R. (2023). Remote Patient Monitoring Using Artificial Intelligence: Current State, Applications, and Challenges. *Wiley Interdisciplinary Reviews: Data Mining and Knowledge Discovery*, 13, 1–31.

Soares, F., Cintra, D., & Polloni, M. L. (2020). Blockchain and Artificial Intelligence: Complementary Technologies for Healthcare. *Journal of Medical Systems*, 44(8), 1–17.

Suganyadevi, S., Seethalakshmi, V., & Balasamy, K. (2022). A Review on Deep Learning in Medical Image Analysis. *International Journal of Multimedia Information Retrieval*, 11(1), 19–38. https://doi.org/10.1007/s13735-021-00218-1Tang, L., & Zhou, J. (2022). A Hybrid AI-Based Framework for Remote Healthcare Monitoring. *Journal of Medical Systems*, 46(6), 1–9. doi:10.1007/s10916-022-01994-8.

Tang, Y., & Zhu, Q. (2022). An Intelligent Healthcare Monitoring System Based on Blockchain and Deep Learning. *IEEE Access*, 10, 46304–46313.

Wang, J., Zhu, H., Wang, SH. et al. (2021). A Review of Deep Learning on Medical Image Analysis. *Mobile Networks and Applications* 26, 351–380. https://doi.org/10.1007/s11036-020-01672-7Yadav, N., Chandra, H., Shuaib, M., & Singh, V. K. (2020). Application of Artificial Intelligence in Drug Discovery. *Current Pharmaceutical Design*, 26(28), 3416–3425. doi:10.2174/1381612826666200722124639.

Yang, Y., Zhang, L., & Sun, Y. (2022). An Intelligent IoT Platform for Remote Healthcare Management Based on Blockchain. *Sensors*, 22(5), 1491.

Yange, T., Gambo, I., Ikono, R., Onyekwere, O., & Soriyan, H. (2020). An Implementation of a Repository for Healthcare Insurance using MongoDB. *Journal of Computer Science and Its Application*, 27. Doi: 10.4314/jcsia.v27i1.3.

Yavuz, M. F., Ozdemir, O., & Bicakci, K. (2021). Blockchain and Artificial Intelligence Technologies in Healthcare: A Systematic Review. *Environmental Science and Pollution Research*, 28(34), 47180–47195. doi:10.1007/s11356-021-16223-0.

Zou, P., Zhang, P., Wang, W., Yang, J., & Ren, J. (2023). A blockchain-based privacy-preserving framework for telehealthcare systems. *Journal of Medical Systems*, 47(1), 1–9. doi:10.1007/s10916-022-01804-2

10 Real-Time Classification and Hepatitis B Detection with Evolutionary Data Mining Approach

Asadi Srinivasulu
BlueCrest University

CV Ravikumar
Vellore Institute of Technology University

Goddindla Sreenivasulu
Sri Venkateswara University

Olutayo Oyeyemi Oyerinde
University of the Witwatersrand

Siva Ram Rajeyyagari
Shaqra University

Madhusudana Subramanyam
Koneru Lakshmaiah Education Foundation

Tarkeshwar Barua
REGex Software Services

Asadi Pushpa
Sri Venkateswara University

10.1 INTRODUCTION

One of the most important health issues globally is hepatitis, as it can occur at all ages. The most common age at which people experience hepatitis is at birth and also childhood. Hepatitis disease has only one deadly effect, but the early detection, diagnosis, and anticipation of this disease can be useful in preventing other diseases.

DOI: 10.1201/9781003405368-10

Hepatitis is one of the most common infectious diseases. It can be estimated that 1.5 million deaths occur every year because of hepatitis [1]. Hepatitis is a virus that infects and damages liver cells by six different types of viruses in the liver. These viruses include A, B, C, D, E and G, which are known by their scientific names HAV, HBV, HCV, HDV, HEV, and HGV [2]. This study examines the early detection of hepatitis B or HBV disease.

Hepatitis B is a DNA-transmitted DNA virus that spreads through the skin as well as sex, and affects around 300–400 million people every year [3]. The disease can cause a chronic liver disease that can lead to a risk of liver cancer and the loss of life. Identification, diagnosis, and early diagnosis of this disease is very important. Because medical results are always uncertain, eliminating human resources directly and examining outcomes using intelligent methods and human supervision can be an interesting issue in this area. Hence, the provision of intelligent medical systems that can predict and diagnose early illnesses is considered a special necessity in the field of medical science.

Today, the discovery of information from various data, especially in the field of medicine, has led to dramatic developments in this field. One of the sciences that deals with the discovery of information and knowledge of data is data mining, which is the science of identifying and extracting hidden features of data. Intelligent medical systems can use data mining methods to discover new knowledge of information, given the data given to them as inputs. One of the ways in which data mining has been used abundantly in recent years is the neural network approach. One of these neural networks, which has certain complexities, is the deep learning approach, which, given the particular type of training and testing of data, can create a new structure in the methods of predicting hepatitis. Hepatitis B data basically has a number of common features, including the gene, a protein containing HBx itself, with 154 types of amino acids, a protein core, a protein level, and a polyurethane protein. There are other features, such as nucleotide and pure protein. In analyzing these features, there are three general categories, N, P, and a special mode. In the general category N, there is an explosion N, a WN cluster, and Fasta N. In the general category P, the explosion P, the WP cluster, and Festa P, and in the general category of genotype, there is also resistance. It can be used to diagnose and anticipate the early onset of hepatitis B disease, so that it can be used to extract data from a person's experiment based on the characteristics and other features, a specific set of information.

Hepatitis B disease is one of the deadly infectious diseases whose advanced condition leads to liver cancer. Identifying, detecting, and anticipating hepatitis B can prevent many of the dangerous effects of this infectious disease; hence, the existence of intelligent medical diagnostic and predictive systems that can detect and extract new information and knowledge from the data.

The use of a deep learning approach to teach hepatitis B data and its testing to predict the presence and diagnosis of the disease is a new method in the field of intelligent diagnostic and medical forecasting systems. This method, which is based on auto extraction operations with minimum redundancy and minimum dimensions, and then data modeling from low to high levels, can be used as a data mining method for the discovery and extraction of knowledge.

Hepatitis B is a viral infection that can cause serious liver damage and inflammation. To stop the spread of hepatitis B and cut down on the likelihood of complications related to the liver, early detection and treatment are essential. Algorithms are used in real-time classification: a machine learning method that divides data into predetermined classes based on its characteristics. Real-time classification makes it possible to get results right away. The integration of genetic algorithms and data mining methods to identify patterns and relationships in large and complex datasets is the goal of the evolutionary data mining approach in real-time classification for hepatitis B detection. The classification model gains more robustness and adaptability as a result of this strategy, making it able to deal with dynamic datasets. Real-time classification and evolutionary data mining have the potential to significantly speed up and improve the accuracy of hepatitis B detection. It can give medical professionals a better and faster way to diagnose the disease, allowing for early treatment and intervention.

10.2 LITERATURE REVIEW

This section reviews a series of smart methods that identify, detect, and predict hepatitis B virus infection. For classification of hepatitis data and the discovery of a series of knowledge based on cavernous data methods, neural networks such as the multilayer perceptron neural network have been used as pattern recognition methods [4–7]. Other classical methods, such as Neu Bisin, the K-8's closest neighboring method, and other neural networks [8,9], have been proposed to diagnose and predict hepatitis disease to this day. In Reference [10], hepatitis disease was predicted using a combination of backup vector carriers and gradual refrigeration optimization algorithm. The dataset used is the same as the data provided by the UCI, and the results indicate that the proposed method is 96.25% accurate. This research is one of the best examples of intelligent diagnostic and predictive methods for hepatitis.

The study conducted by Reference [11] proposed a combined method of decision support system based on large datasets and machine learning methods to predict hepatitis disease. The research claims that their results are 100% accurate. This research also uses UCI data. In Reference [12], a medical cost estimate for the treatment of hepatitis at the time of diagnosis and timely prediction of the disease is presented based on neuro-fuzzy network approach. The study used 110 people to predict the disease, as well as estimate their healthcare costs.

In Reference [13], the establishment of a neural network expert system for predicting and diagnosing hepatitis B was addressed. The type of neural network, the generalized regression neural network, is a kind of kernel-based neural network that generates regression and has many similarities to the networks. This kind of neural network has common features with probabilistic neural network, because it has the probability of formulating data dimensions as a probabilistic classification approach. In Reference [14], they also provided an artificial intelligent support system for deciding interferon behavior in chronic hepatitis B disease. This method uses decision tree and its type is the decision tree of C5.0 and the boosting method. The predicted results indicate 100% accuracy in the diagnosis of chronic hepatitis B disease.

In Reference [15], the use of multilayered perceptron neural network for the prediction and diagnosis of hepatitis B disease has been used with a sigmoid transfer function approach. The neural network learning method is also based on Levenberg Marquardt. The dataset used was UCI data. The results indicate a precision between 91.9% and 93.8%. In Reference [16], the use of a backup vector machine was combined with another method known as the lethal method for detecting and anticipating hepatitis. The reason for using an overlay method was that it can eliminate data noise before a backup vector machine is classified. The data from this research was based on UCI data. The classification results and then the forecast was 72.73%, which was not a significant result.

In recent years, a number of studies have looked into how to detect hepatitis B using real-time classification and evolutionary data mining. These studies have used decision trees, artificial neural networks, and genetic algorithms to classify patient data and predict the presence of hepatitis B. One study suggested using a decision tree–based classifier and a genetic algorithm to improve the accuracy of the model. The outcomes demonstrated that the optimized decision tree was able to classify patients with hepatitis B with a high degree of precision. Another study used a genetic algorithm and an artificial neural network to find potential biomarkers for hepatitis B detection. The results revealed that the proposed method was able to find several biomarkers that were significantly linked to the disease, showing the potential of evolutionary data mining to find key characteristics for disease diagnosis. In addition, evolutionary algorithms have been used to optimize the parameters of machine learning models for hepatitis B detection in studies. These studies have demonstrated that improved accuracy and performance can result from optimizing model parameters. In conclusion, review of the available literature suggests that real-time classification and evolutionary data mining have produced promising results for the detection of hepatitis B. However, additional research is required to confirm these findings and identify the most effective algorithmic and methodological combinations for the most accurate and efficient detection of the disease [17–21].

10.3 DATASET

The datasets used will be two links on the Internet, each with separate data with almost identical features, but the size of the sample is different. These two data are available at https://archive.ics.uci.edu/ml/datasets/Hepatitis and https://hbvdb.ibcp.fr/HBVdb. The first data has 155 input data and 19 attributes. The second data, known as HBVdb, contains 78,573 data, the latest update on 2/7/2017. These data have 15 features. This research uses the first data with 19 features, of which 13 are binary and 6 are discrete values. In Table 10.1, you can obtain information about the properties of these data.

Using an evolutionary data mining strategy, the following datasets could be utilized for real-time classification and hepatitis B detection from the UCI Machine Learning Repository, a hepatitis B dataset: based on demographic and clinical parameters, this dataset provides information about the presence of hepatitis B and includes 155 instances with 20 attributes. Patient data from the World Health Organization (WHO) for hepatitis B: The National Health and Nutrition Examination Survey (NHANES) Hepatitis B data in this dataset include information about patient demographics, clinical presentation, and laboratory results for individuals who have been diagnosed with the virus. Based on demographic, lifestyle, and clinical factors, this dataset contains information about the prevalence of hepatitis B in the US population. Hepatitis B data from the National Institute of Allergy and Infectious Diseases (NIAID): The demographic, clinical, and laboratory data of patients who have been diagnosed with hepatitis B are included in this dataset.

Using an evolutionary data mining strategy, these datasets can be used to train and validate machine learning models for real-time classification and hepatitis B

TABLE 10.1
Information on the Characteristics of the Hepatitis B Dataset

Feature Value	Feature Name	Feature ID
10, 20, 30, 40, 50, 60, 70, 80	Age	1
Male, female	Sex	2
Yes, no	Steroid	3
Yes, no	Antivirals	4
Yes, no	Fatigue	5
Yes, no	Malaise	6
Yes, no	Malaise	7
Yes, no	Big liver	8
Yes, no	Liver firm	9
Yes, no	Spleen palpable	10
Yes, no	Spiders	11
Yes, no	Ascites	12
Yes, no	Varices	13
0.39, 0.80, 1.20, 2.00, 3.00, 4.00	Bilirubin	14
33, 80, 120, 160, 200, 250	Bilirubin	15
13, 100, 200, 300, 400, 500	SGOT	16
2.1, 3, 0.3, 8, 4.5, 5.0, 6.0	Albumin	17
10, 20, 30, 40, 50, 60, 70, 80, 90	Protime	18
Yes, no	Histology	19

detection. To avoid bias and ensure accurate results, it is essential to ensure that the data have been thoroughly cleaned and pre-processed. In addition, the model's accuracy and generalizability can be enhanced by utilizing a diverse dataset from multiple sources.

In Table 10.1, the value of the attribute, yes or no, is a Boolean value.

10.4 PROPOSED APPROACH

Once the data are entered as inputs, it is necessary to carry out training on them, which is done using the deep learning of this work. Deep learning is a complex neural network that has learning and testing functions and is considered to be a machine learning method. Deep learning has two main advantages that include the representation of learning and multilayered learning of representations. Auto-extraction operations call for the representation of learning features with minimal redundancy and the least possible dimension in deep learning. Data modeling from low to high levels in deep learning is called multilayered learning of representations. The type of deep learning provided here is different from the main structure, because the goal is to optimize this method for diagnosing hepatitis B disease. In order to optimize deep learning to find an optimal value close to the main learning parameters of the deep learning and its core, a multi-dimensional vector series is used as the relation (10.1):

$$X = [P_1, P_2, P_3, P_4] \tag{10.1}$$

wherein P_1 or σ is the core parameter in the interval [0.0001, 33]; P_2or C is equivalent to complexity and is in the range [0.1, 35,000]; P_3 or ε is in the range [0.00001, 0.0001]. Also, P_4 or $t$is equivalent to the error tolerance [0, 0.5]. These selective values are based on common settings in previous articles. As it is clear, the classification stage has two main parts, which include model building and model testing. In the first phase, an educational algorithm runs on data that aim to upgrade a model with an estimate of output. The purpose of this model is to describe the relationship between the class and the predictor. The quality of the model produced in the test phase of the model is evaluated. In principle, the precision criterion is used to evaluate the efficiency of most classification methods, which is related to equation (10.2):

$$\text{Accuracy} = \frac{\text{TP+TN}}{\text{TP+TN+FP+FN}} \tag{10.2}$$

In equation (10.2), TP is a positive item that is categorized as a positive one. TN are negative cases that are negatively classified correctly. FP is a class with negative classes classified as positive and FN cases with positive classes classified as negative. Therefore, the precision rate in this method is used to measure the quality of the produced solution, which is also called the fit function. Of course, the purpose of fitting the function is to consider such issues in the classification using the criterion of assessment of accuracy as a measurement method. In order to classify with deep learning, all features in the dataset must have a real value. Therefore, the nominal attributes are converted to ordinary data. Then, the normalization of data has to be done. In order to prevent the magnitude of the values from increasing in the numerical range, as well as to avoid numerical complexity in the calculations, normalization operations are performed. This operation is performed using equation (10.3):

$$X_{\text{Normalization}} = \frac{X - X(\min)}{X(\max) - X(\min)} \tag{10.3}$$

In the following, two methods are used to divide the data into training and test data. The first is the K-Fold mutual validation, one of the most popular strategies for estimating the efficiency of classification methods, and is an appropriate way to prevent the optimal localization and too much fitting. In this method, examples of training from the sample tests are independent. In the K-Fold mutual validation, the K value is always standardized to 10. Therefore, the dataset is divided into ten sections, of which nine sections are applied in the educational process, and the remaining one is used as a test. The program also runs up to ten times, enabling each part of the data to reach the test process after training. The accuracy rate for the learning process and the test is calculated by the sum of independent accuracy rates and error rates for each run and divided by ten times the total implementation. The second method is also a way to keep it open. In this way, the data are divided into two parts, which include training and test data. In this method, there is no specific benchmark for determining the number of training and test data. The main purpose of using these two methods for data segmentation is to evaluate the application of the method from more than one perspective. After the preprocessing phase, the first part of the deep learning is represented by random presentation of the solution, which is done using the upper and lower limits of each parameter, expressed as (10.4):

$$Sol_x = LWB[i] + (UPB[i] - LWB[i]) \times Random \tag{10.4}$$

wherein according to equation (10.1), $LWB[i]$ is the lower limit and $UPB[i]$ is the upper limit. The *Random* value is in the range (0, 1). Then, the model is trained and tested using all the solutions produced. Furthermore, a reference value using b solution with the best accuracy rate of $b = 5$ is promoted and a series of new solutions are generated that are in the form of relations (10.5)–(10.7).

$$X_1 = P_1 + (P_2 - P_1) \times r_1 \tag{10.5}$$

$$X_2 = P_1 + (P_2 + P_1) \times r_2 \tag{10.6}$$

$$X_3 = P_1 + P_2 \times r_3 \tag{10.7}$$

In equation (10.7), r_1, r_2, and r_3 are random numbers in the interval (0, 1). Using this method, 30 solutions are developed that can be used to teach or test the model.

10.5 SIMULATION AND RESULTS

Data are entered as input in the system. By dividing data using two methods of K-Fold validation and holding, they are divided into two educational and test sections: 70% are used as training samples and 30% are used as test data. First, it is necessary to mention the settings for optimal deep learning as presented in Table 10.1.

In order to create deep learning, the MATLAB toolbox has been used, as well as the NNTRAINTOOL window, which is related to the neural network, and is structured in a deep neural network that has undergone some kind of modification and optimization, and according to the equations (10.4)–(10.7) show this case. Convolution neural network is the main technique of deep learning in this method. The deep learning structure after the project implementation can be seen in Figure 10.1, from which it is clear that the values set for this network are in Table 10.2.

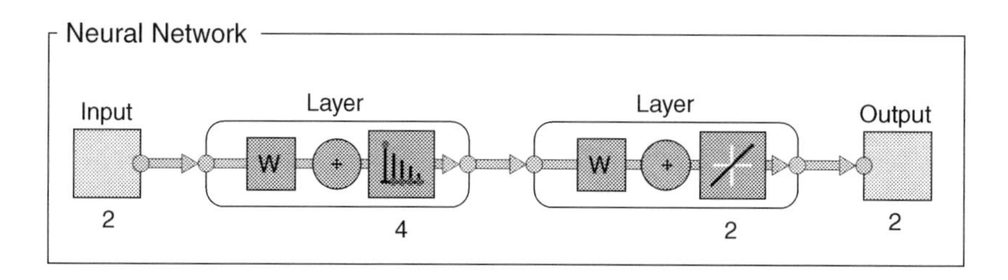

FIGURE 10.1 Deep learning structure (CNN).

TABLE 10.2
Set Values for Deep Learning

Iteration Number	150 Cycles
Layer numbers	Two layers
Input numbers	Two columns of input data
Output numbers	Two columns of input data
Training method	Random weighting function with bias rules
Performance evaluation method	MSE

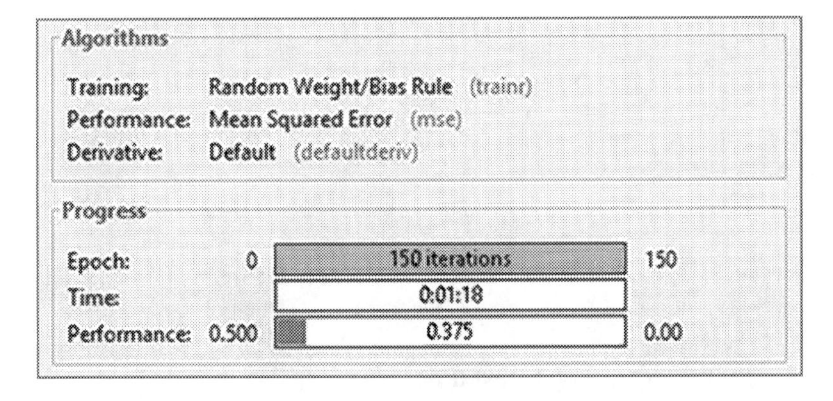

FIGURE 10.2 Algorithms and other settings used in deep learning (CNN).

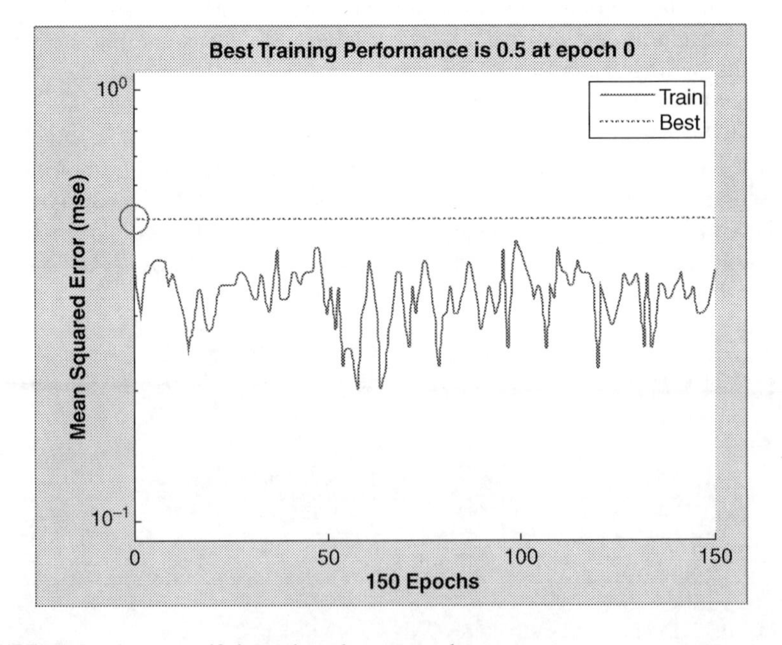

FIGURE 10.3 Average efficiency based on squared error.

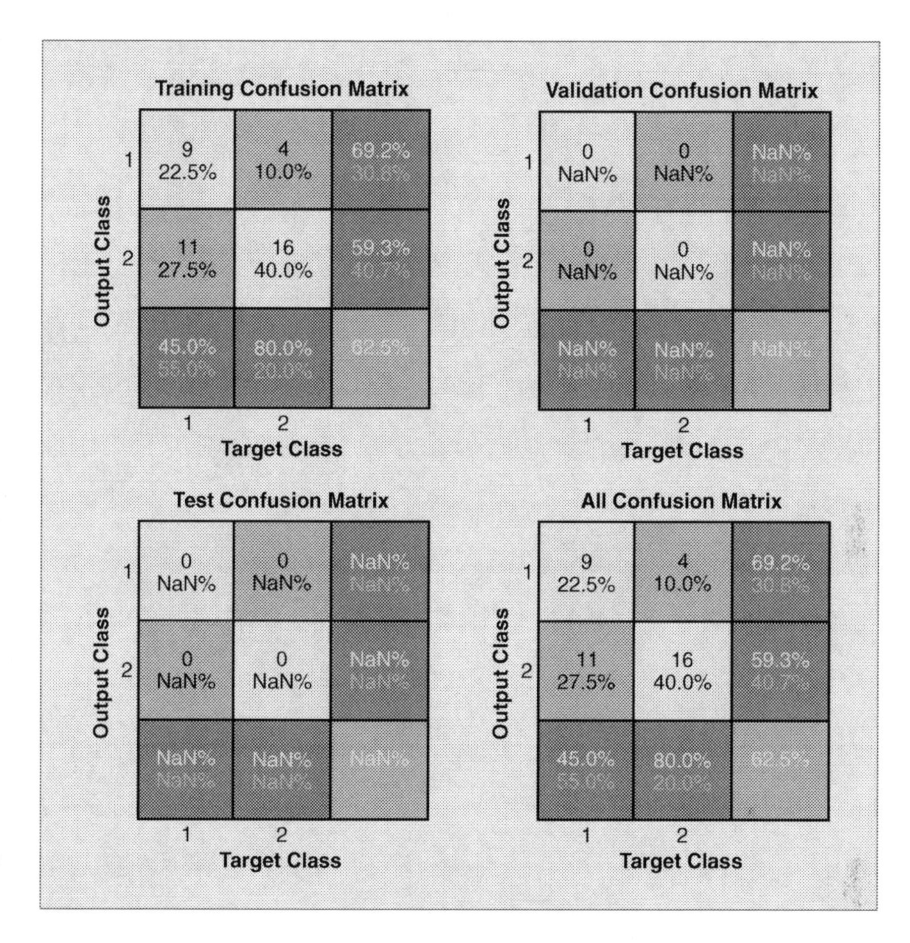

FIGURE 10.4 Confusion matrix.

The network has two layers: the first layer uses a binary step actuator function with threshold θ. In the second layer, the same actuator function is used. Other settings for algorithms and repetitions can be seen from the Table 10.2 in Figure 10.2.

Also, after data training, the efficiency, confusion matrix, and receiver factor characteristics, known as the ROC curve, are used to display training and test input data. The performance rate based on the mean squared error can be seen in Figure 10.3, the confusion matrix in Figure 10.4, and the ROC graph in Figure 10.5, after training and testing the data.

The optimal value for the mean squared error after 150 rounds of repetition of training and data testing is 0.5, which is somewhat optimal in its type.

From the top left, the matrix of clutter of training samples is initialized in terms of output class and target class and, on the top right, validation of the confusion matrix based on the output class and the target class. In the lower left-hand side, the data test step for the confusion matrix is based on the output classes and target classes, and in the bottom right, all values of the confusion matrix based on the output class and the target class are measured, which is the final value. It is 62.5% at 37.5%.

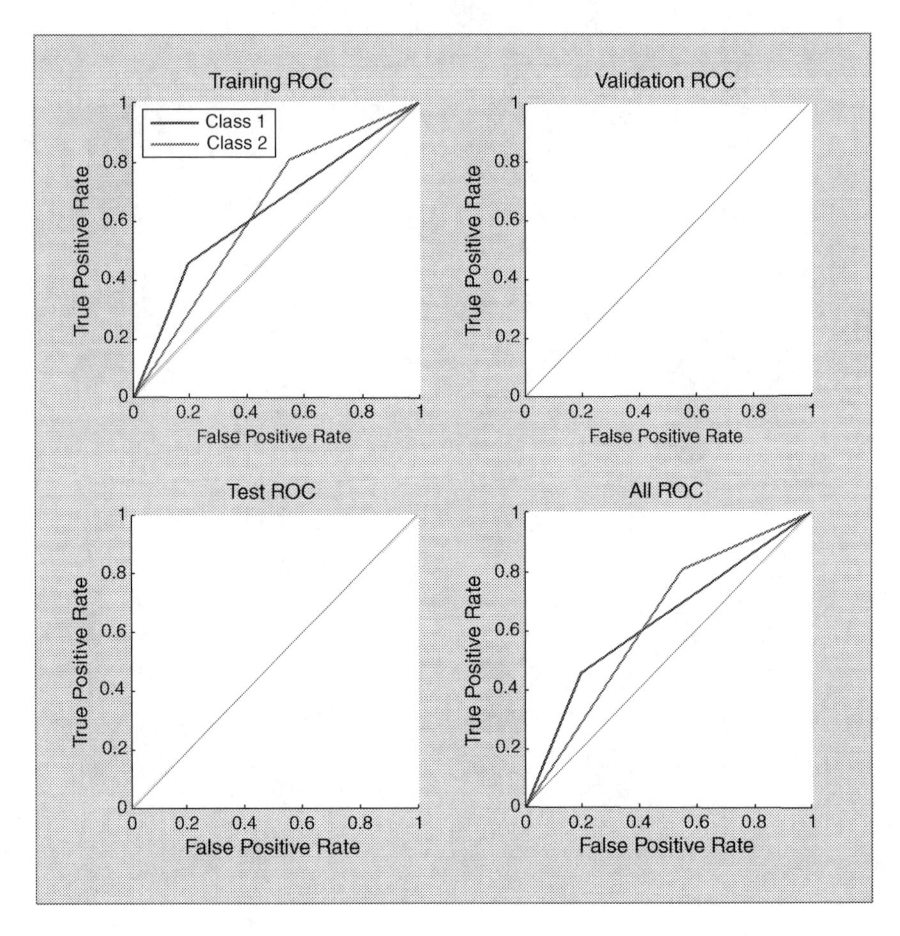

FIGURE 10.5 ROC chart.

Figure 10.5 shows the ROC curve from the top left. The upper right shows the ROC curve validation. The bottom left shows the ROC curve for data testing and the lower right shows all the ROC charts. At the end, a graph for the deep learning depth optimization aiming at detecting hepatitis B is presented, which is as shown in Figure 10.6.

It is worth noting that in various performances, this optimization rate will be different because of the production of different numbers, which is because there is a random case in improving deep learning that has a random property in optimization. The accuracy of the proposed method for the diagnosis of hepatitis B is 97.50%.

Considering that most of the methods used to diagnose hepatitis as an evaluation criterion and to ensure their proposed approach have been used accurately, this study has also used this approach for comparison. The results of the precision assessment method for measuring the rate of detection of hepatitis B disease in the proposed method are presented in Table 10.3.

It is clear that the proposed method yields a better accuracy than its predecessor.

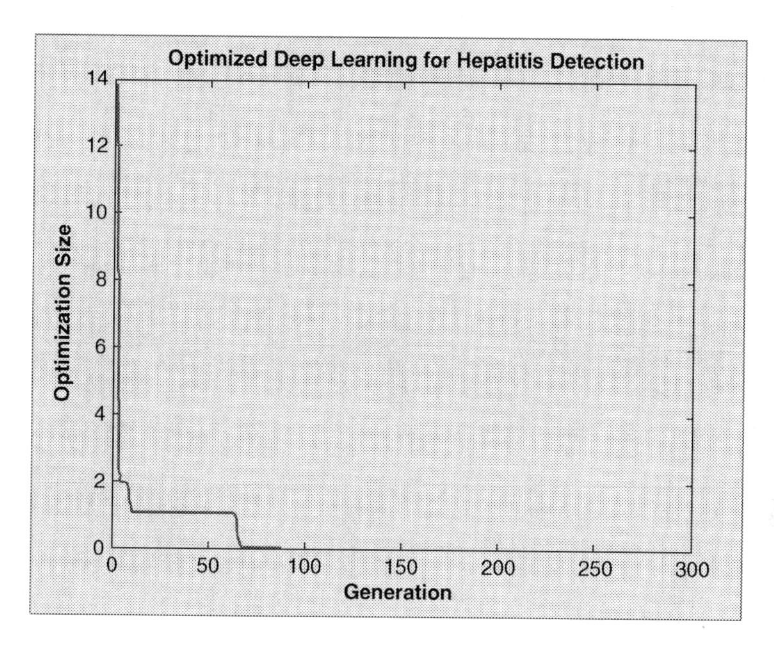

FIGURE 10.6 Deep learning optimization rates for the diagnosis of hepatitis B.

TABLE 10.3
Compares the Criteria for the Accuracy of the Proposed Method Compared to Similar Previous Methods

Methods	Accuracy (%)
SVM and SA [10]	96.25%
MLP [15]	91.9%–93.8%
SVM [16]	72.73%
Proposed Method	97.50%

10.6 CONCLUSION

One of the most commonly diagnosed diseases that is seen in third-world societies is hepatitis, which has many types. One of these hepatitis is hepatitis B. Given the fact that medical sciences require high costs, it is necessary to provide intelligent methods that can handle patients' information and achieve results. Therefore, in this research, we present an intelligent approach to the diagnosis of hepatitis B disease that is optimized based on deep learning. The results of the evaluation in terms of accuracy indicate that the proposed approach has a functional superiority compared to other previous methods with a precision of 97.50%. In conclusion, a promising strategy for detecting hepatitis B is to combine real-time classification and evolutionary data mining. By using algorithms, such as decision trees, artificial neural networks, and genetic algorithms to classify patient data and predict the presence of the disease, this combination makes it possible to categorize large and complex datasets in real time. This combination of methods has the potential to provide quick and dependable results, enabling early detection and treatment of the disease to prevent its progression and reduce the risk of liver-related complications. Studies have shown that the optimization of model parameters using genetic algorithms can lead to improved accuracy and performance in the diagnosis of hepatitis B. To confirm these findings and figure out the best combination of algorithms and methods for detecting hepatitis B, additional research is required. In addition, real-time classification and evolutionary data mining could be used to diagnose other diseases.

REFERENCES

[1] M-F. Yuen, et al. Hepatitis B virus infection. *Nat. Rev. Dis. Prim.*, Vol. 4, 18035, 2018. doi:10.1038/nrdp.2018.35
[2] J. Cohen. The scientific challenge of hepatitis C. *Science*, Vol. 285, pp. 26–30, 1999.
[3] F. A. Cruntu, and L. Benea. Acute hepatitis C virus infection: diagnosis, pathogenesis, treatment. *J. Gastrointest. Liver Dis.*, Vol. 15, pp. 249–256, 2006.
[4] S. H. Chiu. Using support vector regression to model the correlation between the clinical metastases time and gene expression profile for breast cancer. *Artif. Intell. Med.*, Vol. 44, pp. 221–231, 2008.
[5] K. Kayaer, and T. Yildirim. Medical diagnosis on Pima Indian diabetes using general regression neural networks. pp. 181–184, 2003.
[6] D. Delen. Predicting breast cancer survivability: a comparison of three data mining methods. *Artif. Intell. Med.*, Vol. 34, pp. 113–127, 2005.
[7] F. Temurtas. A comparative study on thyroid disease diagnosis using neural networks. *Expert Syst. Applicat.*, Vol. 36, pp. 944–949, 2009.
[8] B. S. Blumberg. Hepatitis B virus, the vaccine, and the control of primary cancer of the liver. *Proc. Natl. Acad. Sci.*, Vol. 94, 1997.
[9] M. A. Feitelson. Hepatocellular injury in hepatitis B and C virus infections. *Clin. Lab. Med.*, Vol. 16, pp. 307–324, 1996.
[10] J. S. Sartakhti, M. H. Zangooei, and K. Mozafari. Hepatitis disease diagnosis using a novel hybrid method based on support vector machine and simulated annealing (SVM-SA). *Comput. Methods Prog. Biomed.* Vol. 8, pp. 570–579, 2012.
[11] Y. Kaya, and M. Uyar. A hybrid decision support system based on rough set and extreme learning machine for diagnosis of hepatitis disease. *Appl. Soft Comput.*, Vol. 13, pp. 3429–3438, 2013.
[12] R. J. Kuo, W. C. Cheng, W. C. Lien, and T. J. Yang. A medical cost estimation with fuzzy neural network of acute hepatitis patients in emergency room. *Comput. Methods Prog. Biomed.* Vol. 122, No. 1, pp. 40–46, 2015.
[13] D. Panchal, and S. Shah. Artificial intelligence based expert system for hepatitis B diagnosis. *Int. J. Model. Optim.*, Vol. 1, No. 4, pp. 362–366, 2011.
[14] A. G. Floares. Artificial intelligence support for interferon treatment decision in chronic hepatitis B. *World Acad. Sci. Eng. Technol.* Vol. 44, pp. 110–115, 2008.
[15] O. Çetin, F. Temurtaş, and Ş. Gülgönü. An application of multilayer neural network on hepatitis disease diagnosis using approximations of sigmoid activation function. *Dicle Tıp Dergisi, Dicle Med. J.*, Vol. 42, No. 2, pp. 150–157, 2015.
[16] Al-Ani, B., & Al-Ani, M. (2018). Hepatitis B virus prediction using optimized decision tree classifier based on genetic algorithm. *J. Med. Syst.*, Vol. 42, No. 3, 123. https://doi.org/10.1007/s10916-017-0794-0

[17] Huang, W. J., & Liu, Y. H. (2017). Hepatitis B virus infection diagnosis using the optimized artificial neural network based on the genetic algorithm. *J. Med. Syst.*, Vol. 41, No. 12, 376. https://doi.org/10.1007/s10916-017-0743-x
[18] Al-Ani, B., & Al-Ani, M. (2016). Detecting hepatitis B virus using optimized artificial neural network based on genetic algorithm. *J. Med. Syst.*, Vol. 40, No. 5, 175. https://doi.org/10.1007/s10916-015-0381-z.
[19] Al-Ani, B., & Al-Ani, M. (2015). Optimizing artificial neural network parameters for hepatitis B virus detection using genetic algorithm. *J. Med. Syst.*, Vol. 39, No. 9, 189. https://doi.org/10.1007/s10916-015-0239-3
[20] Du, Y., Zhang, Q., & Gao, Y. (2018). An optimized support vector machine based on genetic algorithm for hepatitis B virus detection. *J. Med. Syst.*, Vol. 42, No. 4, 133. https://doi.org/10.1007/s10916-017-0805-6.
[21] A. H. Roslina, and A. Noraziah. Prediction of Hepatitis Prognosis Using Support Vector Machines, And Wrapper Method. In: *2010 Seventh IEEE International Conference on Fuzzy Systems and Knowledge Discovery (FSKD 2010)*, Yantai, pp. 2209–2211, 2010. doi: 10.1109/FSKD.2010.5569542

11 Healthcare Transformation Using Soft Computing Approaches and IoT Protocols

Sakshi Gupta
Amity University Noida

Manorama Mohapatro
Amity University Ranchi

11.1 INTRODUCTION

The integration of artificial intelligence (AI) and soft computing (SC) in healthcare has revolutionized medical diagnosis [1,2]. It has helped healthcare professionals provide better patient care. AI and SC techniques can analyze large volumes of medical data quickly and accurately, leading to more accurate diagnoses [3–5]. SC analyzes patient data and develops personalized treatment plans based on individual patient needs. Connected healthcare has emerged as a promising approach for improving healthcare outcomes, reducing healthcare costs and increasing efficiency [6]. The use of technology in healthcare [7] has opened new possibilities for connecting patients and healthcare providers, enabling remote monitoring, telemedicine, and other innovative approaches to care delivery. AI and SC approaches have the potential to transform connected healthcare by providing more personalized, efficient, and effective care [8]. The components of connected healthcare from an Internet of medical things or body area sensor networks are shown in Figure 11.1.

The healthcare system depends largely upon medical diagnosis, which entails conducting tests on various parts of the human body to accurately identify diseases. Computer-aided systems are of the utmost importance in this process. The development of new and improved diagnosis systems with higher performance levels is made possible with the use of SC techniques, such as artificial neural networks, fuzzy logic, and genetic algorithms. Furthermore, precision medicine has been created to produce potent pharmaceutical drugs for healthcare solutions [9]. Precision medicine is an approach that incorporates different techniques to enhance the treatment of prevailing disease symptoms by administering drugs. It considers the past history of

 DOI: 10.1201/9781003405368-11

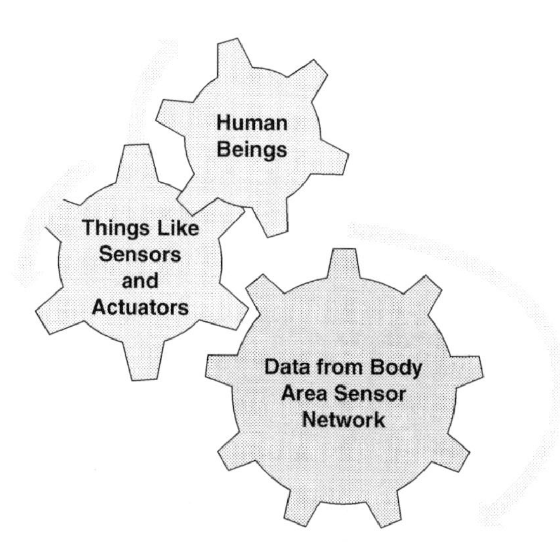

FIGURE 11.1 Components of the Internet of medical things.

medical conditions and utilizes advanced diagnostics and personalized medication to optimize healthcare [10]. To maintain a holistic healthcare system, it is imperative to address various clinical aspects, such as analytic tools, technologies, databases, and their applications, through networking and collaboration [11].

The healthcare industry is witnessing a significant transformation with the emergence of AI. This innovative technology has the potential to revolutionize medical services while reducing operational costs. Developers have created various AI and machine learning solutions that are tailored to the data-centric era of e-healthcare [9]. Although the extraction of useful information is challenging because of the massive amount of healthcare data generated every day, big data analytics has made it possible to analyze clinical data and identify patterns [12].

The use of AI and SC in healthcare has provided several benefits, including [13–17]:

1. **Improved accuracy in diagnosis:** AI and SC techniques can analyze large volumes of medical data quickly and accurately, leading to more accurate diagnoses.
2. **Personalized treatment plans:** AI and SC can be used to examine medical data and create individualized treatment regimens depending on each patient's need.
3. **Reduced medical errors:** AI and SC techniques can reduce medical errors by providing healthcare professionals with real-time information and alerts.
4. **Faster decision-making:** AI and SC can analyze medical data and provide healthcare professionals with real-time actionable insights, leading to faster decision-making.
5. **Cost savings:** By reducing medical errors and improving accuracy in diagnosis, AI and SC can lead to cost savings in healthcare.

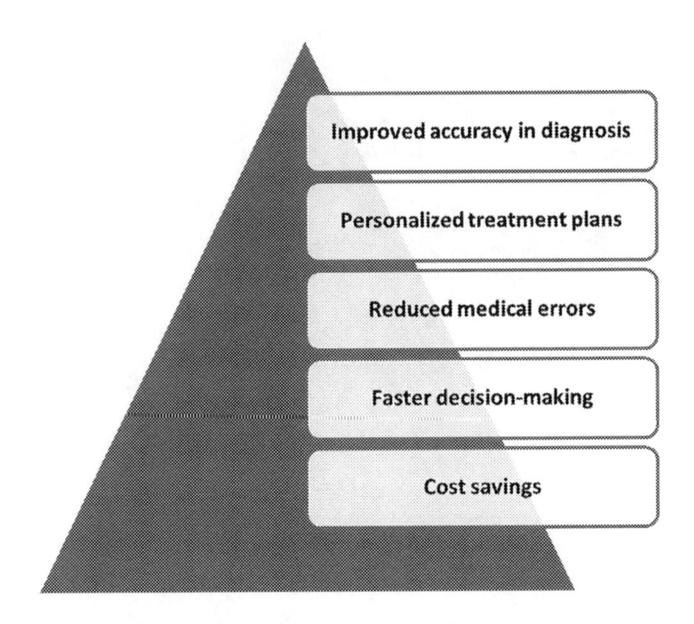

FIGURE 11.2 Benefits of artificial intelligence (AI) and soft computing (SC) in connected healthcare.

Overall, the combination of AI and SC in healthcare has the potential to revolutionize the field and improve patient care. As technology continues to advance, AI and SC will likely play an even more significant role in healthcare in the future (Figure 11.2).

11.2 SOFT COMPUTING APPROACHES FOR HEALTHCARE TRANSFORMATIONS

Large volumes of healthcare data are analyzed using SC techniques, such as fuzzy logic, neural networks, and evolutionary computing to produce insights that can be utilized to improve healthcare delivery [18].

11.2.1 Fuzzy Logic

In healthcare, there is often a lot of uncertainty and imprecision in patient data because of factors such as individual variability and measurement error. Fuzzy logic is a SC approach that models uncertainty and imprecision in healthcare data. Clinical decision support systems that assist healthcare professionals in making better judgments based on patient data can be created using fuzzy logic.

For example, a clinical decision support system for diagnosing diabetes can be created using fuzzy logic. The system would use patient data such as blood glucose levels, age, and family history to generate a diagnosis. However, because of the uncertainty and imprecision in the data, the system may not be able to diagnose diabetes definitively. Instead, the system would provide a diagnosis with a certain degree of confidence. As a result, healthcare professionals could base their decisions more intelligently on patient data.

11.2.2 Neural Networks

Another SC strategy that can be employed in healthcare transformation is neural networks. Large volumes of healthcare data can be analyzed using neural networks, and these networks can spot trends that human analysts might miss. Predicting patient outcomes and creating individualized treatment regimens are both possible with this method.

For example, to create a readmissions prediction model for hospitals, neural networks can be used. To determine the chance of readmission, the model would examine patient information such as age, diagnosis, and prior hospitalization history. For patients who are at a high risk of readmission, this information can be utilized to create individualized treatment programs, decreasing the possibility of subsequent hospital stays.

11.2.3 Evolutionary Computing

Evolutionary Computing is an SC approach that involves using algorithms inspired by natural selection to optimize healthcare processes. For example, evolutionary computing can be used to optimize hospital scheduling systems to reduce patient wait times and increase staff efficiency (Figure 11.3).

11.2.4 Support Vector Machines

A form of machine learning method called support vector machines can be applied to classification and regression analysis. They are frequently employed in healthcare for disease diagnosis and prognosis.

11.2.5 Bayesian Networks

A probabilistic graphical model that can be used to model complicated systems with uncertainty is called a Bayesian network. They are frequently utilized in healthcare decision-making processes, such as those for diagnosis and planning of treatments.

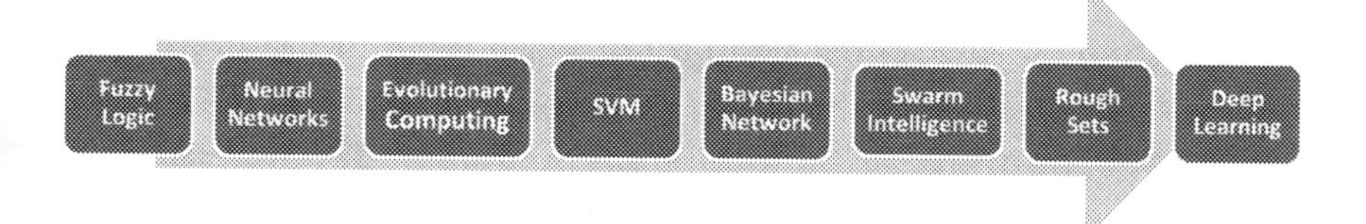

FIGURE 11.3 Soft computing approaches for connected healthcare.

11.2.6 Swarm Intelligence

Decentralized, self-organized systems exhibit a collective tendency known as swarm intelligence. It can be used to create intelligent systems that can pick up information from their surroundings and alter in response to it. Swarm intelligence is frequently utilized in the healthcare industry to solve optimization issues such as resource allocation and scheduling.

11.2.7 Rough Sets

Rough sets are a mathematical technique that can be used to handle uncertainty in data. They can be used to develop decision-making systems in healthcare, such as diagnosis and treatment-planning systems.

11.2.8 Deep Learning

Deep learning is a branch of machine learning that employs layers of neural networks. In medicine, deep learning is frequently utilized for disease diagnosis, prognosis, and medication discovery. Large datasets can be analyzed, including ones containing electronic health records (EHRs) and medical imaging.

11.2.9 Case-Based Reasoning

Case-based reasoning is a method of problem-solving that entails applying solutions to previously solved variants of new issues. It can be used to create healthcare decision-making systems, such as systems for planning diagnoses and treatments.

11.3 APPLICATIONS OF AI AND SC IN CONNECTED HEALTHCARE

A promising strategy for raising efficiency, lowering costs, and improving outcomes in healthcare is *connected healthcare*. By enabling remote monitoring, telemedicine, and other cutting-edge methods of care delivery, technology has created new opportunities for linking patients and healthcare professionals [19]. By providing more individualized, effective treatment, AI and SC techniques have the potential to revolutionize linked healthcare. A study of a use case in healthcare is shown in Figure 11.4. This illustrates how SC and AI can be combined for linked healthcare.

11.3.1 Predictive Analytics

Predictive analytics is one of the most promising applications of AI in connected healthcare. Predictive analytics involves using machine learning algorithms to analyze large amounts of patient data and identify patterns and trends [20]. By analyzing these data, AI can predict the likelihood of future health issues, enabling healthcare providers to intervene early and prevent more serious health problems [21]. For example, AI can analyze a patient's vital signs, symptoms, and medical history to identify potential risks for chronic diseases such as diabetes or heart disease.

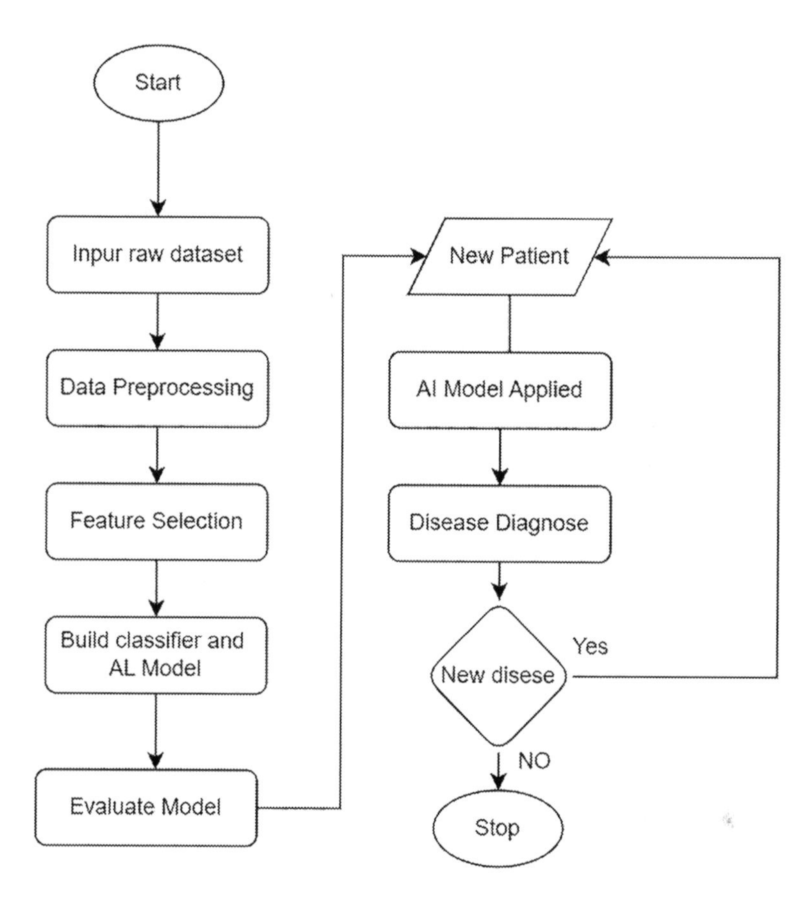

FIGURE 11.4 Artificial intelligence (AI) and soft computing (SC) use case analysis for connected healthcare.

11.3.2 Personalized Medicine

Another promising application of AI in connected healthcare is personalized medicine. AI can analyze a patient's unique health data, including genetics, medical history, and lifestyle factors, and provide tailored treatment recommendations. For example, AI can analyze a patient's genetic data to determine the best course of treatment for a particular type of cancer. This approach can potentially improve patient outcomes by providing more personalized, effective care.

11.3.3 Remote Patient Monitoring

Remote patient monitoring is another area where AI can be beneficial. By using sensors and wearable devices, healthcare providers can monitor patients remotely and provide real-time feedback on their health status. For example, AI can analyze data from a patient's wearable device to determine whether they are experiencing heart attack symptoms and alert healthcare providers if necessary. This approach can benefit patients with chronic conditions who require frequent monitoring but may not need to visit a healthcare provider in person.

11.3.4 Medical Imaging

AI can also be used to analyze medical images, helping to detect diseases and abnormalities more quickly and accurately. For example, AI can be trained to analyze X-rays and CT scans to identify potential tumors or other abnormalities that a human radiologist might miss. This approach can be particularly useful in areas with a shortage of radiologists, enabling healthcare providers to provide more efficient and effective care.

11.3.5 Drug Discovery

The analysis of huge volumes of data and the identification of prospective medication candidates can be accomplished using SC techniques such as neural networks. AI, for instance, can be used to analyze molecular data to find prospective therapeutic targets. This strategy may speed up the discovery of novel drugs, cutting down on the time and expense involved in bringing them to market.

11.3.6 Electronic Health Records

AI can analyze EHRs and find patterns and trends that can help doctors make better decisions about patient care. Healthcare professionals can spot potential health hazards and create better treatment options by analyzing EHRs. AI can, for instance, analyze information from EHRs to identify patients who are at a high risk of contracting sepsis and to conduct preventative measures.

11.4 IoT TECHNOLOGIES IN HEALTHCARE

Kevin Ashton's concept, the "Internet of Things," gives devices the ability to connect with one another without the need for human involvement over the internet. IoT is a paradigm that converts the physical world to the virtual world. IoT has become an important part of daily life in almost every application. Healthcare is a trending IoT application as of now. IoT communication technologies and protocols provide seamless connectivity among medical devices, patients, and healthcare providers [22]. Hospitals, doctors, nursing homes, and communities are just a few of the stakeholders that IoT healthcare applications can manage to evaluate data collection. In addition, they can accurately track individuals, objects, samples, and supplies.

The development of IoT-based healthcare systems must guarantee and improve patient safety, life quality, and other healthcare-related activities [23]. In healthcare applications, sensors are deployed near the patient's body (wireless body area network [WBAN]) to collect and share real-time biosignals. And, network layer provides connectivity to transmit bio-signals to the upper layer for further processing. Sensors have revolutionized healthcare by supplying precise and timely data that help medical personnel make knowledgeable decisions and enhance patient outcomes. Timely

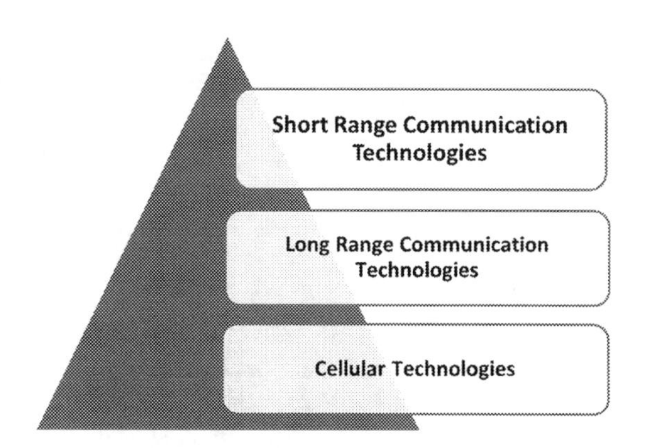

FIGURE 11.5 Types of communication in the Internet of Things.

data delivery in healthcare applications is crucial to improve patient outcomes, lower healthcare expenses, and boost facility efficiency. To deliver data on time, the selection of IoT communication technology plays an important role. Various IoT communication technologies and protocols can be chosen as per application requirements. IoT communication technologies can be divided into three categories—short-range communications, long-range (LoRa) communications, and cellular technologies. This chapter gives a vision of the importance of each communication technology in the healthcare sector (Figure 11.5).

11.4.1 Short-Range Communication

The IoT relies on short-range communication to allow objects to connect in proximity. Near field communication (NFC), Bluetooth, Zigbee, and Wi-Fi Direct are examples of short-range communication technologies.

11.4.1.1 Bluetooth

The most popular short-range communication technique in IoT applications is Bluetooth. This protocol allows devices to communicate wirelessly and with low power over short distances. Several IoT applications, including smart homes, wearable technology, and medical equipment, employ Bluetooth.

Recent years have seen more Bluetooth advancements in commercial settings, including retail, hospitality, tourism, and healthcare.

The market for linked devices has been dominated by Bluetooth-enabled wearables, which range from fitness trackers and smart watches to glucose monitors and pulse oximeters.

The main application of Bluetooth has long been for wirelessly connecting a mobile phone and a headset. A recent advancement in Bluetooth core technology is the Bluetooth 5.1 specification, which was introduced in 2019 and included enhanced localization services with the so-called direction-finding capabilities.

The patient's body can be fitted with Bluetooth-enabled devices to create a body area network and monitor vital signs, including blood pressure, oxygen saturation, and heart rate [24].

11.4.1.2 Zigbee

In essence, ZigBee wireless technology acts as a publicly available, all-encompassing standard to satisfy the unique needs of low-power, low-cost wireless M2M (machine-to-machine) networks and IoT devices. It can operate on unlicensed frequencies, including 2.4 GHz, 900 MHz, and 868 MHz, and employs the IEEE 802.15.4 physical radio specification to do so. There are three ZigBee devices in the ecosystem: a coordinator, a router, and end devices.

Using body sensors, wireless sensor networks (WSNs) are used to monitor a person's biological characteristics. Wireless communication technology is known as ZigBee. This mode can be used in healthcare facilities to observe parameter variations in a patient under evaluation. ZigBee is, among them, the real network connectivity solution. In Reference [25], Zigbee monitors the health in assisted ambient living for elders.

ZigBee is like Bluetooth but is less complicated, uses a lower data rate, and is mostly dormant, making it a very efficient technology that can operate for up to 2 years on an AA battery. ZigBee devices have a maximum data rate of 250 kbps, but Bluetooth devices can even reach 1 Mbps. ZigBee supports up to 254 nodes in master–slave networks, while Bluetooth only supports 8 slave nodes.

11.4.1.3 Wi-Fi Direct

Wi-Fi direct is a short-range communication technology that does not require a centralized network or wireless router; instead, devices in the network can behave as an access point or hotspot. Even without a network connection, data can be transferred immediately between neighboring devices once this connection is made.

As long as the devices are in range of one another, Wi-Fi direct allows rapid connectivity in any situation. This provides significant mobility in business applications. Wi-Fi direct technology can be helpful in many gaming and file-sharing applications. However, Wi-Fi direct is also useful when there is little or no network connectivity. In Reference [26], Wi-Fi direct–based mobile application for early response to persons in emergency situations has been proposed by the authors. This software helps establish links with the assistance workers who help provide relief to the people in an emergency.

Wi-Fi monitoring enables nurses to constantly be aware of the condition of their patients and avert potential injury. Bed alarms, for instance, can alert medical workers if patients try to leave their beds (Table 11.1).

TABLE 11.1
Short-Range Wireless Communication Technologies Comparison

Criterion	Wi-Fi Direct	Bluetooth v4.0	BLE	NFC
Range	Up to 180 m	Up to 1 00 m	Up to 10 m	Up to 4 cm
Data rate	Up to 250 Mbps	Up to 25 Mbps	Up to 200 kbps	Up to 424 kbps
Security	High	High	High	Medium
Power consumption	High	Medium	Low	Low
Communication	Unidirectional	Bidirectional	Bidirectional	Unidirectional

NFC, Near field Communication; BLE, Bluetooth Low Energy

11.4.1.4 NFC and RFID

NFC is a short-range wireless networking technology that is based on standards and enables a quick and secure two-way communication between electronic devices. Consumers may connect electronic gadgets with only one touch, conduct contactless transactions, and access digital content.

In addition, radio frequency identification (RFID) has helped hospitals track patients, medications, and medical equipment more effectively and safely by digitizing these processes. In Reference [27], a basic architecture for tracking medical supplies and medications was proposed using UHF RFID hardware. This design is simple to integrate with most hospitals' network infrastructure and medical information services already in place. Ultra-high-frequency UHF RFID testing in a hospital has demonstrated that it can be utilized successfully.

11.4.2 Long-Range Technology

11.4.2.1 Low-Power Wide-Area Network

IoT devices and M2M applications benefit from the low-power wide-area coverage offered by LPWANs (low-power wide-area networks). The development of LPWAN solutions is becoming more and more popular than traditional short-range and cellular communication technologies because it can manage long-distance communication, cost-effective connection, and large-scale communication at a lower power level.

11.4.2.2 LoRa

One of the newest technologies for long-distance, low-power communication, LoRa employs LoRaWAN as the MAC layer standard. The standard LoRa topology uses single-hop star communication, which uses more energy and has a slower data rate [28]. LoRa differs from other LPWAN technologies in that it gives users the option to customize physical layer characteristics (transmission power, bandwidth, coding rate, spreading factor, and carrier frequency) for their applications [29].

In Reference [30], novel Medical IoT designs for homecare and hospital services based on LoRa technology and edge computing have been suggested. The coverage range for various LoRa settings is examined by the authors.

In Reference [31] the authors suggested a system for monitoring the health of old people in Portugal that is powered by LoRa technology and connected to The Things Network. They discovered a strong enough signal for communication, but LoRa networks are not intended for sending huge amounts of data.

In Reference [32], data from the patient was sent to the gateway using the LoRa (HEADR) protocol. In this study, the authors also provided a healthcare model based on machine learning to monitor a patient's health, in particular diabetes symptoms, and to assess the severity of the patient's sickness.

11.4.2.3 NB-IoT

A narrowband radio technology called NB-IoT (Narrowband-IoT) is used by M2M and IoT devices. It is a form of cellular LPWAN that uses current cellular networks to facilitate wireless communication. Narrowband IoT, sometimes referred to as LTE Cat NB1, was created by the 3GPP for IoT applications. NB-IoT employs a variant of the LTE standard but limits the bandwidth to a single narrow band of 200 kHz.

Elderly persons with disabilities and patients at risk of falling can be helped using NB-IoT devices. Being a member of the LPWAN family of low-power wide-area networks, NB-IoT enables the connection of gadgets with modest data requirements, low bandwidth, and extended battery life. Because of its simpler waveform and capability to “sleep” while not in use, it has higher power efficiency. In Reference [33], an IoT device based on NB-IoT shield is created to immediately notify about the occurrence of a fall of elderly people and to notify up to four contacts of the patients automatically.

The authors offered remote wellness monitoring and smart health services using NB-IoT in WBAN. To send and receive data between wireless sensors in these systems, long-range wireless links are used. With NB-IoT, a cellular base station that is already in place could provide coverage for the entire hospital as well as everyone in the neighborhood. NB-IoT guarantees that end-user terminals will have a long lifespan even when powered by batteries [34].

11.4.2.4 Sigfox

Another wireless network, Sigfox, falls under the LPWAN proprietary technology group and strives to fulfill IoT criteria for wireless systems, including WBANs. Sigfox was designed with long-distance communication, straightforward connectivity, extended battery life, higher network capacity, and less expensive hardware (i.e., Sigfox module) aims in mind. High QoS standards that are less susceptible to interference issues are delivered. Low-cost gadgets that can transport data over a long distance while consuming minimal power are ideal candidates for Sigfox technology. In addition, Sigfox works in Europe and the United States on the sub-1 GHz ISM unlicensed frequency bands of 915 and 868 MHz, respectively, at a very low data rate of roughly 100 bps. By employing a line-of-sight communication method, Sigfox is able to attain great sensitivity over long-distance communication. For instance, Sigfox can achieve communication coverage of around 50 km utilizing frequency hopping; however, barriers reduce the range to about 10 km in cities [35,36]. The authors of Reference [33] proposed a routing protocol (PLTAAR) for path discovery and selection in the WBAN of the Sigfox network.

11.4.3 Cellular Technologies

The many cellular network generations, including 2G, 3G, 4G, 5G, and 6G, have played and will continue to play a crucial role in the healthcare industry. The following are some examples of the contributions each generation has made to healthcare:

2G: Text message sending and receiving became possible with the introduction of 2G cellular networks. This technology has been deployed in the healthcare industry to remind patients to take their meds, show up for appointments, and complete other medically related chores.

3G: The third generation of cellular networks, or 3G, allowed for quicker data transmission than 2G. With the aid of this technology, medical pictures like X-rays and CT scans may be sent from one place to another, enabling quicker diagnosis and treatment.

4G: With the advent of the fourth generation of cellular networks, data transmission rates surpassed those of 3G. Telemedicine, remote patient monitoring, and virtual doctor consultations have all been made possible by using this technology in the healthcare industry.

5G: The fifth generation of cellular networks, or 5G, provides high speed, high data rates, incredibly low latency, ubiquitous device connectivity, reliability, and increased network capacity. This technology has the potential to transform healthcare by providing real-time monitoring of patient health and vital signs, remote surgery, and other cutting-edge applications. In Reference [37], the authors reviewed 5G technologies in a number of applications, including wearable and implantable medical devices, mobile-connected ambulances, and remote robotic surgery, and assessed key performance indicators such as availability, accessibility, reliability, data rate, and retainability for 5G in the aforementioned use cases. The authors discussed the function of 5G in the COVID-19 period in Reference [38]. Real-time contact tracing, robot-assisted tela-ultrasound, telementoring during ocular surgery, and the exchange of enormous amounts of data in Fangfang (cabin) hospitals were all benefits of employing 5G as the key technology for COVID-19 applications. In other instances, COVID-19-related applications such as patient monitoring used 5G as a supporting technology.

6G: Although still under construction, 6G networks are anticipated to offer even higher speeds and lower latency than 5G. This might make it possible to perform remote surgeries and telemedicine in ever more sophisticated ways. 6G technology can overcome barriers like time and space in healthcare. Telesurgery, pandemics, and epidemics can all benefit from the use of 6G services. New techniques such as intelligent wearable devices (IWD), intelligent Internet of medical things (IIoMT), and hospital-to-home (H2H) services could result from the integration of 6G facilities with the healthcare industry [39].

11.5 METHODOLOGIES OF AI AND SC IN CONNECTED HEALTHCARE

11.5.1 Structured Data Classification

Data are categorized using the classification process into a predetermined number of classes [40]. As seen in Figure 11.6, classification can be done on both structured and unstructured data. Finding the category or class that fresh data will belong to is the basic objective of a classification challenge.

Following are some terms associated with the classification of medical datasets:

- **Classifier:** Classifiers are algorithms that map the input data to a specific category.
- **Classification model:** Some conclusion is drawn by the input values of the training set in the classification model. This always predicts the class categories/labels for generating new data.
- **Feature:** The phenomenon of observing an individual's property is a feature.

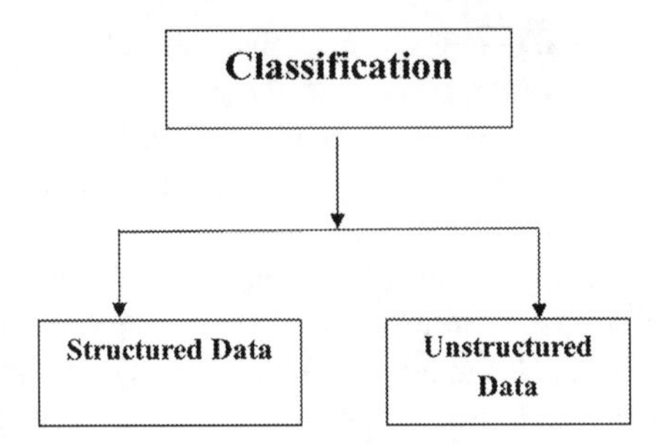

FIGURE 11.6 Types of classification in machine learning.

- **Binary classification:** In the binary classification task, there may be two possible outcomes, such as the classification of gender (male/female).
- **Multiclass classification:** More than two classes are classified in multiclass classification. Each sample of the multiclass classification has a single target label attached to it. For instance, a person can be either male or female, but not both at once.
- **Multilabel classification:** In multilabel, each sample is mapped to a collection of target labels (from more than one class). For instance, a news story could concurrently be about sports, a person, and a place.

The following are the steps involved in the building of a classification mode:

- **Initialize:** The classifier must be used for initialization in this phase.
- **Train the classifier:** The fit (*X*, *Y*) method is used by all classifiers during training to fit the models to the provided train data (*X*) and train label (*Y*).
- **Predict the target**: For an unlabelled observation X, predict (X) provides the predicted labels Y.
- **Evaluates:** the classifiers models.

11.6 MACHINE LEARNING TECHNIQUES

In AI-based SC, machine learning algorithms are widely utilized for identifying any abnormalities in the network, and their effectiveness in providing accurate detection rates has been established [41]. These algorithms are trained with relevant data to create a predictive model [38], which is then employed to analyze incoming data for malicious behavior [42]. This technique is commonly referred to as supervised learning.

Different algorithms are used in machine learning, including support vector machines, Naive Bayes (NB) classifiers, decision tables, and decision trees. Compared to unsupervised learning algorithms, supervised learning algorithms require labeled data. Unsupervised algorithms include clustering, K-means, and deep neural networks, among others. Semi-supervised algorithms, on the other hand, sit in the middle between the two and do not demand that all of the data be labeled.

Semi-supervised machine learning algorithms include generative, graph-based, and self-training models. To obtain the best results, it is essential to use an appropriate algorithm.

Following are the machine learning techniques applied for anomalies detection and intrusion detection [43];

11.6.1 Logistic Regression

Definition: A machine learning algorithm for classification is logistic regression (LR). In this approach, a logistic function is used to model the probabilities describing the potential outcomes of a single experiment [44–49].

Advantages: LR is a statistical method that is specifically designed for classification purposes. It is particularly useful when determining how various independent variables impact a single outcome variable. This approach is highly valued professionally for its ability to provide valuable insights into complex datasets.

A discriminative model, LR is dependent on the caliber of the dataset. The characteristic weights $w = w1$, $w2$, $w3,\ldots,wn$, where $F = F1$, $F2,\ldots,Fn$ (where $F1$–$Fn =$ distinct features). Bin and classes $C = C1$, $C2$, $C3$, and $B = B1$, $B2$, $B3\ldots$ The following is the posterior estimation equation for LR (in the current instance, we only have two classes).

$$\text{Predicted values}: P\left(y = C / F; W, b\right) = \langle I\rangle \frac{1}{1 + \exp\left(-W^{\text{transpose}} F - b\right)} \langle /I\rangle\langle /B\rangle \quad (11.1)$$

Disadvantages: Works only when the predicted variable is binary, presupposes that all predictors are unrelated to one another, and presupposes that the data are missing no values.

11.6.2 Decision Tree

Definition: A decision tree generates a set of rules that may be used to categorize the given data of characteristics and its classes [50–52].

Advantages: The decision tree method is straightforward to comprehend and visualize and demands only a small amount of data. This technique generates a collection of regulations that can help group data, given a dataset that contains attributes and their corresponding classes. The classification process guarantees the preciseness of the provided information and can handle both numerical and categorical data.

When using a decision tree, it is important to consider all potential actions and their benefits, costs, and likelihoods. This tool serves as a visual representation of the possible outcomes that can result from different decisions. The decision tree begins with a single node and branches out into multiple child nodes for each possible outcome. For example, the data in the parent node Pd are separated into two child nodes, LCd and RCd, as illustrated in Figure 11.6. The main motive is to maximize information gain (IG) that calculates the reduction in entropy or disclosure from transforming a dataset somehow (Figure 11.7).

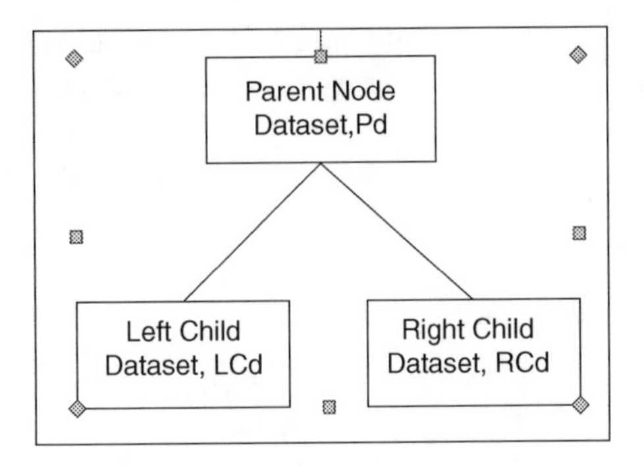

FIGURE 11.7 Splitting of decision tree.

The amount of "information" an element provides us about the class is measured by IG. The primary element that Decision Tree Algorithms use to construct a decision tree is IG. Consistently, decision tree calculations aim to increase IG. The initial split will occur for the attribute with the highest Information increase. Gain in information can be expressed as:

$$\text{Information gain}(Pd,F) = I(Pd) - \frac{Lcn}{Pn} I\ (Lcd) - \frac{Rcn}{Pn} I\ (Rcd) \quad (11.2)$$

Gini Index "*Ig*," Entropy "*Ih*," and classification error "*Ie*" are three different techniques that measure the impurity of data "*I*," where "*c*" specifies labels or classes, "*n*" specifies any node, and "*p*(*c*/*n*)" represents the ratio between "*c*" with "*n*".

$$Ih(n) = -\sum_{i=1}^{c} p\left(\frac{c}{n}\right) \log 2p\left(\frac{c}{n}\right) \quad (11.3)$$

$$Ig(n) = 1 - \sum_{i=1}^{c} p\left(\frac{c}{n}\right)^2 \quad (11.4)$$

$$Ie(n) = 1 - \max\{p(c/n)\} \quad (11.5)$$

Disadvantages: Decision trees can yield complex, hard-to-generalize trees, and they can also be unstable because even minor changes in the data could result in the formation of a completely different tree.

11.6.3 Random Forest

Definition: The random forest (RF) classifier is a type of meta-estimator that utilizes multiple decision trees trained on different subsets of the data to enhance the accuracy of predictions and prevent overfitting. The subsamples used for training are the same size as the original input data but are selected randomly with replacement. This method helps improve the model's performance by averaging the results from multiple decision trees [53].

Advantages: In most cases, the RF classifier is more precise than decision trees because of its ability to minimize over-fitting. It is essential to note that no information should be left out when paraphrasing the statement. The tone of voice should remain professional.

This algorithm, which falls under supervised classification, constructs a forest comprising numerous decision trees. Its quick execution time makes it a highly sought-after machine-learning classification algorithm. Compared to a solitary decision tree, it performs significantly better in predictive accuracy, thanks to the presence of multiple decision trees within the forest, enhancing its robustness.

Disadvantages: This shows a slow real-time prediction, challenging implementation, and demanding algorithm.

11.6.4 Support Vector Machine

Definition: Support vector machines (SVMs) organize training data into distinct categories by creating a gap in a space as large as possible. This gap is then used to map new examples into the same space and predict which category they belong to based on which side of the gap they fall on. It is a professional method of data representation and prediction. No information has been omitted in this paraphrased text [54–59].

Advantages: Effective in high-dimensional spaces and memory-efficient because of the employment of a small fraction of training points in the decision functions.

SVM is another straightforward calculation in the discriminative classification model that machine learning has. Numerous individuals profoundly favor SVM as it produces critical exactness with less calculation power. SVM can be utilized for non-linear data. However, it is generally used in calculating weight vector W as is given in:

$$W = \sum_{i=1}^{m} \alpha_i c_i x_i \tag{11.6}$$

Targeted optimized SVM is calculated as:

$$Maximize_{\alpha i} \sum_{i=1}^{m} \alpha i - \sum_{i=1}^{m} \sum_{j=1}^{m} \alpha_i \alpha_j c_i c_j \left\langle x_i, x_j \right\rangle \tag{11.7}$$

where $\left\langle x_i, x_j \right\rangle$ can be brought by different kernels such as radical and Sigmoid kernels.

Disadvantages: The algorithm does not output probability estimates directly. These are computed with a pricey five-fold cross-validation method.

11.6.5 Naive Bayes

The NB method is an effective and straightforward way to create linear models. This approach is especially beneficial when dealing with a massive number of samples. For categorization purposes, it also supports a variety of loss functions and penalties [60–61].

Advantages: The ability to operate effectively and with minimal difficulty in execution is of utmost importance in professional settings.

Disadvantages: This model encounters the "zero-frequency problem," because it assigns zero to any missing categorical value in categorical dataset that is designed for traning purpose.

NB classifiers are a set of "probabilistic classifiers" used in machine learning that rely on Bayes' theorem and assume strong independence between features. These models are some of the simplest Bayesian network models.

11.6.6 K-Nearest Neighbor

Definition: K–nearest neighbor (K-NN)-based classification falls under the lazy learning category, which means it does not aim to create a comprehensive internal model. Instead, it merely holds onto examples of the training data. The classification is determined by taking a majority vote of the k-nearest neighbors for each data point [62–64].

Advantages: The implementation of this algorithm is straightforward, and it is resilient to training data that are contaminated with noise. It proves to be efficient when the training data are extensive.

A machine learning method that utilizes supervised learning is the K-NN algorithm. The algorithm organizes fresh data points based on their similarity and stores all of the already available data. It functions by comparing new data with cases already in existence and classifying the new data into the category that closely resembles those already in existence. This implies that the K-NN algorithm can quickly classify newly introduced data. K-NN is primarily used for classification problems, while it can be used for regression as well. The K-NN algorithm is nonparametric, which means it makes no assumptions about the underlying data [65].

Disadvantages: It is essential to calculate the distance between each occurrence and each training example in order to establish the value of K, which results in a large computational cost.

11.6.7 XGBoost

XGBoost offers additional benefits, such as fast dataset training that can be distributed among multiple classes. It is a specialized implementation of the Gradient Boosting technique, which uses more precise estimations to find the optimal tree model. The algorithm employs various techniques that greatly enhance efficiency, mainly when working with structured data [66].

11.7 EXPLORATORY DATA ANALYSIS IN AI-BASED SC

11.7.1 Accuracy

The precision of a model is just a component of its overall performance. Accuracy is one of the metrics used to evaluate classification models and can be calculated using equation (11.8) to determine single-class precision.

$$\text{Accuracy} = \frac{\text{True positive} + \text{True negative}}{\text{True positive} + \text{True negetive} + \text{false positive} + \text{False negetive}} \tag{11.8}$$

11.7.2 Precision

In the field of data analysis, precision refers to the ability of a model to predict positive outcomes correctly. It is determined by comparing the number of true positives in the model's results to the total number of positive outcomes predicted by the model. A formula for calculating precision for a specific class can be found in the following equation. This information is crucial for maintaining a professional standard in data analysis:

$$\text{Precision} = \frac{\text{True positive}}{\text{True positive} + \text{False positive}} \tag{11.9}$$

11.7.3 Recall

The real positive rate, also called recall, measures the number of positive cases in the model compared to the actual number of positive cases in the data. The formula for calculating the recall value of a single class is provided in the following equation. This is a technical aspect that requires attention to detail and precision:

$$\text{Recall} = \frac{\text{True positive}}{\text{True positive} + \text{False negetive}} \tag{11.10}$$

11.7.4 F1 Score

The performance of a model can be evaluated through the F1 score, which is a combination of precision and recall. This score is calculated using a weighted average formula and is represented by equation (11.11). It is a valuable metric for assessing the performance of a single class in the model:

$$F1\ \text{score} = \frac{2\ {*}\text{True positive}}{2\ {*}\text{True positive} + \text{False positive} + \text{False negative}} \tag{11.11}$$

11.7.5 Receiver Operating Characteristic Curve

The receiver operating characteristic (ROC) curve is a frequently used graph that lowers the classifier's performance from every viewpoint. Plotting the true positive rate against the false positive rate while adjusting the edge's value for categorizing data results in the curve shown above. This method of measuring a classifier's effectiveness is extensively used [67–69]. The adjacent equations offer the formulas for the true positive rate and false positive rate:

$$\text{False positive rate} = \frac{\text{Number of false positive samples}}{\text{Total number of samples}} \tag{11.12}$$

$$\text{True positive rate} = \text{recall} = \frac{\text{Number of true positive samples}}{\text{Total number of samples}} \tag{11.13}$$

11.8 COMPARISON MATRIX

The formula below calculates a matrix, which is determined after accuracy and F1-score have been estimated. No details are left out in this process.

11.8.1 Accuracy: (True Positive + True Negative)/Total Population

- Accuracy, which is determined by dividing the number of properly predicted observations by the total number of observations, is the most logical way to assess performance [70].
- The true positive metric refers to the accurate predictions of positive occurrences.
- The true negative metric refers to the accurate prediction of an event as negative.

11.8.2 F1-Score: (2 × Precision × Recall)/(Precision + Recall)

- The F1-score is a metric used in classification algorithms that computes a weighted average while taking precision and recall into account. When dealing with unbalanced class distributions, this score is more helpful than accuracy because it considers false positives and false negatives [71–74].
- **Precision:** How often is a prediction accurate when it forecasts a positive outcome?
- **Recall:** What is the frequency of accurate predictions when the real value is positive?

11.9 CHALLENGES AND LIMITATIONS

Despite the potential advantages of SC and AI in linked healthcare, various issues and constraints need to be resolved. Keeping patient data secure and private is one of the major challenges. Large volumes of data are required to train AI systems, which raises questions about accessing and retaining private patient data. Furthermore, there is a chance that AI algorithms could reinforce current biases in healthcare, resulting in the unjust or uneven treatment of particular patient populations.

Although AI can provide valuable insights and recommendations, it is important that healthcare providers remain involved in the decision-making process to ensure that care is personalized and compassionate. Another challenge is the need for human oversight and intervention. In addition, SC techniques offer a wide range of tools and methods for developing intelligent healthcare systems that can improve patient outcomes and reduce healthcare costs. By combining these techniques with domain expertise, healthcare professionals can create personalized and effective healthcare solutions for their patients.

REFERENCES

1. Goldstein, Y. Shahar, E. Orenbuch, and M. J. Cohen, Evaluation of an automated knowledge based textual summarization system for longitudinal clinical data, in the intensive care domain, *Artificial Intelligence in Medicine*, vol. 82, pp. 20–33, 2017.
2. Liang, J., C.-H. Tsou, and A. Poddar, A novel system for extractive clinical note summarization using EHR data, In: *Proceedings of the 2nd Clinical Natural Language Processing Workshop,* Minneapolis, Minnesota, USA, 2019, pp. 46–54.
3. Das, S., S. Biswas, A. Paul, and A. Dey, AI doctor: An intelligent approach for medical diagnosis, In: S. Bhattacharyya, S. Sen, M. Dutta, P. Biswas, and H.Chattopadhyay (Eds), *Industry Interactive Innovations in Science, Engineering and Technology.* Springer Singapore, Singapore, vol. 11, pp. 173–18, 2018.
4. Mohapatro, Manorama, and Itu Snigdh. Security in IoT healthcare. In: *IoT Security Paradigms and Applications.* CRC Press, Boca Raton, FL, pp. 237–259, 2020.
5. Gambhir, Shalini, Sanjay Kumar Malik, and Yugal Kumar. Role of soft computing approaches in healthcare domain: A mini review. *Journal of Medical Systems*, vol. 40, 1–20, 2016.
6. Suwarno, I., A. Cakan, N. M. Raharja, M. A. Baballe, and M. S. Mahmoud. Current trend in control of artificial intelligence for health robotic manipulator. *Journal of Soft Computing Exploration*, vol. 4(1), 2023.
7. Gupta, S., and U. Singh. Ontology-based IoT healthcare systems (IHS) for senior citizens. *International Journal of Big Data and Analytics in Healthcare (IJBDAH)*, vol. 6(2), 1–17, 2021.
8. Bajgain, Bishnu et al. Determinants of implementing artificial intelligence-based clinical decision support tools in healthcare: A scoping review protocol. *BMJ Open*, vol. 13(2), e068373, 2023.
9. Mishra, Ishani, and Sanjay Jain. Soft computing based compressive sensing techniques in signal processing: A comprehensive review. *Journal of Intelligent Systems*, vol. 30(1), 312–326, 2020.
10. Manorama, Snigdh Itu. Anonymity in body area sensor networks-an insight. In: *2018 IEEE World Symposium on Communication Engineering (WSCE).* IEEE, New York, NY, 2018.
11. Kumar, Ayush, and Pooja Jha. Fuzzy logic applications in healthcare: A review-based study. In: *Next Generation Communication Networks for Industrial Internet of Things Systems.* pp. 1–25, 2023.
12. Dionisio, Marcelo et al. The role of digital transformation in improving the efficacy of healthcare: A systematic review. *Journal of High Technology Management Research*, vol. 34, 100442, 2023.
13. Dash, Prasannajit. A novel classifier architecture based on deep CNN neural network for COVID-19 detection using laboratory findings. *Applied Soft Computing*, vol. 107, 107329, 2023.
14. Srivani, M., Abirami Murugappan, and T. Mala. Cognitive computing technological trends and future research directions in healthcare-A systematic literature review. *Artificial Intelligence in Medicine*, 102513, 2023.
15. Loh, Erwin. Medicine and the rise of the robots: A qualitative review of recent advances of artificial intelligence in health. *BMJ Leader*, 2018.

16. Fatani, Faris, N. Binhowaimel, M. Alfaisal, M. Aamer, F. Al Asmari, and A. Almuflih. Machine Learning opportunities and challenges in health system management: Review of the existing literature. *Journal of Health Informatics in Developing Countries*, vol. 16(2), 2022.
17. Selvachandran, Ganeshsree et al. Developments in the detection of diabetic retinopathy: A state-of-the-art review of computer-aided diagnosis and machine learning methods. *Artificial Intelligence Review*, vol. 56(2), 915–964, 2023.
18. Shaik, Thanveer, X. Tao, N. Higgins, L. Li, R. Gururajan, X. Zhou, and U. R. Acharya. Remote patient monitoring using artificial intelligence: Current state, applications, and challenges. *Wiley Interdisciplinary Reviews: Data Mining and Knowledge Discovery*, vol. 13(2), e1485, 2023.
19. Yuen, Kevin Kam Fung. The primitive cognitive network process in healthcare and medical decision making: Comparisons with the analytic hierarchy process. *Applied Soft Computing*, vol. 14, 109–119, 2014.
20. Greco, Luca et al. Trends in IoT based solutions for health care: Moving AI to the edge. *Pattern Recognition Letters*, vol. 135, 346–353, 2020.
21. Sahu, Nilesh Kumar, Manorama Patnaik, and Itu Snigdh. Data analytics and its applications in brief. In *Applications of big data in large-and small-scale systems*, IGI Global, pp. 115–125, 2021.
22. Gupta, S., and I. Snigdh. Analyzing impacts of energy dissipation on scalable IoT architectures for smart grid applications. In: *Advances in Smart Grid Automation and Industry 4.0: Select Proceedings of ICETSGAI4. 0.* Springer, Singapore, pp. 81–89, 2021.
23. Gupta, S., and Snigdh, I. (2022). A review on parameters that impact IoT application-an experimental evaluation. In: *Microelectronics, Communication Systems, Machine Learning and Internet of Things: Select Proceedings of MCMI.* pp. 203–213, 2020.
24. Babburu, Kiranmai et al. Brain MRI image active contour segmentation for healthcare systems. In: *Artificial Intelligence in Cyber-Physical Systems: Principles and Applications.* p. 125, 2023.
25. Al-Qaseemi, S. A., H. A. Almulhim, M. F. Almulhim, and S. R. Chaudhry, IoT architecture challenges and issues: Lack of standardization. In: *2016 Future technologies conference (FTC)*. IEEE, New York, NY, pp. 731–738, 2016.
26. Zhang, Ting et al. Bluetooth low energy for wearable sensor-based healthcare systems. In: *2014 IEEE Healthcare Innovation Conference (HIC)*. IEEE, New York, NY, 2014.
27. Buthelezi, B. E., M. Mphahlele, D. Deon Du Plessis, S. Maswikaneng, & T. Mathonsi, ZigBee healthcare monitoring system for ambient assisted living environments. *International Journal of Communication Networks and Information Security (IJCNIS)*, vol. 11(1), 2022. doi:10.17762/ijcnis.v11i1.3677
28. Gupta, S., Itu Snigdh, and Sudip Kumar Sahana. A fuzzy logic approach for predicting efficient LoRa communication. *International Journal of Fuzzy Systems*, 1–9, 2022.
29. Kumar, Abhishek, Sourab Das, and C. Deepti. A Wi-Fi direct based mobile application for early response to persons in emergency situations.
30. Álvarez López, Y., J. Franssen, G. Álvarez Narciandi, J. Pagnozzi, I. González-Pinto Arrillaga, and F. Las-Heras Andrés. RFID technology for management and tracking: E-health applications. *Sensors (Basel, Switzerland)*, vol. 18(8), 2018.
31. Alliance, L. *White Paper: A Technical Overview of LoRa and LoRaWAN.* The LoRa Alliance, San Ramon, CA, pp. 7–11, 2015.
32. Bor, M., and U. Roedig. Lora transmission parameter selection. In: *2017 13th International Conference on Distributed Computing in Sensor Systems (DCOSS)*, IEEE, New York, NY, pp. 27–34. 2017.
33. Drăgulinescu, Ana Maria Claudia et al. LoRa-based medical IoT system architecture and testbed. *Wireless Personal Communications*, 1–23, 2020.

34. Lousado, José Paulo, and Sandra Antunes. Monitoring and support for elderly people using LoRa communication technologies: IoT concepts and applications. *Future Internet*, vol. 12(11), 206, 2020.
35. Verma, Navneet, Sukhdip Singh, and Devendra Prasad. Performance analysis and comparison of Machine Learning and LoRa-based Healthcare model. *Neural Computing and Applications*, vol. 35(17), 12751–12761, 2023.
36. Manatarinat, Wiraphon, Suvit Poomrittigul, and Panjai Tantatsanawong. Narrowband-internet of things (NB-IoT) system for elderly healthcare services. In: *2019 5th International Conference on Engineering, Applied Sciences and Technology (ICEAST)*. IEEE, New York, NY, 2019.
37. Robinson, Daniel. *NB-IoT (LTE Cat-NB1/Narrowband IoT) Performance Evaluation of Variability in Multiple LTE Vendors, UE Devices and MNOs*. Diss, Stellenbosch University, Stellenbosch, 2020.
38. Ali, A., G. A. Shah, M. O. Farooq, and U. Ghani. Technologies and challenges in developing machine-to-machine applications: A survey. *Journal of Network and Computer Application*, vol. 1, 124–139, 2017.
39. Lavric, A., A. I. Petrariu, and V. Popa. Sigfox communication protocol: The new era of IoT? In: *2019 International Conference on Sensing and Instrumentation in IoT Era (ISSI)*. IEEE, New York, NY, pp. 1–4, 2019.
40. Bakkaiahgari, P. V. D., and K. V. Prasad. Prim based link quality and thermal aware adaptive routing protocol for IoMT using SigFox network in WBAN. *Evolving Systems*, 2022, doi:10.1007/s12530-022-09451-3.
41. Moglia, Andrea et al. 5G in healthcare: From COVID-19 to future challenges. *IEEE Journal of Biomedical and Health Informatics*, vol. 26(8), 4187–4196, 2022.
42. Qureshi, H. N. et al. Communication requirements in 5G-enabled healthcare applications: Review and considerations. *Healthcare*, 10, 293, 2022.
43. Nayak, Sabuzima, and Ripon Patgiri. 6G communication technology: A vision on intelligent healthcare. In: *Health Informatics: A Computational Perspective in Healthcare*, pp. 1–18, 2021.
44. Alsheikh, Mohammad Abu et al. Machine learning in wireless sensor networks: Algorithms, strategies, and applications. *IEEE Communications Surveys & Tutorials*, vol. 16(4), 1996–2018, 2014.
45. Lakshminarayana, Deepthi Hassan. Intrusion detection using machine learning algorithms. East Carolina University. 2019.
46. Singh, Taranveer, and Neeraj Kumar. Withdrawn: Machine learning models for intrusion detection in IoT environment: A comprehensive review. *Computer Communications* 2020.
47. Kumar, D. Praveen, Tarachand Amgoth, and Chandra Sekhara Rao Annavarapu. Machine learning algorithms for wireless sensor networks: A survey. *Information Fusion*, vol. 49, 1–25, 2019.
48. Saeedi, Kubra. Machine learning for ddos detection in packet core network for iot. 2019.
49. Ukil, Arijit et al. IoT healthcare analytics: The importance of anomaly detection. In: *2016 IEEE 30th international conference on advanced information networking and applications (AINA)*. IEEE, New York, NY, 2016.
50. Xiao, Liang et al. IoT security techniques based on machine learning: How do IoT devices use AI to enhance security? *IEEE Signal Processing Magazine*, vol. 35(5), 41–49, 2018.
51. Branch, J. W., C. Giannella, B. Szymanski, R. Wolff, and H. Kargupta, In-network outlier detection in wireless sensor networks, *Knowledge and Information Systems*, vol. 34(1), 23–54, 2013.
52. Narudin, F. A., A. Feizollah, N. B. Anuar, and A. Gani, Evaluation of machine learning classifiers for mobile malware detection, *Soft Computing*, vol. 20(1), 343–357,. 2016.

53. Buczak, A. L. and E. Guven, A survey of data mining and machine learning methods for cyber security intrusion detection, *IEEE Communication Surveys Tutorials*, vol. 18(2), 1153–1176, 2015.
54. Kulkarni, R. V. and G. K. Venayagamoorthy, Neural network based secure media access control protocol for wireless sensor networks, In: *2009 International Joint Conference on Neural Networks*, Atlanta, GA, pp. 1680–1687, June 2009.
55. Strobel, Gero, and Juliane Per!. Health in the era of the internet of things: A smart health information system architecture. *Health*, vol. 6, 20–22, 2020.
56. Z. Tan, A. Jamdagni, X. He, P. Nanda, and R. P. Liu, A system for denial-of service attack detection based on multivariate correlation analysis, *IEEE Transactions on Parallel and Distributed Systems*, vol. 25(2), 447–456, 2013.
57. Nadeem, Adnan, et al. Application specific study, analysis and classification of body area wireless sensor network applications. *Computer Networks*, 83, 363–380, 2015.
58. Agarwal, Sonali, and G. N. Pandey. SVM based context awareness using body area sensor network for pervasive healthcare monitoring. *Proceedings of the First International Conference on Intelligent Interactive Technologies and Multimedia*, pp. 271–278, 2010.
59. Di, Ma, and Er Meng Joo. A survey of machine learning in wireless sensor netoworks from networking and application perspectives. In: *2007 6th International Conference on Information, Communications* and *Signal Processing*. IEEE, New York, NY, 2007.
60. Verner, Alexander, and Dany Butvinik. A machine learning approach to detecting sensor data modification intrusions in WBANs. In: *2017 16th IEEE International Conference on Machine Learning and Applications* (*ICMLA*). IEEE, New York, NY, 2017.
61. Gupta, Sakshi, and Itu Snigdh. Applying Bayesian belief in LoRa: Smart parking case study. *Journal of Ambient Intelligence and Humanized Computing*, vol. 14(6), 7857–7870, 2023.
62. Gupta, Sakshi, and Itu Snigdh. Clustering in Lora networks, an energy-conserving perspective. *Wireless Personal Communications*, vol. 122(1), 197–210, 2022.
63. Restuccia, Francesco, Salvatore D'Oro, and Tommaso Melodia. Securing the internet of things in the age of machine learning and software-defined networking. *IEEE Internet of Things Journal*, vol. 5(6), 4829–4842, 2018.
64. Al-Saud, Khalid Abu, Massudi Mahmuddin, and Amr Mohamed. Wireless body area sensor networks signal processing and communication framework: Survey on sensing, communication technologies, delivery and feedback. *Journal of Computer Science*, vol. 8(1), 121–132, 2012.
65. Özdemir, Ahmet Turan, and Billur Barshan. Detecting falls with wearable sensors using machine learning techniques. *Sensors*, vol. 14(6), 10691–10708, 2014.
66. Nagdeo, Sumit Kumar, and Judhistir Mahapatro. Wireless body area network sensor faults and anomalous data detection and classification using machine learning. In: *2019 IEEE Bombay Section Signature Conference* (*IBSSC*). IEEE, New York, NY, 2019.
67. Ohri, Kriti, R. Vijaya Saraswathi, and L. Jai Vinita. Performance analysis of wireless body area sensor analytics using clustering technique. In: *2019 International Conference on Communication and Signal Processing* (*ICCSP*). IEEE, New York, NY, 2019.
68. Singh, Shweta, A. Bhardwaj, I. Budhiraja, U. Gupta, and I. Gupta. Cloud-based architecture for effective surveillance and diagnosis of COVID-19. In: *Convergence of Cloud with AI for Big Data Analytics: Foundations and Innovation*, pp. 69–88, 2023.
69. Gupta, Umesh et al. Sign language detection for deaf and dumb students using deep learning: Dore Idioma. In: *2022 2nd International Conference on Innovative Sustainable Computational Technologies (CISCT)*. IEEE, New York, NY, 2022.

70. Gupta, Madhuri, D. Srivastava, D. Pantola, and U. Gupta Brain tumor detection using improved otsu's thresholding method and supervised learning techniques at early stage. In: *Proceedings of Emerging Trends and Technologies on Intelligent Systems: ETTIS 2022*. Springer Nature Singapore, Singapore, pp. 271–281, 2022.
71. Gupta, Deepak, A. Choudhury, U. Gupta, P. Singh, and M. Prasad. Computational approach to clinical diagnosis of diabetes disease: A comparative study. *Multimedia Tools and Applications*, 1–26, 2021.
72. Dwivedi, Shri Prakash, Vishal Srivastava, and Umesh Gupta. Graph Similarity Using Tree Edit Distance. In *Joint IAPR International Workshops on Statistical Techniques in Pattern Recognition (SPR) and Structural and Syntactic Pattern Recognition (SSPR)*. Springer International Publishing, Cham, pp. 233–241, August 2022.
73. Gupta, Umesh et al. Next-generation networks enabled technologies: Challenges and applications. In: *Next Generation Communication Networks for Industrial Internet of Things Systems*, pp. 191–216, 2023.
74. Gupta, Umesh and Deepak Gupta. Least squares structural twin bounded support vector machine on class scatter. *Applied Intelligence*, vol. 53(12), 15321–15351, 2023.

12 Automated Detection and Classification of Focal and Nonfocal EEG Signals Using Ensemble Empirical Mode Decomposition and ANN Classifier

C. Ruth Vinutha, S. Thomas George, J. Prasanna, Sairamya Nanjappan Jothiraj, and M. S. P. Subathra
Karunya Institute of Technology and Sciences

12.1 INTRODUCTION

Epilepsy is characterized by recurrent seizure activity [1,2], which is brought on by a disruption in the normal balance between the brain's neuronal excitation and inhibition. The electroencephalogram (EEG) is a therapeutic tool for gaining an insight into a person's physiological and psychological state by measuring electrical impulses in the brain. Traditionally, medical professionals have spent hours poring over EEG signals, looking for signs of epileptic spikes [3]. As a result, this changes how neurologists treat patients with epilepsy. Hence, eliminating human subjectivity in diagnosis makes the development of an automated diagnostic tool for recognizing epileptic EEG patterns all the more urgent. According on the signal processing techniques used, automated epileptic signal classifications can be loosely placed into one of the four buckets: frequency domain, time domain, time-frequency domain, and nonlinear approach.

More than 50 million people are analyzed every year globally [4]. Epileptic seizures range in severity from humorous to fatal in a matter of seconds (SUDEP). Epileptic seizures, which occur when an abrupt surge of electricity in the brain disrupts the normal balance of excitation and inhibition among neurons (leading to a seizure), can produce this condition, which can last anywhere from a few seconds to several minutes [5–7]. There are two types of epileptic seizures: focal and generalized. Impact on a localized region of the brain can produce focal seizures [8,9].

 DOI: 10.1201/9781003405368-12

Convulsions affect the whole brain from the force of the blow. Medical professionals with expertise are required to visually evaluate epileptic spikes of EEG signal [10]. As a result, the challenges of manual diagnosis can be ameliorated through the use of automatic diagnosis of EEG signals.

Independent component analysis (ICA) is used for the purposes of source separation and source localization [11,12] of EEG data in scientific applications such as generalized seizure [12]. ICA, combined texture pattern (CTP), and closed neighborhood gradient pattern (CNGP) are used to extract important features from brain maps and conduct subsequent analyses (CNGP). The obtained data are fed into the least square support vector (LS-SVM) classifier [13], which is trained using examples of brain anomaly patterns such as poly spikes, sharp waves, spikes, and slow waves [14]. Five distinct signals are categorized by means of empirical mode decomposition using data provided by Bonn University. An EMD-derived IMF is utilized to train an LS-SVM classifier [15] that can distinguish between interictal and ictal EEG signals. The support vector machine (SVM) classifier has made EMD a practical possibility [16]. Here, we apply EMD to artificial neural network (ANN) classification to identify and differentiate between normal, localized, and diffuse EEG activity [17]. EMD with ANN classification is useful for distinguishing between focal and normal, normal and generalized, and normal and generalized.

A noise-assisted data analysis strategy, EEMD, has recently been proposed [18] to address the EMD method's mode mixing problem. To break down the first signal with noise into a series of IMFs, the EEMD iteratively applies the first EMD procedure. Means of associated IMFs are considered in each iteration toward a final EEMD decomposition result. As noise is introduced all the way through the breakdown of the signal, mode mixing is essentially nullified. This study presents a novel approach to the classification problem, using ANN as a classifier to distinguish between focal, generalized, and normal epilepsy based on EEG signals, samples, and fuzzy entropies.

An IMF is a capacity or function characterized by having an equal number of minimum and maximum points and no junctions or crossing points. Hence, the definition of the IMF ensures that the Hilbert transform of the IMF is well-behaved. Hilbert spectrum analysis (HSA) of the instantaneous frequencies of each IMF can provide a frequency-time distribution of the signal's amplitude or energy, allowing us to home in on its specific characteristics. These papers were chosen to inspire more study in this field because, in many circumstances, techniques based on this domain outperform those in the popular wavelet domain. The results are particularly dominant when variables from multiple domains are combined (IMFs + Frequency, in particular). In this section, we introduce the time-frequency domain known as EMD and the famous and very important transformation known as the Hilbert transform through the presentation of eight works. Each of the eight publications takes a unique tack in attempting to deduce seizure activity from EEG recordings. It has also been shown that there are connections between different methods. EMD-based techniques have shown promising results, outperforming those of wavelet and Fourier domain methods, prompting academics to dive more into this area. Readers can also use papers to determine the best classifier and the optimal number of IMF levels.

Eftekhar et al. [19] examined the drawbacks of common time-frequency approaches like spectrograms and wavelet analysis, including the requirement of an a priori signal interpretation and the assumption of linearity. By combining the two well-known methods of signal processing, the Hilbert transform and the Huang transform, Eftekhar et al. applied a novel time-frequency methodology to the problem of seizure detection using EEG and ECG output called the Hilbert–Huang technique or empirical time-frequency technique (Hilbert–Huang). Huang et al. first proposed it (1996, 1998, 1999, 2003, 2009, 2012, 2013). Using EEG and ECG data, they examined this empirical time-frequency method (CHB-MIT). After comparing their findings to those of other time-frequency methods, they concluded that further research and comprehension of their methods was necessary and complemented those already in use. The findings encouraged the researchers to further explore the seizure detection approach they had developed.

Tafreshi et al. [20] defined the IMF using the average absolute values of the Hilbert transform. They compared their method to another that used a wavelet transform to extract features. Algorithmically, MLP classifiers were found to be more effective than SOM neural networks. The Freiburg database is used for validation: it contains data from five patients collected over 128 channels and sampled at 256 Hz. Each patient has both an "ictal" and an "interictal" seizure dataset. A window size of 1,500 samples was determined to be optimal. Based on their findings, the EMD technique is superior to the wavelet technique, and only four empirical modes are required for best results. The MLP networks outperform the SOM networks, with an accuracy of 90.69% over the same four empirical modes.

Seizure and nonseizure activities in an EEG signal are distinguished from one another using the energies of IMFs, as proposed by Orosco et al. [21]. They applied this strategy by comparing the IMF energy to thresholds they had set for themselves. The proposed technique is put to the test and evaluated using data from nine patients in the invasive Freiburg database, with only six channels (three focal and three extrafocal). Totaling 90 minutes across both channels, the recordings feature 51 periods that do not have seizures and 39 that show epileptic activity. To begin preprocessing the EEG data, a notch filter was used. Next, a second-order, bidirectional Butterworth filter with a bandwidth of 0.5–60 Hz was applied. The researchers determined a sensitivity of 56.41% and a specificity of 74.86%. The method relies on a single threshold to identify seizures. The findings are low compared to the method provided in Reference [22], and no classifier is used.

All EMD components are characterized by higher-order statistics in addition to Shannon's entropy, as stated by Guarnizo and Delgado [23]. Relevant features are selected using mutual information (MI), a metric that measures the degree to which features are similar to one another. In this chapter, we utilize a linear Bayes classifier and run cross-validation with five iterations to better predict clinical outcomes. They Guarnizo and Delgado combined the results of the EMD procedure with four IMFs for classification. Using the Bonn database, the researchers found that the best results were obtained by integrating all attributes across all datasets. On average, the algorithm was 98% accurate. It demonstrates that as little as four IMFs can produce satisfactory results, and it compares favorably to methods proposed in References [23,24].

Sabrina et al. [25] presented an unsupervised learning-based strategy for seizure detection (clustering techniques). The proposed method differentiated between seizure and nonseizure activity by employing the rapid potential-based hierarchical agglomerative (PHA) clustering algorithm and the empirical mode decomposition (EMD)EMD strategy. Data including metrics of dissimilarity between the IMFs (e.g., Euclidean, Bhattacharya, and Kolomogorov distances) was supplied into the PHA cluster algorithm. The overall accuracy of the suggested method was 98.84%, according to tests and evaluations conducted on CHB-MIT.

Dattaprasad et al. [26] utilized empirical mode decomposition (EMD) to extract characteristics from the EEG data and an ANN for classification to assess whether the given EEG epochs constitute seizure or nonseizure activity. The signal was first decomposed into intrinsic mode functions (IMFs), and then the Hilbert transform was applied to each IMF to extract feature coefficients that characterize the instantaneous frequencies of interest. The suggested technique was evaluated on the Bonn database, where it achieved a 96% success rate.

Alam and Bhuiyan [27] used a variety of statistical and chaotic properties of EEG signals, including kurtosis, skewness, the maximum Lyapunov exponent, variance, approximation entropy, and correlation dimension. In this case, a 256-by-256 rectangular window was used to divide an IMF into 16 equal-sized sections. For every window, we calculated the variance, skewness, and kurtosis together with the LLE, CD, and ApEn chaotic features. The classifiers used in this study were based on ANNs. When applied to the (D,E) set with IMF3 and IMF4, validation on the Bonn database verified the algorithm's results with a sensitivity of 100%, a specificity of 100%, and an accuracy of 100%. They also demonstrated that the computational complexity of their approach is lower than that of competing time-frequency algorithms by researchers such as Liang et al., Tzallas, etc.

Patients with temporal lobe epilepsy can be identified using a method proposed by Bajaj and Pachori [28], which uses EMD to detect seizures. For use in the algorithm, the IMFs produced using EMD were converted. Then, the analytical tracing of IMFs in the EEG data is used to determine the instantaneous area, which is then used to detect epileptic seizures. IMF2 was shown to be sufficient for detecting the onset of seizures when the signal was broken down to IMF3. The local mean of EMD is the statistic of choice when trying to detect seizures. This algorithm's efficiency was measured using data collected from the University of Freiburg. Sensitivity is 90%, specificity is 89.31%, and the error rate is 24.25%. In other words, this is an algorithm tailored to each individual patient.

When compared to various time-frequency domains, we found that the one involving EMD and the Hilbert transform performed the best. The commencement of a seizure can be determined with as few as four IMFs. Researchers look into numerous classifiers and find that ANNs excel in many circumstances. It's a completely novel technique to seizure identification that uses unsupervised classifiers and puts them to the test. The transformational properties of EMD include time frequency and adaptability. IMFs can be employed as features in and of themselves because their constituent parts reliably differentiate between seizure and nonseizure regions. Based on our examination of the aforementioned techniques, we hypothesized that a hybrid strategy combining EMD and wavelet transformations would result in a novel and superior transformation for signal processing. However, a great deal of research is required before EMD can be used effectively in signal processing (Table 12.1) [29–37].

TABLE 12.1
Analysis of Multiple EEG Methods for Detecting Seizures Demonstrates their Efficacy

	Database	No. of Patients	Single/Multi Channel Data	Channe1 Selection	Methodology	Feature	Accuracy (%)
Samiee et al. [29]	Bonn	10	Single	NIA	Rational discrete short-time Fourier transform	Power Spectrum	98.1
Wang et al. [30]	Own	10	Multiple	No	Partial directed coherence analysis	Inflow and outflow information	98.3
Kumar et al. [31]	Bonn	10	Single	NIA	Artificial neural network	Entropies	94.5
Martis et al. [32]	Bonn	10	Single	NIA	Empirical mode decomposition	Spectral peaks, spectral entropy, and spectral energy	95.3
Yuan et al. [33]	Freiburg	21	Multiple	No	Extreme learning machine	Fractal geometry and relative fluctuation index	94.9
Fan et al. [34]	CHB-MIT	23	Multiple	No	Complex recurrence network	Spectral graph theoretic features	98.5
Aldana et al. [35]	Own	14	Multiple	No	CPD and block term decomposition	Three-way tensor with frequency*time*channels	98.7
Yuan et al. [36]	CHB-MIT	23	Multiple	Yes	Multi-view deep learning framework	Spectrogram fragments and channels in formation	94.05
Wen et al. [37]	Bonn + CHB-MIT	10 +2 3	Single + multiple	NIA + Yes	Deep convolution network and autoencoders-based model	Unsupervised features using multiple convolution kernels in the network	98.9

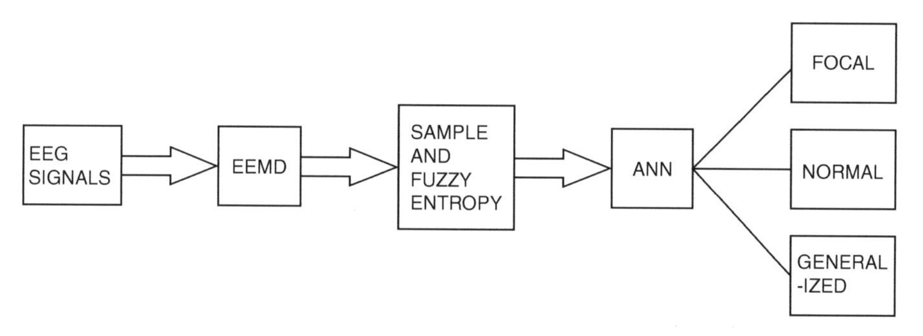

FIGURE 12.1 Block diagram of EMD and ensemble empirical mode decomposition (EEMD).

In Section 12.2, we explain the datasets used in this investigation. In Section 12.3, we introduce the classifier and briefly explore the feature extraction technique. In Section 12.3, we report and discuss the results of our experiments. The conclusions and recommendations are presented in Section 12.4 (Figure 12.1).

12.2 METHODS AND MATERIALS

12.2.1 A Data Base

The first steps in EEG-based seizure detection are often data gathering and pre-processing. Many electrodes are implanted on the scalp to collect the raw signals, and when the data are recorded, denoising utilizing conventional filtering and artifact removals are performed as part of the pre-processing. An epileptic EEG database [38] compiled by researchers at India's Karunya University was used to examine the method's efficacy. The Karunya database, which includes a large number of patients across a wide age range, was chosen as the case study of seizure detection in the experimental verification because it has not been well studied. In addition, there are 174 patients aged 1–84 years in the database (95 males and 79 females). In addition, a standard 10–20 EEG system was used, along with 18 channels (16 scalp channels and 2 periocular channels referencing right and left mastoids). The information was pre-processed with an analog pass band of 0.01–100 Hz and captured at 256 Hz. Seizure abnormalities, such as spikes and sharp waves, were annotated in 10-second chunks of EEG data for each patient. The participants in this study were divided into five age categories: children (aged 1–14), adolescents (aged 15–28), adults (aged 29–44), middle-aged adults (aged 45–65), and the elderly (aged 66 and up) (over 65 years). Afterward, 93 patients were selected at random for studies, yielding 93 sets of EEG readings, each measuring 16 scalp channels for 10 seconds (Table 12.2).

12.2.2 Ensemble Empirical Mode Decomposition

As the majority of the papers we've chosen to highlight here are founded on EMD and the Hilbert transform, let's begin with some background on both concepts. Equal-magnitude decomposition (EMD) is another time-frequency analytical method that does not rely on the Fourier or wavelet domains. In contrast to the fixed base of the

TABLE 12.2
Summarizes the Data, with the Group Shown in the First Column and the Number of Patients Tested in the Remaining Columns

Group	Female	Male	Total
Child	10	7	17
Youth	8	10	18
Adult	8	9	17
Middle-age	11	13	24
Elderly	8	9	17
All	45	48	93

Fourier and wavelet transforms, EMD is well suited for evaluating nonlinear and nonstationary signals like EEG because of its adaptive nature. The novel method proposed by Huang et al. [39] focuses on nonlinear data decomposition. While maintaining the signal's original properties, EMD breaks it down into a set of functions or components called IMFs in the field of signal processing. When the level of decomposition increases, the IMFs get simpler and the signal becomes weaker. When brain signals go awry, these IMFs respond with specific alterations in behavior. As a result, there is a plethora of data to be mined from IMFs, and the IMFs themselves can serve as identifying features for locating potential seizures. The following is the algorithm for EMD:

Given a signal $x(t)$,

1. Determine the signal's peak and valley.
2. Use interpolation to determine the upper and lower limits.
3. The local mean, $m(t)$, is calculated as $m(t)=(e\ \min\ (t)+e\ \max\ (t))/2$, where $e\ \min\ (t)$ and $e\ \max\ (t)$ are the minimum and maximum envelopes, respectively.
4. Find $d(t)$ by subtracting $x(t)$ by $m\ (t)$.
5. If the input $d(t)$ does not meet the requirements for an IMF, steps 6 and 7 are skipped, and the process is restarted from the beginning with the new input $d(t)$.
6. Value preserved is g $I(t)=d$ if and only if $h(t)$ is an IMF (t) signal, $r(t)=x(t)-gI\ (t)$, where I is the ith IMF removed.
7. With the revised signal $r(t)$, we can now revert to step 1 and store $gI\ (t)$ as an IMF.

This method generates a set of IMFs and a residual (represented by the letter r) as output (t). There are two ways to determine when this algorithm has run its course: (1) counting the number of IMFs produced or (2) using some other criterion. The frequency range of the signal analysis window is intrinsically linked to the concept of an IMF. The number of maximum and minimum values and zero crossings were compared as the first criterion, using narrowband frequency parameters (that they must differ at most by 1). If we interpolate between the two extremes, how many IMFs are possible? The traditional Hermite cubic spline method is limited to a maximum of three data points.

12.2.2.1 Advantages of EMD

1. The EMD generates a sequence of IMFs that can be used to compress or limit a signal's dynamic range.
2. Using the empirical time-frequency obtained from an IMF Hilbert spectral analysis, the instantaneous frequency may be easily determined.
3. In contrast to the static nature of a simple harmonic component, the amplitude and frequency of an IMF are free to vary with time.
4. There is absolutely no need for any form of predefined base functions.

Mode mixing is a disadvantage of the standard EMD method despite its usefulness in time-frequency analysis applications for signal processing. The mode mixing problem refers to situations where two or more modes exhibit highly similar oscillations. In order to address this issue, ensemble empirical mode decomposition (EEMD) was developed. Before the signal of interest is decomposed into its IMFs, a layer of Gaussian white noise is overlaid on top of the signal using the EEMD method (IMF). By creating signal continuity across several frequency domains, the statistical features of Gaussian white noise help to alleviate the problem of mode mixing.

The suggested EEMD is carried out as follows:

1. Putting the intended data in the white noise order.
2. When white noise is added to IMFs, the data get deconstructed.
3. To achieve a new white noise effect, simply repeat steps 1 and 2.
4. As a result, you should acquire the ensemble IMFs.

12.2.2.2 Fuzzy Entropy (FuzzyEn)

Randomness is represented by FuzzyEn, which is defined as the entropy of a fuzzy set in which the individual components have different enrolment levels. The enrolment function varies between 0 and 1. The definition of FuzzyEn is:

$$H[B] = \int_{-\infty}^{\infty} S\left(C_r\{B \geq t\}\right) dt \tag{12.1}$$

Credibility is denoted by *Cr*, where *B* is the fuzzy variable [40]. The fundamental benefit of this entropy is its insensitivity to noise and its great sensitivity to changes on the instructional material [41].

FuzzyEn compares two time series signals using a fuzzy function to calculate their similarity. The FuzzyEn is commonly used to classify EEG signals as either epileptic or nonepileptic:

$$y_0 = \frac{1}{m}\sum_{j=0}^{m-1} y(1+j) \tag{12.2}$$

In this case, an exponential function, which is described here, is chosen as the fuzzy function:

$$\mu\left(d_{ij}^{m}, n, r\right) = \exp(-\left(d_{ij}^{m}\right)^{n} / r) \tag{12.3}$$

where d_{ij}^{m} is the farthest possible distance between any two scalars.

The function ϕ^{m} for Y_i^{m} and ϕ^{m+1} for Y_i^{m+1} is given as:

$$\phi^{m}(n,r) = \frac{1}{N-m}\sum_{i=1}^{N-m}\left(\frac{1}{N-m-1}\sum_{j=1\ j\neq i}^{N-m}\mu\left(d_{ij}^{m},n,r\right)\right) \tag{12.4}$$

$$\phi^{m+1}(n,r) = \frac{1}{N-m}\sum_{i=1}^{N-m}\left(\frac{1}{N-m-1}\sum_{j=1\ j\neq i}^{N-m}\mu\left(d_{ij}^{m+1},n,r\right)\right) \tag{12.5}$$

12.2.2.3 Sample Entropy (sampEn)

The sampEn standard, for instance, was developed by Richman and Randall [42] to evaluate the complexity and uniformity of time series data. Self-similarity can also be quantified using SampEn. A low SampEn score indicates that the data are highly similar to themselves, whereas a high number indicates that the data were collected irregularly. When paired with approximation entropy, SampEn becomes a more accurate indicator [43]:

$$SampEn(m,p,N) = -\ln\left(\frac{A^{m}(p)}{B^{m}(p)}\right) \tag{12.6}$$

To determine how similar two vectors are to one another, SampEn employs a Heaviside function, which takes the form:

$$\theta(z) = \left\{1,\text{if } \mid \begin{array}{c} z\geq 0 \\ z<0 \end{array}\right\} \tag{12.7}$$

This results in a binary classifier with just two possible states—whether the vectors are close or not. Unfortunately, this may not be able to capture the borders between classes, which may be more confusing in real-world biomedical data. As a result, FuzzyEn was developed as an alternative to the Heaviside function for determining the level of similarity between vectors because of this restriction.

Research comparing FuzzyEn to ApEn and SampEn indicates that it is superior. Since its inception, FuzzyEn has been put to use characterizing EMGs, gait, and HRV data, to name just a few types of biomedical information. For missing data analysis of biomedical signals, FuzzyEn has also been demonstrated to be a trustworthy entropy estimator.

With N observations in a time series, $x(n)=x(1), x(2),\ldots, x(n)$, FuzzyEn can be calculated as follows (N):

1. For $1 \leq i \leq N-m+1$, form m-vectors $\boldsymbol{X}_m(1),\ldots, \boldsymbol{X}_m(N-m+1)$ defined as:

$$Xm(i) = \left[x(i),\ x(i+1),\ \ldots,x(i+m-1)\right] - x0(i) \tag{12.8}$$

Starting at the ith position, these vectors indicate m successive x values without the baseline $\left(x0(i) = 1mm1j = 0x\left(i+j\right)\right)$.

2. When two vectors, $Xm(i)$ and $Xm(j)$, are compared, the distance, dij,m, between their scalar components is the largest absolute difference between them.
3. Given the values n and r, use a fuzzy function to calculate the similarity Dij,m between the vectors $Xm(i)$ and $Xm(j)$.

$$Dij,m = \mu(dij,m,r) = exp(-(dij,m)nr\,/) \tag{12.9}$$

4. We repeat steps 2–4 until we reach $m+1$ dimensions, at which point we have the function $m+1$.
5. FuzzyEn estimates for time series with a constant sample size, N, can be calculated using the following equation:

$$FuzzyEn(m,n,r,N) = ln\phi m(n,r) - ln\phi m + 1(n,r) \tag{12.10}$$

Based on the original SampEn algorithm proposed by Richman and Moorman in Reference [44], FuzzyEn can be written as the negative logarithm of the conditional probability that two sequences that are similar for m points remain similar when the size of the vectors being considered is increased by one, where the Heaviside function is replaced by the fuzzy function introduced in equation (12.4). Like SampEn, the algorithm does not factor in self-matches when determining the aforementioned probability. As a result, it does not exhibit the bias that is typical of ApEn [44]. In addition, a higher degree of self-similarity in the time series is indicated by a smaller value for FuzzyEn.

Because the values of FuzzyEn change depending on the input parameters' actual values, comparisons should only be conducted using fixed values for m, n, r, and N. The length of the time series, N, is set by the parameters of this analysis: 256 Hz sampling and 5 s epochs. With ApEn and SampEn long sequences will be compared. The width and gradient of the fuzzy exponential function, however, are determined by the values of r and n.

Increasing m could, in theory, allow for a more accurate reconstruction of the system's dynamics. Thus, if there are more matches of vectors of lengths m and $m+1$, the entropy estimate will be more accurate and trustworthy. Thus, it is suggested that m be kept relatively low [11].

12.2.2.4 Artificial Neural Network

In biological signal processing, subject classification is crucial for making diagnoses. In this study, an ANN classifier is utilized to sort data. ANNs are made up of many individual artificial neurons, each of which serves as a processing element and is linked to its neighbors through weighted connections. The neurons receive their initial weight assignments at random. There are three layers that make up the ANN: input, hidden, and output. There are 256 features taken from each channel. The ANN classifier's input layer receives the signal's extracted features. Because it uses the supervised learning technique, it will automatically assign the desired value. Its training function may take in the input and the randomly initialized weights and spit out a value. In this research, the ANN is taught using a training function based on conjugate gradient backpropagation. The ANN classifier training algorithm entails the following procedures:

Step 1: The first step involves the input of the retrieved features.
The weights, the objective, and the learning rate are all set to their initial values.
Step 2: Add up all the inputs and the weights to get the final value.
Step 3: Determine the final result by applying the activation function training algorithm.
Step 4: Halt the procedure if the output value is the same as the desired value.
Step 5: If the results are not comparable, proceed to Step 5 and use the training rule to readjust the weights.
Step 6: Do it again and again, until the result equals the intended value.

ANN is a type of counting system made up of a large number of simple, highly interconnected processing components that, in theory, may compete with the network and functioning of the biological nervous system [45] (called nodes or artificial neurons). The learning rules used in ANN training algorithms are meant to mimic the training processes used by biological systems. In-depth descriptions of several neural network types and their respective learning architectures may be found in the literature [26–31]. In this context, "neural" refers to the brain-stimulated mechanisms that define each person's unique approach to learning. Layers of input, hidden, and output make up a neural network. ANN is a powerful tool for discovering patterns too intricate for a human coder to understand. In this study, ANN is employed because of the benefits it provides when classifying cases as either "normal," "focal," or "generalized." The methods offered made use of a number of different measures of precision, including accuracy, specificity, sensitivity, positive predictive value (PPV), and negative PPV (NPV). These values can be expressed numerically (Figure 12.2):

$$\text{Accuracy} = \frac{\text{TP+TN}}{\text{TP+TN+FP+FN}} * 100\% \tag{12.11}$$

$$\text{Sensitivity} = \frac{\text{TP}}{\text{TP+FN}} * 100\% \tag{12.12}$$

$$\text{Specificity} = \frac{\text{TN}}{\text{TN+FP}} * 100\% \tag{12.13}$$

$$\text{PPV} = \frac{\text{TP}}{\text{TP+FP}} * 100\% \tag{12.14}$$

$$\text{NPV} = \frac{\text{TN}}{\text{TN+FN}} * 100\% \tag{12.15}$$

$$\text{NPV} = \frac{TN}{TN + FN}. * 100\%$$

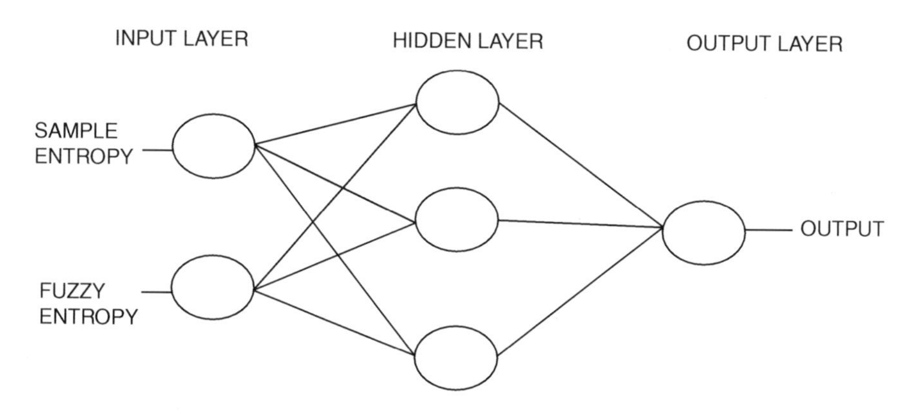

FIGURE 12.2 Block diagram of the artificial neural network(ANN).

12.3 RESULTS AND DISCUSSION

In this research, we employed EEMD and ANN to examine how various types of EEG signals can be categorized. The proposed method is validated by putting signals from the Karunya EEG database through its paces on normal, generalized, and focused EEG signals. The 86 EEG signals found in the Karunya database fall into three categories: standard, focused, and generalized. Here, we pick and evaluate 12 signals from each of the 12 different scenarios. Figure 12.3 displays the EEG data for 16 channels, including normal, generalized, and localized EEG signals. In Figure 12.3, we can see how abnormal, global, and localized EEG signals vary from one another (Figure 12.4).

In order to compute the fluctuations in the IMFs of the signals, the EEMD approach decomposes each channel into IMFs. Typical, global, and local EEG signals are depicted on the IMF plot in Figure 12.5. It is clear from the diagram that the IMF components are sorted from the highest to the lowest frequency. Widespread and localized EEG signals have a faster amplitude modulation rate than average EEG [46].

For each dataset, we estimate its focal entropy, normal entropy, generalized entropy, sample entropy, and fuzzy entropy, and then compare them to the other measures. EMD and EEMD with ANN have been shown to have high levels of accuracy, sensitivity, specificity, PPV, and NPV, as summarized in Tables 12.3–12.8. Table 12.9 shows that our suggested method achieved the highest level of accuracy (100%) in EEMD compared to the other methods (Figures 12.6–12.10) [48].

Using mean and standard deviation information, the maximal overlap discrete wavelet transform (MODWT) was used to categorize epileptic episodes in Reference [47]. The support vector machine classifier achieved an accuracy of 98%, sensitivity of 92.3%, and specificity of 100% when applied to normal focal EEG data, and an accuracy of 91.7%, sensitivity of 85.7%, and specificity of 100% when applied to normal and generalized EEG signals. We used an ANN as a classifier to decompose focused and normal EEG signals into approximation and detailed coefficients, with a resulting sensitivity of 98.39%, specificity of 99.16%, PPV of 99.16%, and NPV of 98.33. Using PSO and ANN as classifiers, the Tunable-Q wavelet transform (TQWT) achieved perfect accuracy (100%) [49]. This was achieved by using log

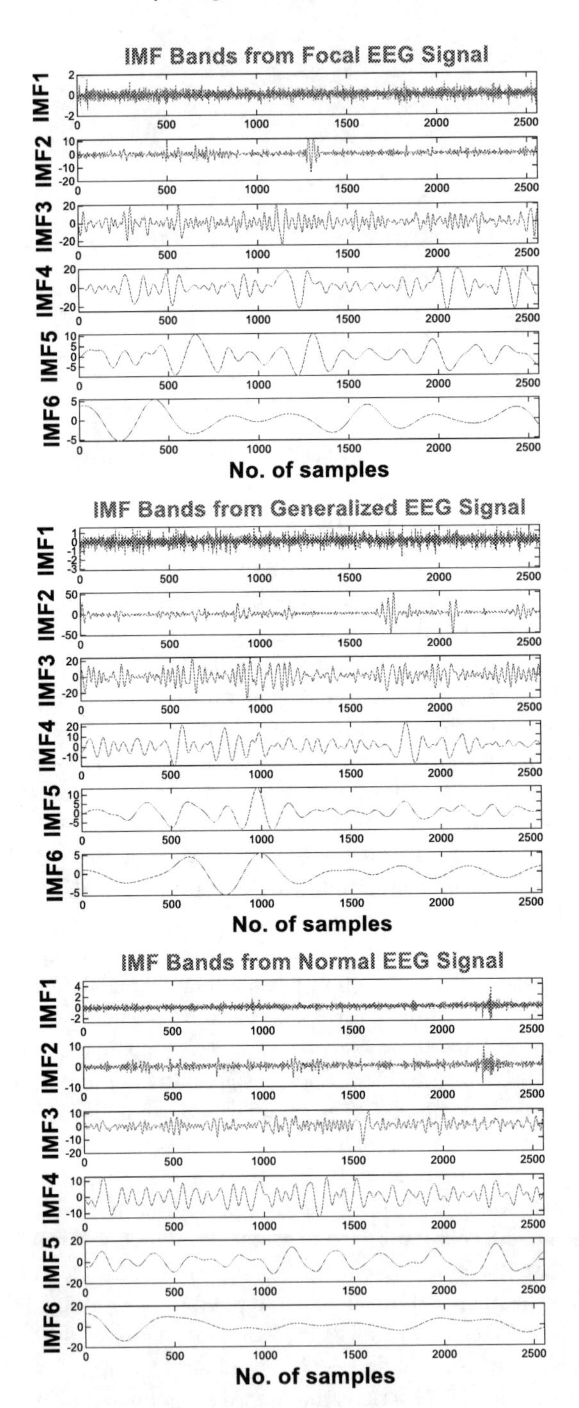

FIGURE 12.3 A sample electroencephalogram (EEG) signal for normal, generalized, and focal subjects.

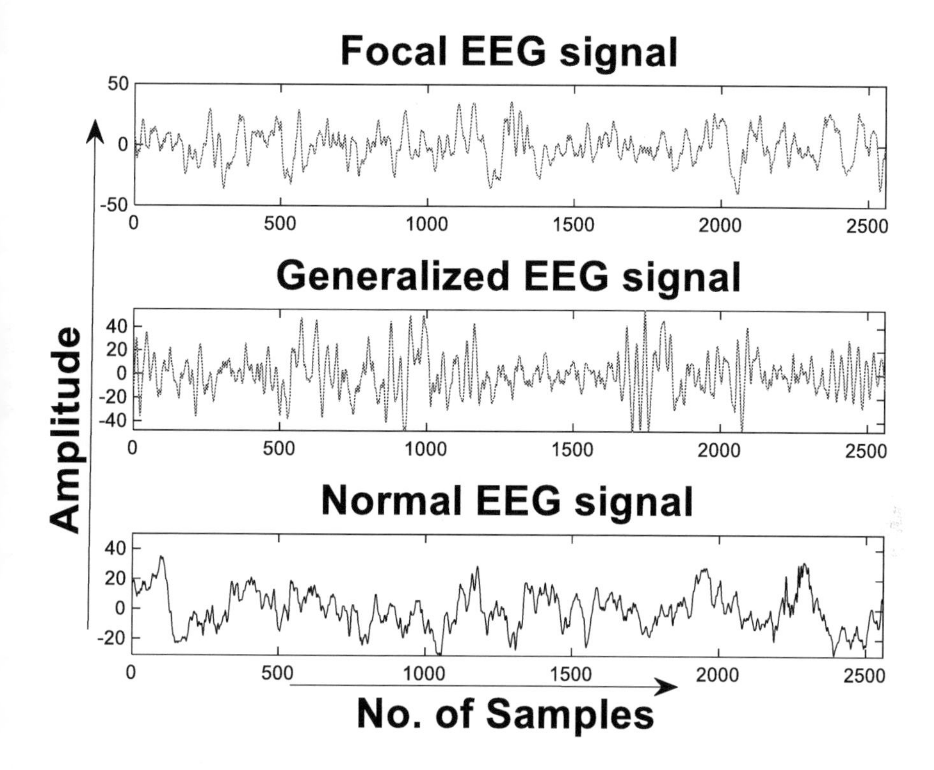

FIGURE 12.4 Example of electroencephalogram (EEG) from normal, generalized, and focal and with one signal.

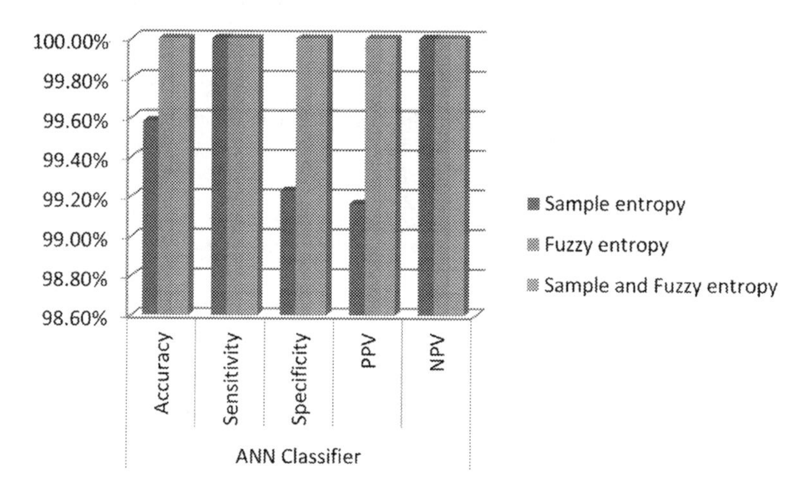

FIGURE 12.5 Analyzing normal and focused electroencephalogram (EEG) data using sample and fuzzy entropies.

TABLE 12.3
The Empirical Mode: Experimental Findings Focal and Normal Decomposition with ANN

	ANN Classifier				
S. No.	**Accuracy (%)**	**Sensitivity (%)**	**Specificity (%)**	**PPV (%)**	**NPV (%)**
Sample entropy	99.5833	100	99.230	99.166	100
Fuzzy entropy	100	100	100	100	100
Sample and Fuzzy entropy	100	100	100	100	100

TABLE 12.4
The Empirical Mode: Experimental Findings Normal and Generalized Decomposition Using Artificial Neural Networks

	ANN Classifier				
S. No.	**Accuracy (%)**	**Sensitivity (%)**	**Specificity (%)**	**PPV (%)**	**NPV (%)**
Sample entropy	100	100	100	100	100
Fuzzy entropy	100	100	100	100	100
Sample and Fuzzy entropy	100	100	100	100	100

TABLE 12.5
Experimental Results for Empirical Mode Decomposition Using ANN for Focal, Normal, and Generalized

	ANN Classifier				
S.no.	**Accuracy (%)**	**Sensitivity (%)**	**Specificity (%)**	**PPV (%)**	**NPV (%)**
Sample entropy	97.77	97.85	98.05	91.66	99.166
Fuzzy entropy	99.44	99.23	100	91.66	100
Sample and Fuzzy entropy	100	100	100	100	100

TABLE 12.6
Experimental Results for Ensemble Empirical Mode Decomposition Using ANN for Focal and Normal

	ANN Classifier				
S. No.	**Accuracy (%)**	**Sensitivity (%)**	**Specificity (%)**	**PPV (%)**	**NPV (%)**
Sample entropy	100	100	100	100	100
Fuzzy entropy	100	100	100	100	100
Sample and Fuzzy entropy	100	100	100	100	100

TABLE 12.7
Experimental Results for Ensemble Empirical Mode Decomposition Using ANN for Generalized and Normal

S. No.	ANN Classifier				
	Accuracy (%)	Sensitivity (%)	Specificity (%)	PPV (%)	NPV (%)
Sample entropy	99.5833	100	99.2307	99.166	100
Fuzzy entropy	95.8333	92.3076	100	100	99.166
Sample and Fuzzy entropy	100	100	100	100	100

TABLE 12.8
Experimental Results for Ensemble Empirical Mode Decomposition Using ANN for Focal, Normal, and Generalized

S. No.	ANN Classifier				
	Accuracy (%)	Sensitivity (%)	Specificity (%)	PPV (%)	NPV (%)
Sample entropy	99.166	98.461	100	100	98.333
Fuzzy entropy	99.5833	100	99.230	99.166	100
Sample and Fuzzy entropy	99.166	99.166	99.166	99.166	99.166

energy, Shannon, and sure entropies. After transforming EEG data into 2D brain maps and altering them with symmetric-weighted scale-invariant local ternary pattern (SWSILTP), ANN was able to discriminate between epileptic and artifact brain maps with a 99.53% success rate, as shown in Reference [50]. Reassignment smoothed pseudo Wigner–ville distribution was used by Jia et al. [51] to forecast the most important brain rhythms with high accuracy. In Reference [52], the authors applied the discrete wavelet transform and wavelet packet decomposition in four distinct scenarios—normal, normal generalized, normal focal, and normal focal generalized—to achieve maximum accuracy for rbio 1.1. The signal-to-noise ratio for epilepsy was improved by 98.42% using a scalar quantizer based on the discrete wavelet transform and an immune feature weighted support vector machine as a classifier [53]. The participants were tasked with recognizing emotions and mental diseases such as joy, melancholy, fear, epilepsy, aneurysm, and sclerosis.

12.4 CONCLUSION

To distinguish between normal, focused, and generalized EEG data, characteristics are acquired using an EEMD and an ANN-based recognition system. With EMD and EEMD, the signal is separated into IMFs. Each IMF has its fuzzy entropy and sample entropy calculated. The ANN classifier used to make the distinction between EEG signals also incorporates the retrieved features. Extracted features were found to be highly effective and precise in experiments. Future work will include assessing the suggested method on a larger database and expanding it to detect the pre-ictal stage in seizure patients.

TABLE 12.9
Comparison with Other Works

Feature Extraction	Data Base	Features	Classifier	Classification Accuracy
MODWT [47]	Karunya Data Base	Mean and standard deviation	SVM	95.8%
MODWT [47]	Karunya Data Base	Mean and standard deviation	SVM	91.7%
SWT [48]	Karunya Data Base	–	ANN	98.5%
Tunable-Q wavelet transform (TQWT) [49]	Karunya Data Base	Log energy, Shannon, Sure	PSO and ANN	100%
Symmetric-weighted scale-invariant local ternary pattern (SWSILTP) [50]	Karunya Data Base	–	ANN	99.5%
Reassigned SPWVD [51]	Karunya Data Base	Brain rhythm sequence data	–	98.9%
DWT [52]	Karunya DataBase	Min, Max, Mean, STD, Kurtosis, Skewness, Energy, Nstd, nEnergy	Cubic SVM	Normal-generalized: 100 Normal-focal: 93.7 Normal-generalized −focal: 93 Normal-generalized + focal: 96.1
WPD [53]	Karunya Data Base	Min, Max, Mean, STD, Kurtosis, Skewness, Energy, Nstd, nEnergy	Cubic SVM	Normal-generalized: 100 Normal-focal: 91.66 Normal-generalized −focal: 91.2 Normal-generalized + focal: 92.77
DWTSQ [54]	Karunya DataBase	–	Immune feature weighted support vector machine (IFWSVM)	98.42%
Proposed method EEMD	Karunya DataBase	Sample and fuzzy entropy	ANN	100%

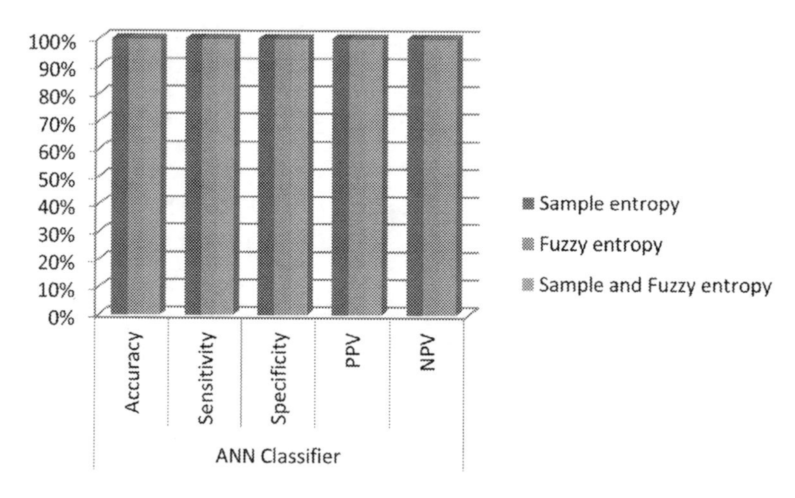

FIGURE 12.6 Sample and fuzzy entropies with generalized and normal electroencephalogram (EEG) signals.

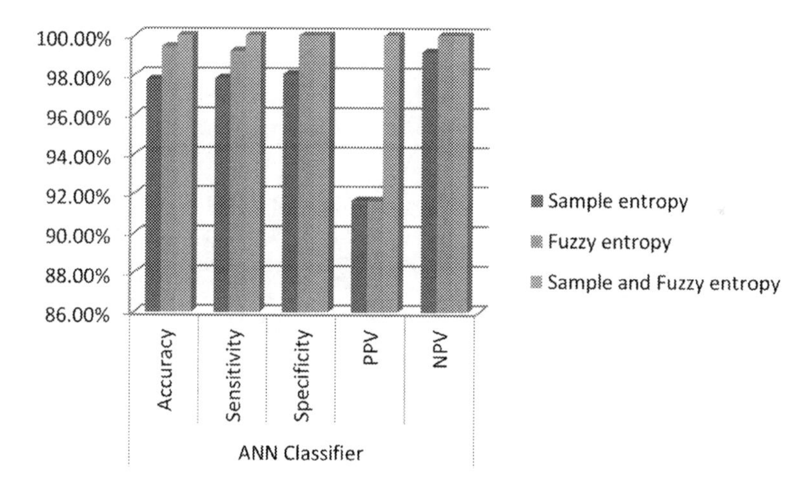

FIGURE 12.7 Sample and fuzzy entropies with focal, normal, and generalized electroencephalogram (EEG) signals.

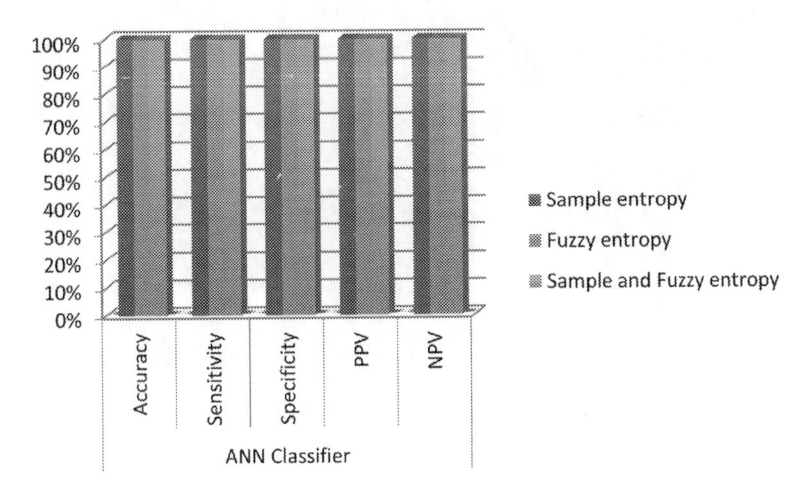

FIGURE 12.8 Sample and fuzzy entropies with focal and normal electroencephalogram (EEG) signals.

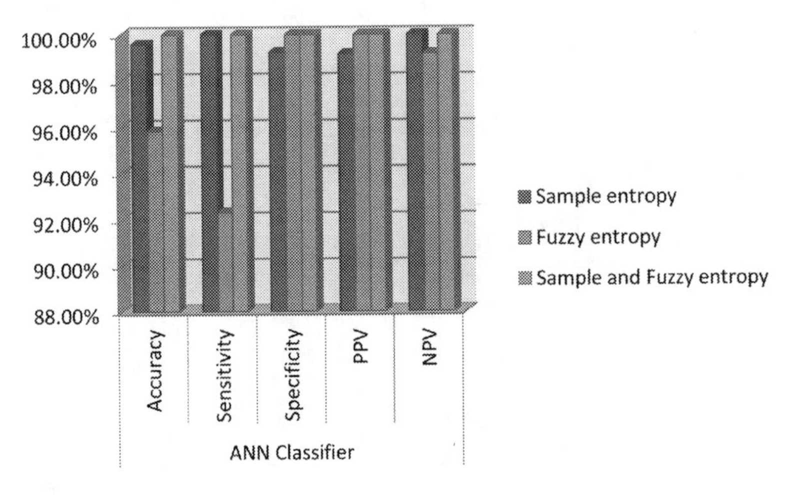

FIGURE 12.9 Sample and fuzzy entropies with generalized and normal electroencephalogram (EEG) signal.

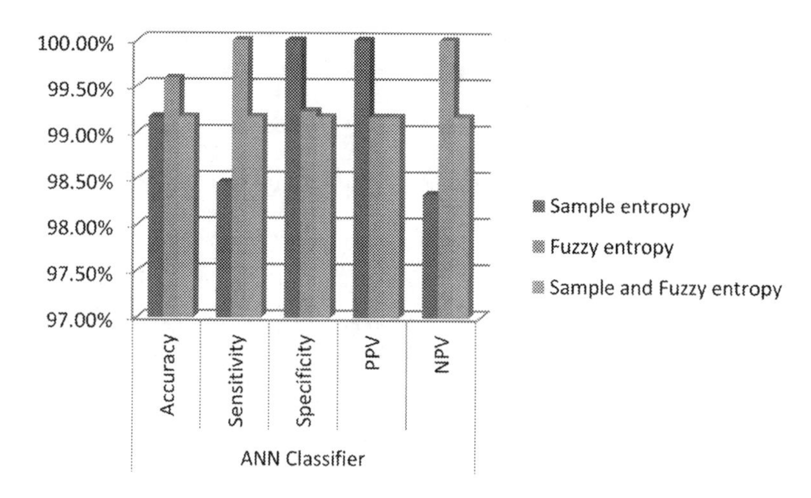

FIGURE 12.10 Sample and fuzzy entropies with focal, normal, and generalized electroencephalogram (EEG) signals.

REFERENCES

1. Fisher, R. S., Boas, W. V. E., Blume, W., Elger, C.,Genton, P., Lee, P., & Engel, J. (2005). Epileptic seizure and epilepsy: Definitions proposed by the International League against Epilepsy (ILAE) and the International Bureau for Epilepsy (IBE). *Epilepsia*, 46(4), 470–472.
2. Yuan, Q., Zhou, W., Li, S., & Cai, D. (2011). Epileptic EEG classification based on extreme learning machine and nonlinear features. *Epilepsyresearch*, 96(1), 29–38.
3. Noachtar, S., & Rémi, J. (2009). The role of EEG in epilepsy: A critical review. *Epilepsy & Behavior*, 15(1), 22–33.
4. Mansouri, A., Singh, S. P., & Sayood, K. (2019). Online EEG seizure detection and localization. *Algorithms*, 12(9), 176.
5. Hirtz, D., Thurman, D. J., Gwinn-Hardy, K., Mohamed, M., Chaudhuri, A. R., & Zalutsky, R. (2007). How common are the "common" neurologic disorders? *Neurology*, 68(5), 326–337.
6. Fisher, R. S., Scharfman, H. E., & DeCurtis, M. (2014). How can we identify ictal and interictal abnormal activity? In: Helen E. Scharfman, and Paul S. Buckmaster, eds., *Issues in Clinical Epileptology: A View from the Bench* (Vol. 813, pp. 3–23). Springer.
7. Shorvon, S., & Tomson, T. (2011). Sudden unexpected death in epilepsy. *The Lancet*, 378(9808), 2028–2038.
8. Fisher, R. S., Boas, W. V. E., Blume, W., Elger, C., Genton, P., Lee, P., & Engel Jr, J. (2005). Epileptic seizures and epilepsy: Definitions proposed by the International League Against Epilepsy (ILAE) and the International Bureau for Epilepsy (IBE). *Epilepsia*, 46(4), 470–472.
9. Yuan, Q., Zhou, W., Li, S., & Cai, D. (2011). Epileptic EEG classification based on extreme learning machine and nonlinear features. *Epilepsy Research*, 96(1–2), 29–38.
10. Noachtar, S., & Rémi, J. (2009). The role of EEG in epilepsy: A critical review. *Epilepsy & Behavior*, 15(1), 22–33.
11. George, S. T., Balakrishnan, R., Johnson, J. S., & Jayakumar, J. (2017). Application and evaluation of independent component analysis methods to generalized seizure disorder activities exhibited in the brain. *Clinical EEG and Neuroscience*, 48(4), 295–300.

12. Selvan, S. E., George, S. T., & Balakrishnan, R. (2015). Range-based ICA using a nonsmooth quasi-Newton optimizer for electroencephalographic source localization in focal epilepsy. *Neural Computation*, 27(3), 628–671.
13. Jothiraj, S. N., Selvaraj, T. G., Ramasamy, B., Deivendran, N. P., & Subathra, M. S. P. (2018). Classification of EEG signals for detection of epileptic seizure activities based on feature extraction from brain maps using image processing algorithms. *IET Image Processing*, 12(12), 2153–2162.
14. Selvaraj, T. G., Ramasamy, B., Jeyaraj, S. J., & Suviseshamuthu, E. S. (2014). EEG database of seizure disorders for experts and application developers. *Clinical EEG and Neuroscience*, 45(4), 304–309.
15. Bajaj, V., & Pachori, R. B. (2011). Classification of seizure and nonseizure EEG signals using empirical mode decomposition. *IEEE Transactions on Information Technology in Biomedicine*, 16(6), 1135–1142.
16. Li, S., Zhou, W., Yuan, Q., Geng, S., & Cai, D. (2013). Feature extraction and recognition of ICTAL EEG using EMD and SVM. *Computers in Biology and Medicine*, 43(7), 807–816.
17. Ruth, V. C., Prasanna, J., Thomas, G. S., Sairamya, N. J., & Subathra, M. S. P. (2019, January). Classification of normal. focal, and generalized EEG signals using EMD and ANN. In: 2019 *International Conference on Computer Communication and Informatics (ICCCI)*. IEEE, New York, pp. 1–4.
18. Wu, Z., & Huang, N. E. (2009). Ensemble empirical mode decomposition: A noise-assisted data analysis method. *Advances in Adaptive Data Analysis*, 1(1), 1–41.
19. Eftekhar A., Vohra F., Toumazou C., Drakakis E. M., & Parker K. (2008) Hilbert-Huang transform: Preliminary studies in epilepsy and cardiac arrhythmias. In: *Proceedings of the IEEE Biomedical Circuits and Systems Conference*. BioCAS, Baltimore, MD, pp 373–376.
20. University of Freidberg (2014). *Seizure Prediction Project Freidburg*. University of Freiburg. https://epilepsy.uni-freiburg.de/freiburgseizurepredictionproject/eeg-database.
21. Orosco L., Laciar E., Correa A. G., Torres A., & Graffigna J. P. (2009) An epileptic seizures detection algorithm based on the empirical mode decomposition of EEG. In: *Proceedings of the international conference of the IEEE EMBS*. IEEE, Minneapolis, MN, pp. 2651–2654.
22. Tafreshi A. K., Nasrabadi A. M., & Omidvarnia A. H. (2008) Epileptic seizure detection using empirical mode decomposition. In: *Proceedings of the IEEE International Symposium on Signal Processing and Information Technology*. ISSPIT, Sarajevo, pp. 238–242.
23. Guarnizo C., & Delgado E. (2010) EEG single-channel seizure recognition using empirical mode decomposition and normalized mutual information. In: *Proceedings of the IEEE International Conference on Signal Processing (ICSP)*. IEEE, Beijing, pp. 1–4
24. Orosco L., Laciar E., Correa A. G., Torres A., & Graffigna J. P. (2009) An epileptic seizures detection algorithm based on the empirical mode decomposition of EEG. In: *Proceedings of the International Conference of the IEEE EMBS*. IEEE, Minneapolis, MN, pp. 2651–2654.
25. Belhadj S., Attia A., Adnane B. A., Ahmed-Foitih Z., & Ahmed A. (2016) *Whole Brain Epileptic Seizure Detection Using Un-Supervised Classification*. IEEE, Algiers, Algeria.
26. Torse D. A., Desai V., & Khanai R. (2017) EEG signal classification into seizure and non-seizure class using empirical mode decomposition and artificial neural network. *IJIR*, 3(1), 2454–1362.

27. Alam S. M. S., & Bhuiyan M. I. H. (2011) Detection of epileptic seizures using chaotic and statistical features in the EMD domain. In: *2011 Annual IEEE India Conference*, pp. 1–4. IEEE.
28. Bajaj V., & Pachori R. B. (2013) Epileptic seizure detection based on the instantaneous area of analytic intrinsic mode functions of EEG signals. *Biomedical Engineering Letters*, 3(1), 17–21.
29. Samiee, K., Kovács P., & Gabbouj, M. (2015). Epileptic seizure classification of eeg time-series using rational discrete short-time Fourier transform, *IEEE Transactions on Biomedical Engineering*, 62(2), 541–552.
30. Wang G., Sun Z., Tao R., Li K., Bao G., & Yan X. (2016). Epileptic seizure detection based on partial directed coherence analysis. *IEEE Journal of Biomedical and Health Informatics*, 20(3), 873–879.
31. Kumar S. P., Sriraam N., Benakop P. G., & Jinaga B. C. (2010). Entropies based detection of epileptic seizures with artificial neural network classifiers. *Expert Systems with Applications*, 37(4), 3284–3291.
32. Martis R. J., Acharya U. R., Tan J. H., Petznick A., Yanti R., Chua C. K., Ng E. Y. K., & Tong L. (2012). Application of empirical mode decomposition (EMD) for automated detection of epilepsy using EEG signals. *Expert Systems with Applications*, 22(6), 1250027.
33. Yuan Q., Zhou W., Liu Y., & Wang J. (2012). Epileptic seizure detection with linear and nonlinear features. *Epilepsy & Behavior*, 24(4), 415–421.
34. Fan M., & Chou, C.-A. (2019). Detecting abnormal pattern of epileptic seizures via temporal synchronization of EEG signals. *IEEE Transactions on Biomedical Engineering*, 66(3), 601–608.
35. Aldana, Y. R., Hunyadi, B., Reyes, E. J. M., Rodríguez, V. R., & Van Huffel, S. (2019). Nonconvulsive epileptic seizure detection in scalp EEG using multiway data analysis. *IEEE Journal of Biomedical and Health Informatics*, 23(2), 660–671.
36. Yuan, Y., Xun, G., Jia, K., & Zhang, A. (2019). A multi-view deep learning framework for EEG seizure detection. *IEEE Journal of Biomedical and Health Informatics*, 23(1), 83–94
37. Wen T., & Zhang, Z. (2018). Deep convolution neural network and autoencoders-based unsupervised feature learning of EEG signals. *IEEE Access*, 6, 25399–25410
38. https://www.karunya.edu/research/EEGdatabase/public/index.php
39. Huang, N. E., Shen Z., Long S. R., Wu M. L., Shih H. H., Zheng Q., Yen N. C., Tung C. C., & Liu H. H. (1998). The empirical mode decomposition and Hilbert spectrum for nonlinear and non stationary time series analysis. *Proceedings of the Royal Society of London Series A, Mathematical and Physical Sciences*, 454, 903–995
40. Wu, Z., & Huang, N. E. (2009). Ensemble empirical mode decomposition: A noise-assisted data analysis method. *Advances in Adaptive Data Analysis*, 1(1), 1–41.
41. Kosko, B. (1986). Fuzzy entropy and conditioning. *Information Sciences*, 40(2), 165–174
42. Acharya, U. R., Fujita, H., Sudarshan, V. K., Bhat, S., & Koh, J. E. (2015). Application of entropies for automated diagnosis of epilepsy using EEG signals: A review. *Knowledge-Based Systems*, 88, 85–96.
43. Richman, J. S., & Moorman, J. R. (2000). Physiological time-series analysis using approximate entropy and sample entropy. *American Journal of Physiology-Heart and Circulatory Physiology*, 278(6), H2039–H2049.
44. Richman J. S., & Moorman J. R. (2000). Physiological time-series analysis using approximate entropy and sample entropy. *American Journal of Physiology-Heart and Circulatory Physiology*, 274, 2039–2049. doi:10.1152/ajpheart.2000.278.6.H2039.
45. Acharya, U. R., Sree, S. V., Swapna, G., Martis, R. J., & Suri, J. S. (2013). Automated EEG analysis of epilepsy: A review. *Knowledge-Based Systems*, 45, 147–165.

46. Subasi, A., & Ercelebi, E. (2005). Classification of EEG signals using neural network and logistic regression. *Computer Methods and Programs in Biomedicine*, 78(2), 87–99.
47. Hagan, M. T., & Menhaj, M. B. (1994). Training feedforward networks with the Marquardt algorithm. *IEEE Transactions on Neural Networks*, 5(6), 989–993.
48. Prasanna, J., Thomas, G. S., Subathra, M. S. P., & Sairamya, N. J. (2019, March). Automatic epileptic seizure classification using MODWT and SVM. In: *2019 2nd International Conference on Signal Processing and Communication* (*ICSPC*). IEEE, New York, pp. 113–116.
49. Mercy, F. J., Prasanna, J., Thomas, G. S., Belvina, R. S., Evangelin, G. N., & Mabel, J. (2019, March). Epilepsy seizure detection using SWT features and ANN classifier. In: *2019 2nd International Conference on Signal Processing and Communication (ICSPC)*. IEEE, New York, pp. 335–338.
50. George, S. T., Subathra, M. S. P., Sairamya, N. J., Susmitha, L., & Premkumar, M. J. (2020). Classification of epileptic EEG signals using PSO based artificial neural network and tunable-Q wavelet transform. *Biocybernetics and Biomedical Engineering*, 40(2), 709–728.
51. Dominguez, E. C., Subathra, M. S. P., Sairamya, N. J., & George, S. T. (2020). Detection of focal epilepsy in brain maps through a novel pattern recognition technique. *Neural Computing and Applications*, 32(14), 10143–10157.
52. Li, J. W., Barma, S., Mak, P. U., Pun, S. H., & Vai, M. I. (2019). Brain rhythm sequencing using EEG signals: A case study on seizure detection. *IEEE Access*, 7, 160112–160124.
53. Sairamya, N. J., Premkumar, M. J., George, S. T., & Subathra, M. S. P. (2019). Performance evaluation of discrete wavelet transform, and wavelet packet decomposition for automated focal and generalized epileptic seizure detection. *IETE Journal of Research*, 67(6), 778–798.
54. Balamareeswaran, M., & Ebenezer, D. (2015). Denoising of EEG signals using discrete wavelet transform based scalar quantization. *Biomedical and Pharmacology Journal*, 8(1), 399–406.

13 Challenges and Future Directions of Fuzzy System in Healthcare Systems

A Survey

Manish Bhardwaj and Jyoti Sharma
KIET Group of Institutions

Yu-Chen Hu
Providence University

Samad Noeiaghdam
Irkutsk National Research Technical University

13.1 INTRODUCTION

Because of its unique perspective, fuzzy logic can serve as a bridge between human experience and technological systems. Fuzzy logic helps us relate real-world propositions to machine propositions by accounting for the fact that events in the real world are rarely black-and-white and that we are constantly exposed to a spectrum of right and wrong [1]. In addition, the use of computer-based decision support systems to automate parts of the medical decision-making in difficult clinical areas is becoming increasingly important [2].

Fuzzy systems, a subset of intelligent systems, make substantial use of the knowledge of techno-scientific professionals to solve semi- or ill-structured issues for which there is no clear-cut explanation of the method [3]. Fuzzy systems have been described as an intelligent system that use data and inference procedures to solve complex problems that typically necessitate the assistance of trained engineers [4].

An example of an expert system is the fuzzy expert system (FES), which has its roots in fuzzy systems. The FES comprises fuzzification, judgment rules, an expertise database, and defuzzification, and it uses fuzzy logic alongside the use of Boolean mathematics in the judgment procedure. When describing decision-making situations for which no good algorithm exists, this framework is typically employed. Nonetheless, experts may be relied upon to provide heuristic considerations of

DOI: 10.1201/9781003405368-13

potential solutions to problems in the form of If-Then rules [5]. The issue, which elucidates ambiguity stemming from fuzziness, uncertainty, or both, can be satisfactorily addressed by an FES subjectivity. The parts of an expert system are depicted in Figure 13.1.

The primary goal of this research is to examine the benefits and drawbacks of current approaches and theories for creating and modeling fuzzy systems in various fields such as manufacturing [6], healthcare, business, marketing, and more. About the accomplishing of such goals, a survey of recent international journals and conference publications is conducted [7].

Figure 13.2 shows the example of an MYCIN Expert system. This diagram shows the communication between the user and expert system. The user received the proper advice and decision of the expert system after complete analysis of the problem with inputs.

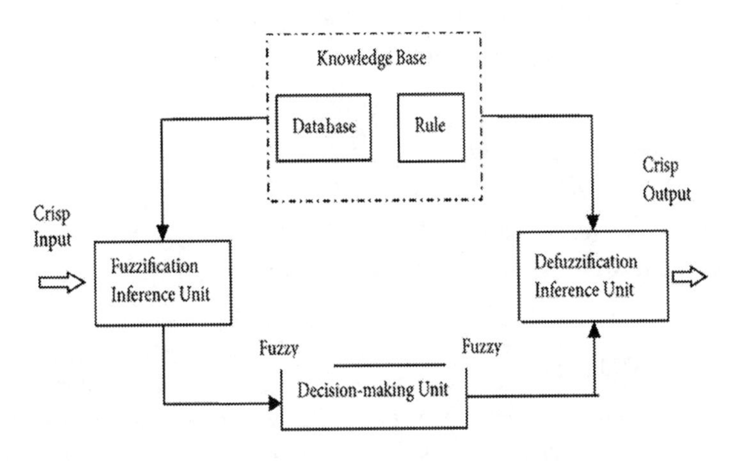

FIGURE 13.1 Basic block diagram of fuzzy-based expert system.

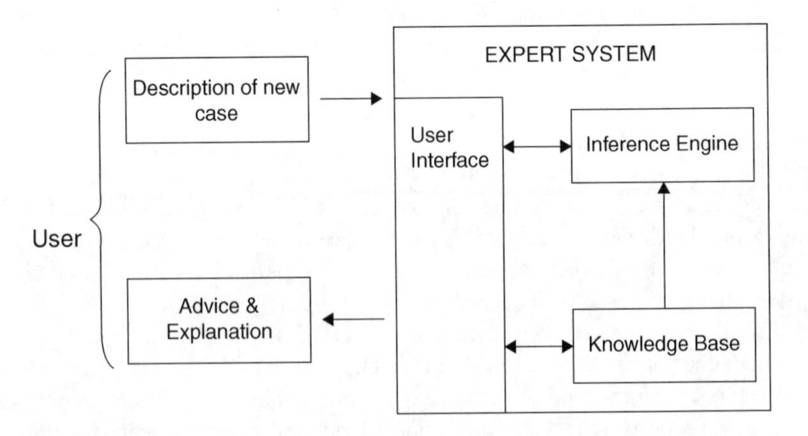

FIGURE 13.2 Example of the MYCIN expert system process.

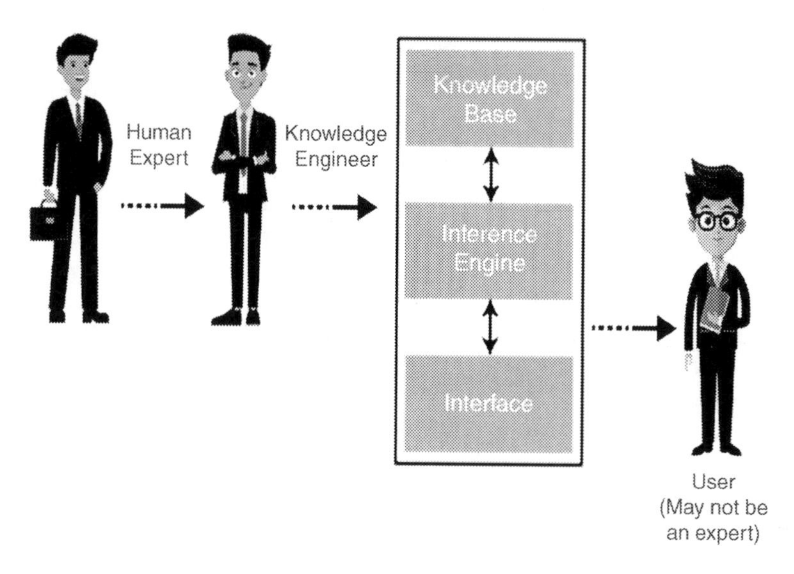

FIGURE 13.3 Expert system representation in artificial intelligence.

13.2 FUZZY SYSTEM APPROACH

To put it simply, Zadeh [8] proposed using the dominant theory of fuzzy logic to make sense of vague and ill-defined expert knowledge. Fuzzification interface uses a variety of membership methods to convert sharp data into fuzzy linguistic values. With an FES, fuzzification is to be expected on a frequent basis because the input values from robust detectors are invariably precise numerical values [9]. The inference dynamo requires fuzzy input and rules, and it only generates fuzzy output in return. The fuzzy rule base, which may include linguistic variables, should ideally be in the form of "IF-THEN" rules. Defuzzification, the final step in an FES, is responsible for enacting crisp yield operations [10].

Computer Science, Engineering, and Mathematics account for the largest percentages (29.4%, 28.8%, and 12.4%) of published research results on FESs across all disciplines, according to the most recent statistical SCOPUS data [11–13]. In Figure 13.3, we see the intricate web of connections between artificial intelligence and the uses of expert systems.

13.3 ADVANTAGE OF FUZZY SYSTEM

Fuzzy logic systems have an easily graspable and intuitive framework. Currently, both businesses and academic institutions make heavy use of fuzzy logic. Fuzzy logic also allows for improved machine control and more efficiency, as well as savings. Although fuzzy logic has been attacked for its inaccuracy, it can be employed with confidence, especially when working with imperfect data [14–16]. Fuzzy logic in the realm of control allows for contingency plans to be made so that production continues even if sensors fail.

Increased performance at low cost is feasible in systems using fuzzy logic if proper control of the system can be achieved through the use of inexpensive sensors. Actually, providing easier to understand and implement answers to difficult situations is a major driving force behind the development of fuzzy logic [17]. The use of fuzzy logic permits the modeling and incorporation of ambiguity into a body of information. In opposition to systems based on regulations, where just one regulation can have far-reaching, negative consequences, and saturate microchip processor-based gadgets, compensatory systems increase system autonomy (the rules in the knowledge base operate autonomously of each other).

Systems with fuzzy logic are useful for reasoning that is approximative or uncertain, especially when applied to a situation for which a precise mathematical model has proven elusive.

Decisions can be made with approximate values using fuzzy logic, even when necessary information is lacking. Fuzzy logic can be used to develop a strategy for foreseeing potential outcomes. Predicting the future is nearly impossible without the help of fuzzy logic [18].

Fuzzy modeling can be either subjective or objective. The prediction system is built using an objective kind of fuzzy modeling. The effectiveness of the prediction is enhanced by the hybrid of a clustering technique and fuzzy system identification. The prediction method is arrived at by mining the web for historical data [19–21]. The data are collected and stored in a format suitable for the intended use. The purpose of keeping these data are to ensure that the prediction system receives just the necessary inputs. The procedure for subtractive clustering is used for this purpose.

The system identification approach has computational benefits and fuzzy rules can be created.

Knowledge-based systems in medicine can use fuzzy logic for tasks, including interpreting groups of medical findings, differentiating syndromes, diagnosing diseases, choosing the most effective treatments, and monitoring patients in real time.

Specifically, for the recognition and cure of lung diseases, practitioners of Oriental indigenous medicine use an evaluation method that incorporates Western disease diagnosis, Oriental syndrome distinctions, a fuzzy system for classifying Western and Eastern medicines, and a fuzzy set theory case-based reasoning approach [22,23]. Figure 13.4 shows the outputs of Boolean logic and fuzzy logic for a particular input.

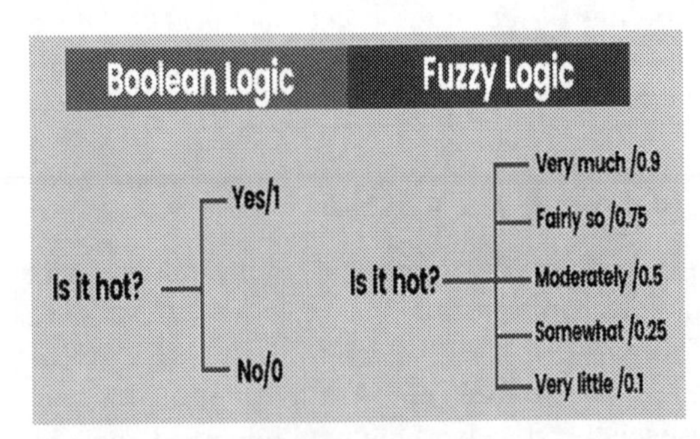

FIGURE 13.4 Output representation for a particular input in Boolean and fuzzy logic.

In case of Boolean logic, it provides only two outputs, but in case of fuzzy logic, it can be divided into various outputs with different choices. Subjective judgments, ambiguous data, and approximative system models are commonplace in reliability analysis. Fuzzy logic is helpful for determining a system's dependability in certain scenarios, because it permits relative analysis without forcing precision where it is not present. Fuzzy logic can be applied to both arithmetic and language variables for the purposes of assessing system configurations, defect trees, activity trees, reliability of chromic systems, and assessing the performance of networks based on the magnitude of an interruption and the probability of it occurring.

Fuzzy logic is used in a new yaw moment control to enhance vehicle control and longevity. The advantages of using fuzzy logic approaches for regulating nonlinear systems include their ease of use and high performance. To keep the vehicle tracking the desired yaw rate and side-slip angle, the enhanced controller calculates the necessary yaw moment by subtracting the brake forces applied to the two front wheels [24].

Results from the simulations are used to evaluate the performance of the suggested control system when the vehicle is instructed to perform various cornering steering maneuvers [25], such as a change line and a J-turn, on both dry and snow-covered roads.

Diabetes is rapidly nearing epidemic proportions in many emerging and newly industrialized countries, and has risen to the fourth leading cause of death in developed nations. Diagnosis models in medicine incorporate both clinical signs and test results [26–29]. The dangers and expenses of the various tests vary widely. The suggested approach solves the issue of selecting the appropriate feature subsets for automatic pattern classification designs. It is the process of determining which patterns are most helpful and choosing which ones to depict. collected attributes from a larger database.

In order to improve classification certainty, the suggested system employs a fuzzy rule-based classification framework. Fuzzy logic is an approach to problem-solving that takes into account the way humans naturally think and behave.

It reveals predicates that occur naturally and are connected to things big and tiny. Fuzzy logic is an extension of traditional (Boolean) logic that allows for the consideration of truth values that fall between "totally true" and "absolutely untrue," thus modeling the way humans make decisions [30].

Hardware, software, or a combination of the two can be used to accomplish it, and its potential applications range from handheld devices to large-scale computerized process control systems.

Fuzzy logic is being accepted by the automotive industry as a means to improve quality, reduce development time [31–35], and save costs in today's highly competitive market. In many control system applications, fuzzy logic has proven to be superior to alternative methods of data classification and processing.

The learner demonstrating section of a web-based instructional tool for instructing in the programming language Pascal makes use of fuzzy logic techniques. The identification of a student's errors of judgment and intellectual abilities is laden with unpredictability; one technique to deal with this is through fuzzy student simulation.

This is why we choose fuzzy logic approaches when describing a student's knowledge and skills. In addition, we employ a method of practices over the fuzzy sets, which is activated after any change in the students' knowledge level of a domain concept and updates the students' knowledge level of all related concepts with this notion [36–38].

The primary mathematical and practical difficulties in doing so are outlined, and demonstrations of fuzzy logic control strategies with affiliations to the Technical University of Denmark are provided.

Two efforts to regulate rotary cement kilns are described in detail, along with the first successful pilot run of the industrial approach [39]. Further studies are required to determine the viability of structural programming and stability issues in the context of fuzzy control systems.

To more accurately model ambiguity and imprecision, type-2 fuzzy sets are used. Zadeh first proposed these type-2 fuzzy sets in 1975; they are often "fuzzy fuzzy" sets in which the fuzzy degree of membership is a fuzzy set: a fuzzy set of type-1 uncertainty. Mendel and Liang's new theories allow for the specification of a type-2 fuzzy set with two membership functions, one superior and one inferior, either of which may be defined by a membership function for a type-1 fuzzy set. The footprint is defined by the range between these two functions [40].

Type-2 fuzzy sets can be defined by this measure of uncertainty, also called the degree of uncertainty (FOU). This study proposes using fuzzy-logic regulation to modify the rotational velocity of an IM. The FLC makes use of fuzzy logic, which is itself dependent on the indirectness of the vector control system. The outer loop includes the fuzzy-logic-based speed regulator. We experimentally implemented the full IM drive vector management system with the FLC using a digital signal: a DS-1102 lab board with a 1 hp squirrel-cage IM processor.

The computational and experimental depictions of the recommended FLC-based IM propel have been contrasted to those taken from the regular proportional-integral (PI) controller-based drive under a variety of evolving functioning instances, including sudden changes in authority speed and demand [41–43].

The FLC is more stable than the usual PI controller, as demonstrated by the experiments, and is therefore recommended as the new standard for the high-performance industrial drive software.

For situations that lack precision by their very nature and for which quantitative data are unavailable or inappropriate, a new methodology is presented. Probability theory and fuzzy logic are implemented, and the approach for reliability and risk assessment is demonstrated through examples. Fuzzy logic is mentioned as a potential method for creating knowledge bases.

To define the use of fuzzy logic and fuzzy set theory to pharmacological issues, the term "fuzzy pharmacology" was coined. Fuzzy logic is the branch of knowledge that studies how humans make sense of the world and puts that understanding to use. It's a development of binary logic that can handle complex systems without requiring fine-grained distinctions between its constituent parts. Fuzzy modeling has been used for the mechanical regulation of drug distribution in surgical settings, and research has begun to evaluate its potential utility in other areas of pharmacokinetics and pharmacodynamics [44]. These preliminary studies show that fuzzy pharmacology is a promising new area for further study. Fuzzy logic is introduced as a potential application in home electronics.

Several forms of Japanese industry have developed fuzzy-controlled equipment. One advantage of fuzzy controllers is that they can incorporate the qualitative understanding of operations that is available from specialists. Yet, experts' heuristic approach to developing fuzzy controllers is inefficient and error-prone. For this reason, we propose two potential uses for "neuro-fuzzy," one involving a refrigerator and the other involving a welding machine, by developing a self-tuning method of fuzzy in which rules are applied using a learning algorithm of neural networks [45]. Fuzzy logic is very useful in the field of picture recognition.

It's important to figure out how to create a comfortable temperature in a room for the specific number and arrangement of people who will be using it. The classification data can be used to verify the room temperature and wind direction using a fuzzy clustering technique and the "neuro-fuzzy in segmentation and locating method of inhabitants from thermal photos."

Open-source software, such as Linux and other free operating systems, has been gaining popularity recently, drawing the attention of many professionals, governments, and organizations. This has led to major software firms such as Google and Oracle developing their own open-source offerings.

Many others have taken the initiative to further the spread of open-source software, particularly Linux. In this research, the LFS system is used to enhance a Linux-based OS. The origins and growth of Linux, and the role it has played in shaping the modern world [46], are the first topics of discussion. Secondly, we break down the components that make up a Linux distribution and show you how you may use those components to make your own distribution better. A different technique named "Remaster" is compared to LFS.

Students and scholars rely heavily on the resources provided by academic libraries. The performance of these libraries can be enhanced and user satisfaction increased by careful assessment of the quality of services provided. This chapter seeks to probe service A hybrid FES employed for quality estimate in the academic library. Experts in the field of library science contributed to the development of this system. A standard survey is also conducted to collect student feedback. After the system was built, it was inspected and evaluated to determine that it is a hybrid of four different kinds of FESs.

With an accuracy of 98.002%, the approach successfully represented the level of user satisfaction with library services. Academic libraries' strengths and weaknesses can be identified and evaluated with the help of the system. In addition, it can foster mastery and raise performance to new heights of satisfaction among users [47].

In today's technologically advanced world, client satisfaction is the most important factor in the success of a business or production. A company's bottom line will greatly benefit from the application of cutting-edge marketing strategies and the careful management of marketing decisions.

This chapter delves deeper into the four primary marketing strategies (price, product, location, and promotion) and how they may be improved with the use of a system based on fuzzy reasoning and the advice of marketing experts.

The system's distinctive features are its reliance on expert knowledge to produce a deduction framework rule, the production of five fuzzy knowledgeable system outputs, the selection of the ultimate outcome, the design of a specific function expert system on management, and the support for executives in their achievement of practiced management.

We used the data in the targeted database to inform the development of a neural network-based medical expert system. BUPA-related liver disease comparison is made between the system's efficiency while using parametric techniques, such as the Bayesian decision-making approach, and when using nonparametric techniques.

By examination of the data, we found that there is a particularly nasty spot in the bank that slows down the system's rate of learning. If you leave this field blank, you'll get great results. The neural network was shown to have the most dependable operation and influence in the diagnosis of liver disorders by comparing the three systems.

The goal of these interconnected experiments is to assess how different "aggregation" and "defuzzification" approaches affect the final product. Different approaches to aggregation, defuzzification, and overlapping within the fuzzy sets are used to produce six fuzzy inference systems (FIS-1, FIS-2, FIS-3, FIS-4, FIS-5, and FIS-6) that share the same rule bases.

The UCI dataset serves as the basis for the method's goal. Several of the input fields of the UCI dataset were swapped out for other, more relevant fields before the systems were built [48].

Water-to-cement ratio, slump, maximum aggregate size, coarse aggregate, fine aggregate, and age are all inputs shared by all systems (day). All of the techniques incorporate the compressive strength of concrete into their resulting fields. Average error in predicted compressive strength for 401 laboratory samples because of these three differences is 6.43% FIS-1, 6.64% FIS-2, 6.48% FIS-3, 5.56% FIS-4, 4.73% FIS-5, and 5.07% FIS-6.

In terms of aggregation and defuzzification, respectively, the experimental findings reveal that the "sum" and "centroid" procedures (used in the FIS-5) present the best outcomes.

In order to be successful in business and attract a large customer base, one must outperform competition and get a larger portion of the market. Satisfied consumers are the first step toward commercial success. Learning about the many agents required to boost client happiness has led to the growth of multiple businesses. A marketing mix model is used to develop an ANFIS and an FIS for inferential purposes in the study. Success with ANFIS was achieved by combining the knowledge of marketing professionals with the P4 principle (price, product, placement, and promotion).

The system's position as a consultant with high assurance can be crucial in assisting company executives in reducing human mistake and making more informed decisions. ANFIS performed better than the alternative method when the results were compared. The goal of this research is to create an FES for diagnosing heart conditions. The Veteran's Administration (VA) Databases at the Long Beach Medical Center and the Cleveland Clinic Foundation serve as the backbone of the new system. There are 11 inputs and a single output in this system. Inputs include gender, age, sex, thallium scan, resting electrocardiogram, exercise, kind of chest discomfort, blood pressure, cholesterol, maximal heart rate, and resting electrocardiogram. The symptoms of cardiac illness are highlighted in the output field. The values range from 0 (no presence) to 4 (distinguishable presence; 1, 2, 3, 4) and are all integers. The system makes use of the Mamdani logic chain. Results from the designed system are 94% accurate when compared to data in the database and results from the intended system are analyzed. Matlab was used to create the system's architecture. The technology under consideration is an alternative to currently available technologies for detecting the presence of heart disease.

The infectious core of the hepatitis B virus, which causes the disease, measures 42 nm in diameter. The Australian antigen, also known as the HB surface antigen (HbsAg), is present on the surface of the virus. As the virus infects the liver, it replicates into similar viruses. The prevalence of (HbsAg) in the blood of an infected person is estimated. PCR and HBNDNA are the most sensitive blood assays for detecting viral spread in the body.

Positive test results indicate HB virus infection; further testing is required to determine the patient's pathology and the rate at which the condition is clearing up. According to the most up-to-date data from the World Health Organization, there are around 400 million people living with HB infection and an additional 50 million new cases each year (WHO).

This study demonstrates the value of researching and building an intelligence system by applying it to the results of 19 different fields selected from the UCI website. The primary driver of forest loss, which has serious repercussions for both the environment and society as a whole.

Early discovery can be helpful in controlling this fatal catastrophe. Using actual data from the University of California's Forest Fire dataset, a hybrid FES was developed to predict the extent of forest fires with high accuracy and efficiency (UCI).

The desired system is a crossbreed of six independently effective fuzzy inference methods. Predictions of the fire's size are 81.2% accurate.

Liver disease is one of the leading causes of death around the world. Because the diagnosis is based on the cumulative experience and knowledge of professionals, it is highly useful to treat this condition through early detection, notwithstanding the possibility of inaccuracy.

It's becoming increasingly common for doctors to use various forms of artificial intelligence to aid in the diagnosis of liver problems. Artificial neural networks (ANNs), expert systems (ESs), fuzzy neural networks (FNNs), classification (ClassNets), fuzzy logic (Fuzzy Logic), and swarm intelligence (SI) are all widely used nowadays.

In this chapter [29], we examine the use of AI and expert systems for the diagnosis and detection of the severity of liver problems.

13.4 APPLICATIONS OF FUZZY SYSTEM IN HEALTHCARE SYSTEM

In this chapter, we explore the applications of fuzzy system designs in the medical profession, focusing on the various ways in which conventional pattern recognition methods might be fused with fuzzy system techniques to achieve better results.

Even though different medical fields are not consistently documented, it is now possible to access real-time approaches for online devices, especially during surgical procedures and in comprehensive care systems.

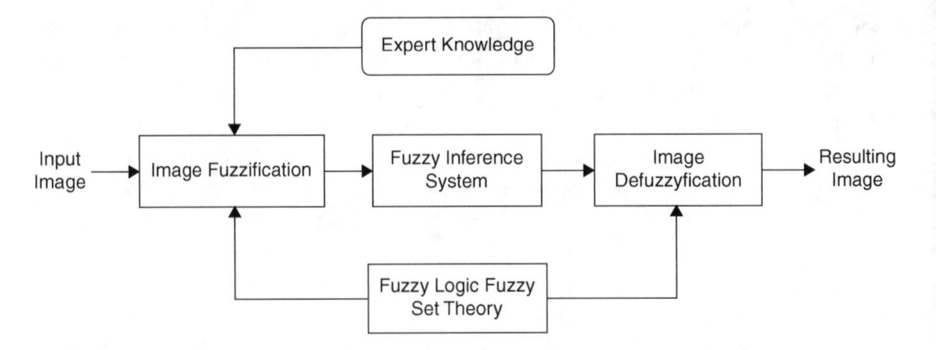

FIGURE 13.5 Fuzzy logic system application in image processing system.

Real-time engineering activities have recently expanded from simple dosage prescription programs to extremely intricate online adaptive healthcare systems. We now recognize that our understanding of the real world is limited, erroneous, and highly contextual.

Fuzzy logic is a hypothesis that describes certain types of medical objects, such as fuzzy sets. Fuzzy logic techniques, as will be assessed in the following sections, have been used in the initial examination of disorders, such as Parkinson, liver, and heart's disease, according to the findings of several studies.

In particular, prime diagnosis has been shown to aid in the development of more effective drug treatment. Patients can benefit much from the early identification of disease, thus any method that can speed up this process is welcome. One possibly useful addition to this area is to evaluate the potential applicability of various models of fuzzy knowledge–based systems in advanced diagnostics of disease groupings. Figure 13.5 shows the structure of image processing system, and it is an application part of the fuzzy logic system.

13.5 CHALLENGES FACED BY FUZZY LOGIC SYSTEM IN TODAY'S ENVIRONMENT

Decisions are made using fuzzy logic because they are based on rules that have already been established. The outcomes may not be satisfactory if these guidelines are flawed. One of the trickiest components of making fuzzy systems is deciding on a membership function and some basic rules. Yet, fuzzy logic is less discussed in Data Science because it is difficult to implement in traditional hardware and because it is not as much effective as a neural setup in machine knowledge in recognizing patterns.

However, there are some drawbacks to using fuzzy logic, i.e., it requires the amendment of participation purposes and it is not necessarily scalable for dealing with enormous or composite complications.

It has been suggested, based on a few widely accepted claims about the foundations of fuzzy logic as proposed by Lotfi Zadeh in 1965, that fuzzy logic differs from traditional bivalent logic in important ways.

The group typically includes the following: fuzzy logic creates results inconsistent with classical logic, fails to meet the standards of classical logic, can't be shown with evidence, is unpredictable, and yields results no human would accept.

The common belief that, at least on a theoretical level, fuzzy logic is too difficult for everyday applications is refuted. We also take a look at fuzzy logic where the accuracy rates are given as intervals smaller than the unit interval itself. Fuzzy logic's truth interval is suggested or proven to be correct, consistent, and comprehensive from a proof-theoretical standpoint.

Significant changes in the way people think of ideas have occurred since the early 1970s, when the classical view of thoughts was closely tested by plausible empirical proof that conceptual levels never had defined boundaries.

Fuzzy set theory and fuzzy logic, as well as the promising results that have arisen from their application, have caught the attention of several scholars. This upbeat perspective dates back to the 1980s, and ever since then, fuzzy organize theory and fuzzy logic have been criticized for being incapable of adequately expressing and managing complex ideas.

In this chapter, we will identify and discuss some of the more well-known applications of fuzzy set theory and fuzzy logic that have resulted from studies in the psychology of theories. We trace the origin and spread of these instances within the relevant discourse. It is shown in great detail why these claims are always wrong and how they stem from a tangle of mistakes and misunderstandings.

13.6 CONCLUSION

Fuzzy and intelligent systems now have many intriguing new applications thanks to numerous and significant developments in computer science and industries. Working at the intersection of computer science and its applications yields the direct and precise output of a fuzzy system. The primary goal of this chapter is to provide a thorough introduction to fuzzy systems and their applications. There was a significant void in the existing literature, which this chapter helped to fill. Various research articles and research papers were carefully selected and other relevant resources were analyzed. From its inception to the cutting edge of the field today, this chapter traces the major developments that have taken place along the way. This chapter's primary contributions can be summed up as follows: It can help researchers in the field of fuzzy and intelligent systems (1) compare their findings to those of other researchers who have used fuzzy systems in their work. (2) It provides a forum where experts from different disciplines can get together to discuss the most pressing problems with fuzzy systems. (3) A specialist in one area of fuzzy systems may see research and development possibilities in other areas. (4) The report draws attention to areas that could use more research, such as handling unclear over-diagnosis issues.

REFERENCES

[1] Mehdi Neshat, Ali Adeli, and Azra Masoumi (2013). A survey on artificial intelligence and expert system for liver disorders.

[2] Mehrbakhsh Nilashi, Hossein Ahmadi, Azizah Abdul Manaf, Tarik A Rashid, Sarminah Samad, Leila Shahmoradi, Nahla Aljojo, and Elnaz Akbari (2020). Coronary heart disease diagnosis through self-organizing map and fuzzy support vector machine with incremental updates. *International Journal of Fuzzy Systems*, 22(4), 1376–1388.

[3] IA Hodashinsky (2020). Fuzzy classifiers in cardiovascular disease diagnostics. *The Siberian Journal of Clinical and Experimental Medicine*, 35(4), 22–31.

[4] Cheng-Hong Yang, Sin-Hua Moi, Ming-Feng Hou, Li-Yeh Chuang, and Yu-Da Lin (2020). Applications of deep learning and fuzzy systems to detect cancer mortality in next-generation genomic data. *IEEE Transactions on Fuzzy Systems* 29 (12), pp. 3833–3844.

[5] Megha Sharma, Shivani Rohilla, and Manish Bhardwaj (2015). Efficient routing with reduced routing overhead and retransmission of manet. *American Journal of Networks and Communications, Special Issue: Ad Hoc Networks*, 4(3–1), 22–26. doi:10.11648/j.ajnc.s.2015040301.15.

[6] M. Bhardwaj (2020). 7 research on IoT governance, security, and privacy issues of internet of things. Agarwal, S., Makkar, S., & Tran, D.-T. (Eds.). (2020).(1st ed.). CRC Press. In: *Privacy Vulnerabilities and Data Security Challenges in the IoT*. p. 115.

[7] A. Kumar, S. Rohilla, and M. Bhardwaj (2019). Analysis of cloud computing load balancing algorithms. *International Journal of Computer Sciences andEngineering*, 7, 359–362.

[8] Lotfi A. Zadeh (1996). Soft computing and fuzzy logic. In: *IEEE Software*, 11(6), pp. 48-56, Nov. 1994, doi: 10.1109/52.329401.

[9] Elzhan Zeinulla, Karina Bekbayeva, and Adnan Yazici (2020). Effective diagnosis of heart disease imposed by incomplete data based on fuzzy random forest. In: *2020 IEEE International Conference on Fuzzy Systems (FUZZ-IEEE)*. IEEE, New York, NY, pp. 1–9.

[10] Zhongyi Hu, Jun Wang, Chunxiang Zhang, Zhenzhen Luo, Xiaoqing Luo, Lei Xiao, and Jun Shi (2021). Uncertainty modeling for multi center autism spectrum disorder classification using takagi-sugeno-kang fuzzy systems. *IEEE Transactions on Cognitive and Developmental Systems* 14(2), pp. 730–739

[11] Aditya Khamparia, Rajat Jain, Poonam Rani, Deepak Gupta, Ashish Khanna, and Oscar Castill (2021). An adaptive neuro fuzzy modelling and prediction system for diagnosis of covid-19. *Applied and Computational Mathematics*, 20, 124–139.

[12] Jovana Nikolic (2021). Expert fuzzy system for estimating risks of hypertension. In: *Ri-STEM-2021: Conference proceedings*, p. 111.

[13] M. Bhardwaj, A. Ahlawat, and N. Bansal (2018). Maximization of lifetime of wireless sensor network with sensitive power dynamic protocol. *International Journal of Engineering & Technology*, 7(3.12), 380–383.

[14] M. Bhardwaj, and A. Ahlawat (2018). Wireless power transmission with short and long range using inductive coil. *Wireless Engineering and Technology*, 9, 1–9. doi:10.4236/wet.2018.91001.

[15] M. Bhardwaj, and A. Ahalawat (2019). Improvement of lifespan of ad hoc network with congestion control and magnetic resonance concept. Bhattacharyya, S., Hassanien, A., Gupta, D., Khanna, A., Pan, I. (eds), In: *International Conference on Innovative Computing and Communications*. Springer, Singapore, pp. 123–133.

[16] M. Bhardwaj, and A. Ahlawat (2017). Optimization of network lifetime with extreme lifetime control proficient steering algorithm and remote power transfer. In: *DEStech Transactions on Computer Science and Engineering*.

[17] Jianfang Wu, Ruo Hu, Ming Li, Shanshan Liu, Xizheng Zhang, Jun He, Jiaxu Chen, and Xiangjun Li (2021). Diagnosis of sleep disorders in traditional Chinese medicine based on adaptive neuro-fuzzy inference system. *Biomedical Signal Processing and Control*, 70, 102942.

[18] Selvaraj Geetha, Samayan Narayanamoorthy, Thangaraj Manirathinam, and Daekook Kang (2021). Fuzzy case-based reasoning approach for finding covid-19 patients priority in hospitals at source shortage period. *Expert Systems with Applications*, 178, 114997.

[19] Sujatha Ramalingam, Kuppuswami Govindan, and Said Broumi (2021). Analysis of covid-19 via fuzzy cognitive maps and neutrosophic cognitive maps. *Neutrosophic Sets and Systems*, 42, 102–116.

[20] Ali Mohammad Alqudah. Fuzzy expert system for coronary heart disease diagnosis in Jordan. *Health and Technology*, 7(2–3), 215–222, 2017.
[21] Anurag Sharma, Arun Khosla, Mamta Khosla, and Yogeswara Rao. Fast and accurate diagnosis of autism (fada): A novel hierarchical fuzzy system based autism detection tool. *Australasian Physical & Engineering Sciences in Medicine*, 41(3), 757–772, 2018.
[22] M. Bhardwaj, and A. Ahlawat (2017). Enhance lifespan of WSN using power proficient data gathering algorithm and WPT. In: *DEStech Transactions on Computer Science and Engineering.*
[23] N. S. Pourush, and M. Bhardwaj (2015). Enhanced privacy-preserving multi-keyword ranked search over encrypted cloud data. *American Journal of Networks and Communications*, 4(3), 25–31.
[24] J. Wu, S. A. Haider, M. Bhardwaj, A. Sharma, and P. Singhal (2022). Blockchain-based data audit mechanism for integrity over big data environments. *Security and Communication Networks.*
[25] M. Bhardwaja, and A. Ahlawat (2019). Evaluation of maximum lifetime power efficient routing in ad hoc network using magnetic resonance concept. *Recent Patents on Engineering*, 13(3), 256–260.
[26] Asha Gowda Karegowda, D. Poornima, and C. R. Lakshmi (2017). Knowledge based fuzzy inference system for diagnosis of diffuse goiter. In: *2017 2nd International Conference On Emerging Computation and Information Technologies (ICECIT).* IEEE, New York, NY, pp. 1–12.
[27] Negar Asaad Sajadi, Shiva Borzouei, Hossein Mahjub, and Maryam Farhadian (2018). Diagnosis of hypothyroidism using a fuzzy rule-based expert system. *Clinical Epidemiology and Global Health*, 7(4), 2019, pp. 519–524.
[28] Alejandro Moya, Elena Navarro, Javier Jaén, and Pascual González (2019). Fuzzy-description logic for supporting the rehabilitation of the elderly. *Expert Systems*, 37(2), 2020, e12464.
[29] Juan Carlos Guzmán, Ivette Miramontes, Patricia Melin, and German Prado-Arechiga (2019). Optimal genetic design of type-1 and interval type-2 fuzzy systems for blood pressure level classification. *Axioms*, 8(1), 8.
[30] Thi Ngoc Mai Nguyen, Quang Chung Tran, Tien Dat Duong, and Ngoc Anh Mai (2018). Design of a medical expert system for consulting tuberculosis diagnosis in vietnam rural areas. Van Toi, V., Le, T., Ngo, H., Nguyen, TH. (eds). In: *International Conference on the Development of Biomedical Engineering in Vietnam.* Springer, New York, NY, pp. 577–583.
[31] A. Sharma, A. Tyagi, and M. Bhardwaj (2022). Analysis of techniques and attacking pattern in cyber security approach: A survey. *International Journal of Health Sciences*, 6(S2), 13779–13798.doi:10.53730/ijhs.v6nS2.8625
[32] A. Tyagi, A. Sharma, and M. Bhardwaj (2022). Future of bioinformatics in India: A survey. *International Journal of Health Sciences*, 6(S2), 13767–13778. doi:10.53730/ijhs.v6nS2.8624.
[33] P. Chauhan, and M. Bhardwaj (2017). Analysis the performance of interconnection network topology C2 Torus based on two dimensional torus. *International Journal of Emerging Research in Management &Technology*, 6(6), 169–173.
[34] Lejla Hadžic, Arnela Fazlic, Osman Hasanic, Nudžejma Kudic, and Lemana Spahic (2019). Expert system for performance prediction of anesthesia machines. Badnjevic, A., Škrbić, R., Gurbeta Pokvić, L. (eds). In: *International Conference on Medical and Biological Engineering.* Springer, New York, NY, pp. 671–679.
[35] Emanuel Ontiveros, Patricia Melin, and Oscar Castillo (2020). Comparative study of interval type-2 and general type-2 fuzzy systems in medical diagnosis. *Information Sciences*, 525, 37–53.
[36] Roan Thi Ngan, Mumtaz Ali, Hamido Fujita, Mohamed Abdel-Basset, Nguyen Long Giang, Gunasekaran Manogaran, MK Priyan, et al. (2019). A new representation of intuitionistic fuzzy systems and their applications in critical decision making. *IEEE Intelligent Systems*, 35(1), 6–17.

[37] Mirza Mansoor Baig, Hamid GholamHosseini, Abbas Kouzani, and Michael J. Harrison (2011). Anaesthesia monitoring using fuzzy logic. *Journal of Clinical Monitoring and Computing*, 25(5), 339.

[38] Abbas Zibakhsh Shabgahi, and Mohammad Saniee Abadeh (2011). Cancer tumor detection by gene expression data exploration using a genetic fuzzy system. In: *2011 Developments in E-systems Engineering*. IEEE, New York, NY, pp. 141–145.

[39] Yun Zou, Zheng Li, Xunsheng Zhu, Jian Yu, and Zheng Gu (2012). Research on the computer-assisted intelligent diagnosis system of traditional chinese medicine. In: *2012 9th International Conference on Fuzzy Systems and Knowledge Discovery*. IEEE, New York, NY, pp. 329–333.

[40] S. Lakkadi, A. Mishra, and M. Bhardwaj (2015). Security in ad hoc networks. *American Journal of Networks and Communications*, 4(3–1), 27–34.

[41] Ishita Jain, and Dr. Manish Bhardwaj (2022). A survey analysis of COVID-19 pandemic using machine learning. In: *Proceedings of the Advancement in Electronics & Communication Engineering 2022*. doi:10.2139/ssrn.4159523.

[42] Maryam Zolnoori, Mohammad Hossein Fazel Zarandi, Mostafa Moin, and Shahram Teimorian. Fuzzy rule-based expert system for assessment severity of asthma. *Journal of Medical Systems*, 36(3), 1707–1717, 2012.

[43] Sachidanand Singh, Atul Kumar, K. Panneerselvam, and J. Jannet Vennila. Diagnosis of arthritis through fuzzy inference system. *Journal of Medical Systems*, 36(3), 1459–1468, 2012.

[44] Mostafa Karimpour, Ali Vahidian Kamyad, and Mohsen Forughipour (2014). Fuzzy modeling of optimal initial drug prescription. Cao, BY., Nasseri, H. In: *Fuzzy Information & Engineering and Operations Research & Management*. Springer, New York, NY, pp. 3–12.

[45] B. M. Gayathri, and C. P. Sumathi (2015). Mamdani fuzzy inference system for breast cancer risk detection. In: *2015 IEEE International Conference on Computational Intelligence and Computing Research (ICCIC)*. IEEE, New York, NY, pp. 1–6.

[46] George-Peter K. Economou, Efrosini Sourla, Konstantina-Maria Stamatopoulou, Vasileios Syrimpeis, Spyros Sioutas, Athanasios Tsakalidis, and Giannis Tzimas (2015). Exploiting expert systems in cardiology: A comparative study. In: Advances in experimental medicine and biology, 820, 79–89. https://doi.org/10.1007/978-3-319-09012-2_6.

[47] Mehdi Neshat, Ghodrat Sepidname, Amin Eizi, and Amanollah Amani (2015). A new skin color detection approach based on fuzzy expert system. *Indian Journal of Science and Technology*, 8, 1–11.

[48] André S. Fialho, Susana M. Vieira, Uzay Kaymak, Rui J. Almeida, Federico Cismondi, Shane R. Reti, Stan N. Finkelstein, and João M. C. Sousa (2016). Mortality prediction of septic shock patients using probabilistic fuzzy systems. *Applied Soft Computing*, 42, 194–203.

14 Perceptual Hashing Function for Medical Images

Overview, Challenges, and the Future

Arambam Neelima and Heisnam Rohen Singh
National Institute of Technology, Nagaland

14.1 INTRODUCTION

Because of the advancement of technology, a huge number of medical images are generated on a daily basis. Different types of medical images available are shown in Figure 14.1. The advent of digital systems has facilitated the accumulation of vast data repositories and empowered medical practitioners to make patient-specific diagnoses by providing them access to all the images associated with a particular patient throughout their medical history. These medical images need to be indexed, registered, and retrieved in an organized manner. Medical image hashing-based content-based medical image retrieval (CBMIR) can be efficiently used to retrieve the images, thereby bridging the gap between medical image and diagnosis.

CMBIR is the process of retrieving images based on the content of the image [1]. CBMIR quantifies the similarity between images by analyzing their underlying visual components. With these systems, it will be much easier to search for similar medical conditions or scenarios that will aid in fast diagnosis. CBMIR has gained significance in computer-aided diagnosis by offering doctors visual access to relevant and existing cases, along with their diagnosis information, which serves as a diagnostic aid. Thus, similar images need to be retrieved in short time. Generalized hashing-based methods have been extensively researched to achieve speedy retrieval without compromising the performance [2].

Hashing is a process of mapping the data into a compact hash value [3]. This hash value is usually fixed-length and shorter value than the original data, which may be in the form of text, image, audio, video, etc. A good hashing function should produce a unique hash value for each unique input. The key properties of a good hashing function include determinism, uniqueness, uniformity, irreversibility, efficiency, and sensitivity to input changes [4]. These properties are important for ensuring the security and integrity of data, detecting changes to data, and efficiently processing large

DOI: 10.1201/9781003405368-14

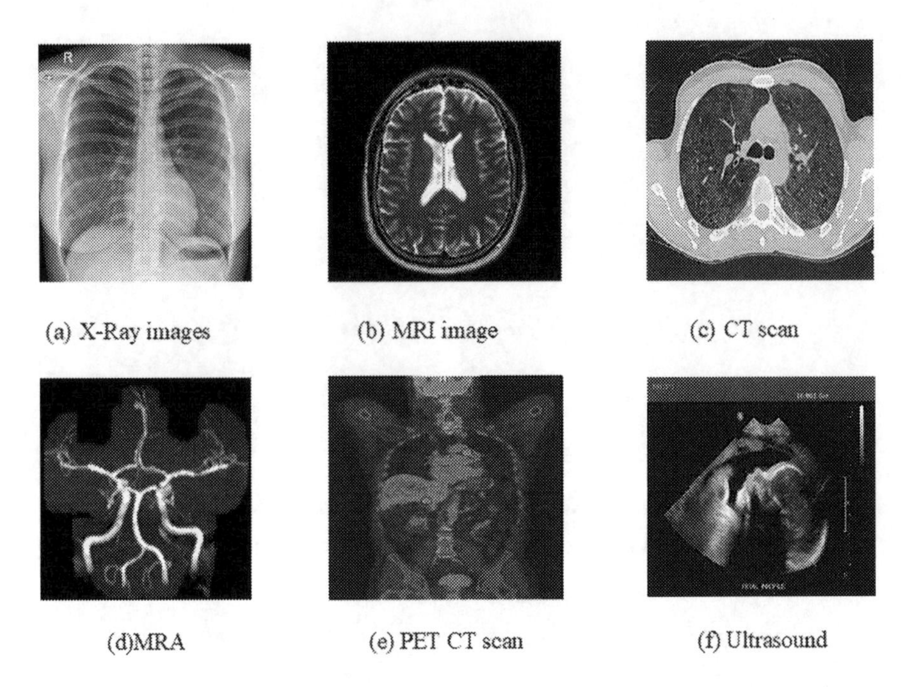

FIGURE 14.1 Sample images of 2D medical scans.

amounts of data. Traditional hashing functions for text, such as SHA [5], MD4, and MD5 [6,7], are not suitable for image hashing in CBMIR because they are too sensitive to each individual bit of input data and do not consider the perceptual structure of the image. Image hashing is the process of converting the images into a compact value called the hash value or hash code. Image hashing functions are designed to be robust to common image processing operations, such as cropping, scaling, rotation, and compression while being sensitive to perceptual changes in the image content [8].

Figure 14.2 shows the various categories of hashing functions. Hashing functions may be keyed or unkeyed. In keyed hash functions, a secret key is used to generate the hash value. For image data, the keyed functions are useful in applications where the integrity and authenticity of data are critical, such as in digital forensics, image verification, and copyright protection [9]. If the secret key used in a keyed hash function is compromised, then the integrity of the data that has been hashed with that key can also be compromised. In unkeyed hash function, secret keys are not required to produce the hash value; only the input data are used to produce the hash value. Unkeyed hash functions are used in a variety of applications where data integrity and authenticity are important, such as in digital signatures, message authentication codes, and file verification, to generate a digital fingerprint of the image [10].

Both the keyed and unkeyed hash functions are based on traditional mathematical principles, while there are also hash functions that use machine learning techniques to generate an efficient hash value. Based on the machine learning techniques used, the hash functions are further divided into supervised hashing, unsupervised hashing, and semi-supervised hashing [11]. Supervised hash functions use labeled data to train a supervised learning algorithm, such as a neural network, to produce

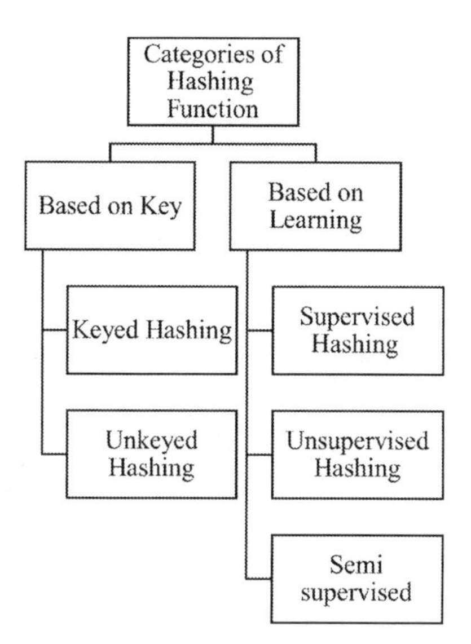

FIGURE 14.2 Categories of hashing.

hash codes for new inputs. Unsupervised hash functions use unsupervised learning techniques, such as autoencoders, to learn compressed representations of the input data that are optimized for hashing. These functions do not require labeled data for training. Semi-supervised hash functions use a combination of labeled and unlabeled data for training a machine learning algorithm to produce hash codes for new inputs. These functions take advantage of the availability of both types of data to improve the accuracy and efficiency of the model. The labeled data are used to train the model to produce hash codes that are similar for inputs that have the same label. On the other hand, the unlabeled data are used to acquire a broader understanding of the input data, which aids in the generation of hash codes for novel inputs.

The objective of this chapter is to analyze various image hashing methodology available for indexing and retrieval and introduce the recent progress, various challenges, and future perspective. Some of the applications of image hashing include image authentication, image integrity, CBMIR, image encryption, etc. [12].

Figure 14.3 illustrates the framework of a medical image retrieval system based on hashing, comprising two primary components: runtime search and offline learning. The feature extraction process is the first step in both components. During offline learning, features are extracted from the image database, and these features are then input into the hashing function to generate the hash codes. The hash codes serve as an index for recording. During runtime search, the input image is taken as a query image and is processed through the feature extraction stage. The resulting features are then input into the hashing function to generate the corresponding hash code. The hash code for the query image is then compared to the different hash codes generated from the database images. The features can be extracted in two ways. It may be extracted as a hand-crafted feature or automatic feature extraction. Different types of feature extraction methods and different types of features are shown in Figure 14.4.

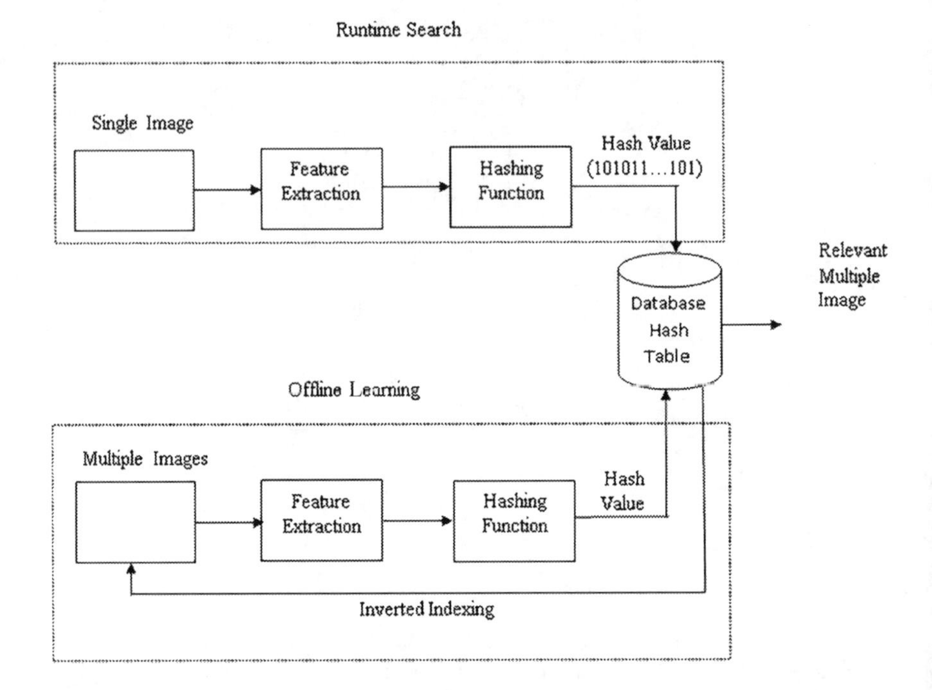

FIGURE 14.3 Framework of a hashing-based medical image retrieval system.

Feature extraction is a crucial step in many machine learning and computer vision applications, including medical image retrieval. There are two main approaches to feature extraction: hand-crafted and automatic feature extraction.

Hand-crafted feature extraction involves manually designing feature descriptors based on domain knowledge and expertise. These features are then extracted from the image using various techniques, such as filtering, edge detection, and feature detection. Hand-crafted feature extraction has been widely used in the past and can produce effective results if the features are carefully designed and chosen. The hand-crafted features can be again divided into two types: low level and high level. Low- and high-level features refer to different levels of abstraction in the visual information extracted from an image.

Low-level features are basic visual elements that can be extracted from an image, such as color, texture, and edges. These features are often extracted using traditional image processing techniques, such as filtering and edge detection. Low-level features are typically simple and represent the raw pixel values of an image. High-level features, on the other hand, are more complex and abstract visual concepts that are derived from low-level features. Examples of high-level features include object shape, object orientation, and object identity [13–15]. These features are typically extracted using more advanced techniques, such as machine learning algorithms like convolutional neural networks (CNNs).

In medical image retrieval, low-level features might include color, texture, and shape, while high-level features could include object detection and recognition, segmentation, and classification of regions of interest in the image.

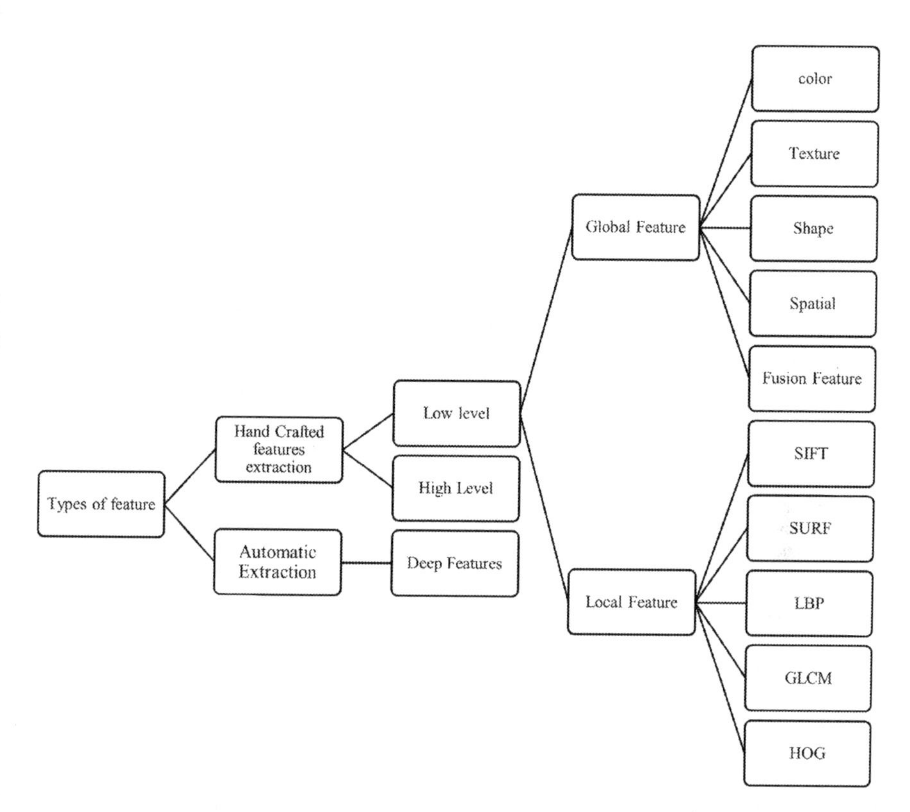

FIGURE 14.4 Different types of features.

In summary, low-level features refer to basic visual elements that can be extracted directly from an image, while high-level features are more complex and abstract visual concepts derived from low-level features. Both types of features are important in medical image retrieval, as they allow for the efficient and accurate identification of relevant images in a large image database.

On the other hand, automatic feature extraction uses machine learning algorithms to learn features directly from the data. This approach does not require domain knowledge or human intervention, and the algorithm learns the best features for a particular task. Popular automatic feature extraction techniques include CNNs and deep learning architectures [16, 17], which have shown impressive results in many computer vision tasks, including medical image retrieval.

Both hand-crafted and automatic feature extraction approaches have their advantages and disadvantages. Hand-crafted features can be designed specifically for a particular task and can be computationally efficient. However, they can be limited in their ability to capture complex patterns and variations in the data. Automatic feature extraction, on the other hand, can learn complex features and patterns, but it can be computationally expensive and may require large amounts of training data [18–20].

In summary, both hand-crafted and automatic feature extraction approaches can be used in medical image retrieval systems, depending on the specific requirements of the task and available resources.

14.2 STATE OF THE ART

The machine learning and computer vision community has extensively investigated hashing methods for large-scale image retrieval [21]. Deep feature extraction is important in hashing-based image retrieval systems because it allows for the creation of compact binary codes that capture high-level semantic information about the images. By using deep neural networks to extract features from the images, the resulting binary codes are more likely to preserve the semantic information required for effective retrieval, while also being computationally efficient to compare. Compared to traditional feature extraction methods, which rely on handcrafted features, deep feature extraction can capture more complex and abstract information about the images, making it more effective for tasks such as medical image retrieval. Previous studies have explored different techniques for feature extraction and hashing in the context of medical image retrieval. Other studies have explored unsupervised hashing approaches for medical image retrieval, which do not require labeled training data. A review of the different techniques such as deep feature extraction and unsupervised hashing are discussed in this section [22, 23].

Anna et al. proposed a medical image retrieval system to address the problem of low-accuracy retrieval [24]. A hash layer was added between the fully connected layer and the classification layer of the DenseNet 121 network. Yibo Tang et al. proposed a deep semantic ranking hashing based on the self-attention (DSHA) approach [25] for medical image retrieval. The authors have tried to consider the issue of lesion context and category-level semantics, which are ignored in most of the works. Yaxiong Chen et al. proposed a multi-scale triplet hashing (MTH) algorithm, which utilizes multi-scale information, convolutional self-attention, and hierarchical similarity to enable simultaneous learning of effective hash codes [26]. The authors have designed a multi-scale dense block, which consists of three parallel convolutions, three parallel dense blocks, and max-pooling operation. Yong Zhang et al. introduced a chest X-ray retrieval system based on supervised cross-modal hashing [27]. A category hashing network is utilized by the model to learn the category hash code. The learned category hash code is then used to supervise the hash code learning of the chest X-ray modality and radiology report modality. A deep balanced discrete hashing method was proposed by Xiangtao Zheng et al. [28]. This method facilitates the direct learning of a discrete hash code by the network. To avoid the continuous relaxation process and reduce the quantization error that results from it, discrete gradient propagation was employed with a straight-through estimator. The discrete hash code generated through this process retains its original appearance throughout the optimization process. A novel CBMIR (CNN) and hash coding framework was proposed by Cai et al. that utilizes a Siamese network to generate images of the same class using image pairs (similar and dissimilar) and a contrasting loss function [29]. Weight sharing is also employed to facilitate the learning process for creating images of the same class. The CNN is utilized to extract features from both network branches, and the dimension of the feature vectors is reduced through hash mapping. A novel loss function [30] has been designed to enhance the inclusion of feature vectors during training and enable the approximation of real value outputs to the required binary values. The trained network generates a compact binary hash code for the image query, which is comparable with the hash code of the database images during retrieval. A novel deep hashing technique is proposed, which involves performing deep extraction, binary code learning, and deep hash function learning under supervision. The learning of the hash code is achieved through the iterative optimization of a discrete and constrained objective function, eliminating the need for relaxation and enabling the solution of the binary code.

A novel approach has been developed to address the semantic gap and attain high retrieval accuracy in the multimodal medical image domain [31]. The features were selected using fast Fourier transform (FFT) [32]. The authors utilized a straightforward linear transformation to convert the feature vector into bits, based on their selection of the required number of frequency components. They then employed the resulting binary codes as hash codes to enable efficient retrieval of large datasets. However, their performance was low when compared with low spherical hashing (SpH) [33] and sensitive hashing (DSH) [34]. Another hashing function based on CNN was employed by Gu and Yang. A reciprocal learning guidance model has been developed, which is specifically designed to handle image data from high–low magnification pairs [35]. A novel framework called multi-magnification correlation hashing (MMCH) [36] was introduced for learning binary codes for histopathologic images. This framework utilizes both low-magnification and high-magnification data to learn discriminative features.

To facilitate similarity learning between samples and ensure the quality of hash learning, the author employed pairwise cross-entropy losses and quantization losses [37]. The authors also utilized bimodal Laplacian prior to eliminating errors caused by the inner product in place of the Hamming distance. Gu et al. developed another MMCH framework that learns the discriminative binary codes for histopathological images [38]. It was mainly deployed for breast cancer diagnosis.

Saban [39] introduced a model that adopts the method of producing hash code through deep features instead of directly generating the codes from images. In addition, the researcher introduced an efficient vector over-sampling technique to address the issue of imbalanced medical image datasets. Saban [40] also proposed a model that effectively decreases the semantic gap without disrupting the high-level feature extraction procedure. It tries to generate the hash code from the features that were extracted using CNN. He also proposes a model [41] that offers an efficient approach for generating hash codes, which involves down-sampling deep features extracted from medical images based on feature selection. The model is trained using the Siamese network in a pairwise manner.

Xiofan et al. suggested a scalable framework for image retrieval that utilizes the supervised hashing technique [42]. The approach employs the supervised kernel hashing technique, which utilizes a limited amount of supervised data to compress image feature vectors into binary bits, while preserving relevant signatures. The resulting binary codes are then indexed in a hash table that enables rapid retrieval of images from a vast database. Zhang introduced a medical hashing retrieval system that will bridge the semantic gap between images and diagnoses [43]. This model is also based on supervised kernel hashing technique. Jamil et al. also introduced a model that employs a fast and efficient technique for compressing selective convolutional features into a sequence of bits, utilizing the FFT. Highly responsive convolutional feature maps are identified for medical images based on their neuronal responses from a pre-trained CNN. Then, the layer-wise global mean activations of the chosen feature maps are transformed into concise binary codes by binarizing their Fourier spectrum [44]. Sailesh et al. proposed a multiple instance (MI) deep hashing technique, which is applicable for large-scale medical image retrieval [45]. Liming et al. developed a method for medical image retrieval called multi-manifold deep discriminative cross-modal hashing (MDDCH), which aims to combine various sub-manifolds from different types of data to preserve the correlation between instances. The correlation was evaluated by measuring the three-step connection on the corresponding hetero-manifold [46].

Chengyuan Zhang and his team put forward a novel approach that utilizes a deep CNN architecture to generate hash codes, while also providing data security and privacy preservation [47]. The CNN is employed to learn the complex visual features of the medical image, which are then transformed into hash codes using the proposed method. This process can facilitate secure and efficient storage and retrieval of medical images while ensuring data privacy.

Yilan Zhjang et al. introduced a novel hybrid dilated convolution spatial attention module aimed at extracting crucial information and suppressing irrelevant details in dermoscopic images based on their complex morphological characteristics [48]. The network architecture has been tailored to effectively capture output variations across different angles and achieve a certain degree of rotation invariance. This approach has the potential to improve the accuracy and robustness of image analysis in the domain of dermatology.

Xiaoshuang Shi et al. proposed a novel approach called the supervised graph-based hashing for medical image retrieval [49]. This model is designed to extract detailed cell-level information, including shape, area, nuclear and cytoplasmic appearance, and uses a group-to-group similarity matching to retrieve the images. By leveraging the graph-based approach, the model can learn and represent the complex relationships between different cells in the images, which may lead to more accurate retrieval results.

Fang and others have presented an innovative approach known as attention-based saliency hashing (ASH) for generating compressed hash codes that represent ophthalmic images [50]. The model leverages the deep network's learning capability to specifically capture the salient features of the images' prominent regions. By focusing on these critical details, ASH can effectively generate compact hash codes while retaining essential information, which may have practical applications in the analysis and storage of ophthalmic images.

The authors integrate a deep convolutional network (CNN) with Hash Coding to improve the efficiency and accuracy of medical image analysis [51]. The integration of these techniques has significant implications for essential healthcare tasks, such as diagnosis and treatment planning, where precise and dependable image analysis is crucial to providing effective care.

Yan et al. proposed a supervised sparse hashing model [52] for medical image retrieval that aims to reduce the storage and computational costs by utilizing a sparse regularizer to generate a sparse projection matrix. This matrix effectively reduces the memory requirements and embedding time while maintaining high accuracy.

By adopting this approach, the model can provide more efficient and cost-effective retrieval of medical images.

A deep learning model for medical sign recognition of lung nodules based on image retrieval was proposed by the authors [53]. This model includes a deep learning framework that extracts semantic features to represent sign information effectively, aiding in the identification of benign and malignant nodules. The high-dimensional image features are then compressed into compact binary codes using principal component analysis [54] and supervised hashing. By leveraging these techniques, the model can efficiently extract essential information from medical images to improve the accuracy and speed of diagnosis.

14.3 DATASET

There are several datasets available for medical image retrieval, each with their own unique characteristics and purposes. Some of the most commonly used datasets for medical image retrieval include:

- **ImageCLEF Medical Retrieval Task [55]**: This dataset is a part of the ImageCLEF challenge and consists of a collection of 10,000 medical images with corresponding textual descriptions. The images cover a variety of modalities and body parts, and the dataset is commonly used for evaluating retrieval methods based on both visual and textual features.
- **OpenI [56]**: This is a large-scale dataset of chest X-rays and corresponding radiology reports, containing over 350,000 images from more than 70,000 patients. The dataset is designed for use in training and evaluating machine learning models for medical image retrieval and analysis.
- **RSNA Pneumonia Detection Challenge [57]**: This dataset consists of chest X-rays labeled for the presence of pneumonia and is commonly used for evaluating retrieval methods focused on a specific disease or medical condition.
- **Digital Database for Screening Mammography (DDSM) [58]**: This dataset contains over 2,500 mammography images and is often used for evaluating retrieval methods for breast cancer detection.
- **Retinal Fundus Glaucoma Challenge [59]**: This dataset contains fundus photographs from patients with and without glaucoma and is used for evaluating retrieval methods focused on the detection and diagnosis of this disease.
- **Medical Segmentation Decathlon [60]**: This dataset includes ten different medical imaging datasets for various tasks, such as brain tumor segmentation, liver segmentation, and lung segmentation. It is designed for benchmarking and comparing the performance of various medical image analysis methods.

These datasets provide a rich source of data for training and evaluating medical image retrieval systems and can be used to assess the effectiveness of different feature extraction and hashing techniques for retrieval of medical images. Some other publicly available datasets for medical image retrieval are listed in Table 14.1.

TABLE 14.1
Details of the Medical Image Retrieval Dataset

Sl. No.	Name of the Dataset	No. of Images	Mode	No. of Class	References
1	MESSIDOR dataset	1,200	Fundus photography	2	[61]
2	Multimodal dataset	7,200	MRI, CT, PET, OPT, X-ray, etc.	22	[62]
3	Kvasir	4,000	Endoscopic	8	[63]
4	IRMA	14,000	X-ray	9	[64]
5	Emphysema CT image	115	CT	3	[65]
6	NEMA CT	1,334	CT	10	[66]
7	NEMA MRI	1,803	MRI	11	[67]
8	TCIA CT	604	CT	8	[68]
10	ILD	128	HRCT	13	[69]
11	Brain tumor	3,063	CE-MRI	3	[70]
12	LIDC-IDRI	144,527 slices	CT	6	[71]
13	MAMMOSET	3,457	Mammogram	2	[72]

14.4 CHALLENGES AND FUTURE TRENDS

Despite the promising results reported in these studies, there are still limitations and challenges in the field of hashing-based medical image retrieval. For example, existing approaches may not be able to handle the large variations in medical image modalities and appearances.

Some of the challenges of hashing-based medical image retrieval are as follows:

1. Multi-scale information plays a crucial role in medical images. Nevertheless, most of the existing algorithms fail to take into account the multi-scale information of medical images, which can lead to substantial information loss and adversely affect the experimental outcomes. Proposed a deep balanced discrete hashing method to reduce the quantization error in traditional supervised hashing.
2. Addressing the problem of data heterogeneity that arises from multiple sources and target domains is crucial. Moreover, as the number of source domains increases, the complexity and training costs may also rise, highlighting the need for exploring a unified model in the future to mitigate the issue of high computational cost.
3. Medical datasets commonly face the challenge of an imbalanced sample distribution across different classes.
4. Advanced cloud computing techniques have made large-scale image retrieval services an essential part of our daily lives by providing storage and computational resources. This enables data owners to share their image data with the public for search and sharing services. However, this trend has brought about

a significant challenge: data security and privacy preservation. As the image data often contain personal privacy information, protecting such data and ensuring their security in the cloud has become a critical issue in the cloud computing and information security community. As a result, an increasing number of researchers are focusing on privacy-preserving image retrieval, recognizing its significance in safeguarding personal privacy information.

5. Cell-level information is crucial for accurate classification or diagnosis of diseases such as cancer, but most models lack this information. Extracting cell information, such as shape, area, and nuclear and cytoplasmic appearance, is a challenging task, as a single-digitized image may contain hundreds of thousands of cells.
6. Multi-scale information plays a crucial role in medical images. Nevertheless, most of the existing algorithms fail to take into account the multi-scale information of medical images, which can lead to substantial information loss and adversely affect the experimental outcomes. Proposed a deep balanced discrete hashing method to reduce the quantization error in traditional supervised hashing.
7. The problem of data heterogeneity is a major challenge that arises when dealing with multiple source and target domains. With an increase in the number of source domains, the complexity and training costs may also increase significantly, making it necessary to explore a unified model in the future. By doing so, we can mitigate the issue of high computational costs and develop more efficient solutions.

In summary, the future of medical image analysis is likely to focus on addressing the challenges discussed above. Multi-scale information integration, imbalanced dataset handling, privacy-preserving image retrieval, and cell information extraction are all areas of research that are likely to receive increasing attention in the coming years. By developing new techniques to address these challenges, researchers can help improve the accuracy and effectiveness of medical image analysis, ultimately leading to better patient outcomes.

14.5 CONCLUSION

In conclusion, the advancement of technology has led to the generation of an enormous number of medical images on a daily basis. Medical image hashing-based CBMIR can be efficiently used to retrieve these images, enabling physicians to perform patient-specific diagnoses by presenting all images related to a particular patient over time. Generalized hashing-based methods have been extensively researched to achieve speedy retrieval without compromising performance. The use of deep learning approaches such as Siamese networks and CNNs have been proposed by many researchers for effective image retrieval. These approaches involve feature extraction, binary code learning, and deep hash function learning under supervision, allowing the solution of binary code without the need for relaxation. These techniques address the semantic gap and ensure the quality of hash learning, allowing for the efficient retrieval of medical images.

REFERENCES

[1] Owais, M., Arsalan, M., Jiho Choi, & Park, Kang Ryoung (2019). Effective diagnosis and treatment through content-based medical image retrieval (CBMIR) by using artificial intelligence. *Journal of Clinical Medicine*, 8(4), 462.

[2] Zhu, L., Shen, J., Xie, L., & Cheng, Z. (2017). Unsupervised visual hashing with semantic assistant for content-based image retrieval. *IEEE Transactions on Knowledge and Data Engineering*, 29(2), 472–486.

[3] Neelima, A., & Manglem Kh., S. (2016). Perceptual Hash Function based on Scale-Invariant Feature Transform and Singular Value Decomposition. *The Computer Journal*, 59(9), 1275–1281.

[4] Neelima, A., & Manglem Kh., S. (2019). Perceptual hash function for images based on hierarchical ordinal pattern. In: Singh, A. K., & Mohan, A., (eds.) *Handbook of Multimedia Information Security*, pp. 267–287. Springer.

[5] Buchmann, J. A. (2004). Cryptographic hash functions. In: Buchmann, J. A. (ed.) *Introduction to Cryptography*, pp. 235–248. Springer, New York, NY.

[6] Rivest, L. (1992). The MD4 message digest algorithm. Request for Comments (RFC) 1320, Internet Activities Board, Internet Privacy Task Force.

[7] Rivest, L. (1992). The MD5 message digest algorithm. Request for Comments (RFC) 1321, Internet Activities Board, Internet Privacy Task Force.

[8] Swaminathan, A., Mao, Y., & Wu, M. (2006). Robust and secure image hashing. *IEEE Transactions on Information Forensics and Security*, 1(2), 215–230.

[9] Neelima, A., & Singh, K. (2014). A short survey on perceptual hash function. *The ADBU Journal of Engineering Technology*, 1, 0011405.

[10] Stalling, W. (2006). *Cryptography and Network Security*, 4th edition, vol. 3. Pearson Education India, Noida.

[11] Ma, Y., Li, Q., Shi, X., & Guo, Z (2022). Unsupervised deep pairwise hashing. *Electronics*, 11(5), 744.

[12] McCarthy, E., Balado, F., Silvestre, G. C. M., & Hurley, N. J. (2004). A framework for soft hashing and its application to robust image hashing. In: *2004 International Conference on Image Processing, ICIP'04*. Springer, Singapore, pp. 397–400.

[13] Jain, R., Kasturi, R., & Schunck, B. G. (1995). *Machine Vision*. McGraw-Hill, New York, NY.

[14] Kim, J. S., Lee, J. J., Han, S. J., & Cha, K. H. (2015). Hand-crafted feature-based medical image retrieval system. *Healthcare Informatics Research*, 21(3), 153–160.

[15] Fathy, M. H., & Mohammed, M. A. (2012). A review of content-based medical image retrieval systems. *Signal & Image Processing: An International Journal*, 3(3), 11–27.

[16] He, K., Zhang, X., Ren, S., & Sun, J. (2016). Deep residual learning for image recognition. In: *Proceedings of the IEEE Conference on Computer Vision and Pattern Recognition*. IEEE, pp. 770–778.

[17] LeCun, Y., Bengio, Y., & Hinton, G. (2015). Deep learning. *Nature*, 521(7553), 436–444.

[18] Islam, S. M. S., Wong, K. W., Siu, W. C., & Liu, D. (2018). Hashing-based medical image retrieval: A survey. *Journal of Medical Systems*, 42(7), 129.

[19] Ciresan, D., Giusti, A., Gambardella, L. M., & Schmidhuber, J. (2012). Deep neural networks segment neuronal membranes in electron microscopy images. In: Pereira, F. and Burges, C.J., Bottou, L., & Weinberger, K.Q., (eds) *Advances In Neural Information Processing Systems*. pp. 2843–2851. ACM.

[20] Kim, J. S., Lee, J. J., Han, S. J., & Cha, K. H. (2015). Hand-crafted feature-based medical image retrieval system. *Healthcare Informatics Research*, 21(3), 153–160.

[21] Wang, J., Liu, W., Kumar, S., & Chang, S.-F. (2016). Learning to hash for indexing big data: A survey. *Proceedings of the IEEE*, 104(1), 34–57.

[22] Zhang, H., Xu, X., Chen, J., & Liu, Q. (2016). Deep hashing for medical image retrieval. In: Yang, Y. (ed) *Proceedings of the 9th International Conference on Intelligent Information Processing*. pp. 297–305. Association for Computing Machinery, New York.
[23] Zhang, J., Liu, L., Li, Y., Zhang, J., & Jiang, Y. (2020). Deep learning-based hashing for medical image retrieval and classification. *Computer Methods and Programs in Biomedicine*, 181.
[24] Guan, A., Liu, L., Fu, X., & Liu, L. (2022). Precision medical image hash retrieval by interpretability and feature fusion. *Computer Methods and Programs in Biomedicine*, 222, 10694.
[25] Tang, Y., Chen, Y., & Xiong, S. (2022). Deep semantic ranking hashing based on self-attention for medical image retrieval. In: *Proceedings of the 26th International Conference on Pattern Recognition (ICPR)*. pp. 4960–4966. IEEE Computer Society.
[26] Chen, Y., Tang, Y., Huang, J., & Xiong, S. (2023). Multi-scale triplet hashing for medical image retrieval. *Computers in Biology and Medicine*, 155, 106633.
[27] Zhang, Y., Ou, W., Zhang, J., & Deng, J. (2023). Category supervised cross-modal hashing retrieval for chest X-ray and radiology reports. *Computers in Biology and Medicine*, 155, 106633.
[28] Zheng, X., Zhang, Y., & Lu, X. (2020). Deep balanced discrete hashing for image retrieval. *Neurocomputing*, 403, 224–236.
[29] Cai, Y., Li, Y., Qiu, C., Ma, J., & Gao, X. (2019). Medical image retrieval based on convolutional neural network and supervised hashing. *IEEE Access*, 7, 51877–51885.
[30] Qi, Y., Gu, J., Zhang, Y., Wu, G., & Wang, F. (2020). Supervised deep semantics-preserving hashing for real-time pulmonary nodule image retrieval. *Journal of Real-Time Image Processing*, 17(6), 1857–1868.
[31] Ahmad, J., Muhammad, K., & Baik, S. W. (2018). Medical image retrieval with compact binary codes generated in frequency domain using highly reactive convolutional features. *Journal of Medical Systems*, 42(2), 1–19.
[32] Proakis, J. G., & Manolakis, D. G. (2007). *Digital Signal Processing: Principles, Algorithms, and Applications*. Prentice Hall, London.
[33] Heo, J. P., Lee, Y., He, J., Chang, S.-F., & Yoon, S.-E. (2012). Spherical hashing. In: *2012 IEEE Conference on Computer Vision and Pattern Recognition*. IEEE, pp. 2957–2964.
[34] Jin, Z., Li, C., Lin, Y., & Cai, D. (2013). Density sensitive hashing. *IEEE Transactions on Cybernetics*, 44(8), 1362–1371.
[35] Gu, Y., & Yang, J. (2018). Densely connected multi-magnification hashing for histopathological image retrieval. *IEEE Journal of Biomedical and Health Informatics*, 23(4), 1683–1691.
[36] Gu, Y., & Yang, J. (2019). Multi-level magnification correlation hashing for scalable histopathological image retrieval. *Neurocomputing*, 351, 134–145.
[37] Sapkota, M., Shi, X., Xing, F., & Yang, L. (2018). Deep convolutional hashing for low-dimensional binary embedding of histopathological images. *IEEE Journal of Biomedical and Health Informatics*, 23(2), 805–816.
[38] Zhu, H., Long, M., Wang, J., & Cao, Y. (2016). Deep hashing network for efficient similarity retrieval. *Proceedings of the AAAI Conference on Artificial Intelligence*, 30, 1.
[39] Öztürk, Ş. (2020). Stacked auto-encoder based tagging with deep features for content-based medical image retrieval. *Expert Systems with Applications*, 161, 113693.
[40] Öztürk, Ş. (2021). Class-driven content-based medical image retrieval using hash codes of deep features. *Biomedical Signal Processing and Control*, 68, 102601.
[41] Öztürk, Ş. (2021). Hash code generation using deep feature selection guided siamese network for content-based medical image retrieval. *Journal of Science*, 34(3), 733–746.

[42] Zhang, X., Liu, W., Dundar, M., Badve, S., & Zhang, S. (2015). Towards large-scale histopathological image analysis: Hashing-based image retrieval. *IEEE Transactions on Medical Imaging*, 34(2), 496–506.

[43] Zhang, X., Liu, W., Dundar, M., Badve, S., & Zhang, S. (2015). Towards large-scale histopathological image analysis: Hashing-based image retrieval. *IEEE Transactions on Medical Imaging*, 34(2), 496–506.

[44] Ahmad, J., Muhammad, K., & Baik, S. W. (2018). Medical image retrieval with compact binary codes generated in frequency domain using highly reactive convolutional features. *Journal of Medical Systems*, 42(2), 24.

[45] Conjeti, S., Paschali, M., Katouzian, A., & Navab, N. (2017). Deep multiple instance hashing for scalable medical image retrieval. In: Descoteaux, M., Maier-Hein, L., Franz, A., Jannin, P., Collins, D., & Duchesne, S. (eds) *Medical Image Computing and Computer Assisted Intervention – MICCAI 2017.* MICCAI 2017. Lecture Notes in Computer Science, vol. 10435. Springer, Cham.

[46] Xu, L., Zeng, X., Zheng, B., & Li, W. (2022). Multi-manifold deep discriminative cross-modal hashing for medical image retrieval. *IEEE Transactions on Image Processing*, 31, 3371–3385.

[47] Zhang, C., Zhu, L., Zhang, S., Yu, W., & Zhang, S. (2020). TDHPPIR: An efficient deep hashing based privacy-preserving image retrieval method. *Neurocomputing*, 406, 386–398.

[48] Zhang, Y., Xie, F., Song, X., Zheng, Y., Liu, J., & Wang, J. (2022). Dermoscopic image retrieval based on rotation-invariance deep hashing. *Medical Image Analysis*, 77, 102301.

[49] Shi, X., Xing, F., Xu, K., Xie, Y., Su, H., & Yang, L. (2017). Supervised graph hashing for histopathology image retrieval and classification. *Medical Image Analysis*, 42, 117–128.

[50] Fang, J., Xu, Y., Zhang, X., Hu, Y., & Liu, J. (2020). Attention-based saliency hashing for ophthalmic image retrieval. In: *2020 IEEE International Conference on Bioinformatics and Biomedicine (BIBM)*. IEEE, pp. 990–995.

[51] Qiu, C., Cai, Y., Gao, X., & Cui, Y. (2017). Medical image retrieval based on the deep convolution network and hash coding. In: *2017 10th International Congress on Image and Signal Processing, BioMedical Engineering and Informatics (CISP-BMEI)*. IEEE, pp. 1–6.

[52] Xu, Y., Shen, F., Xu, X., Gao, L., Wang, Y., & Tan, X. (2017). Large-scale image retrieval with supervised sparse hashing. *Neurocomputing*, 229, 45–53.

[53] Zhao, J. J., Pan, L., Zhao, P. F., Wang, L. H., Liu, X., Zhang, L., & Hu, Y. C. (2017). Medical sign recognition of lung nodules based on image retrieval with semantic features and supervised hashing. *Journal of Computer Science and Technology*, 32, 457–469.

[54] Jolliffe, I. T. (2002). *Principal Component Analysis*. Wiley Online Library.

[55] Müller, H., Michoux, N., Bandon, D., & Geissbuhler, A. (2004). A review of content-based image retrieval systems in medical applications-clinical benefits and future directions. *International Journal of Medical Informatics*, 73, 1–23.

[56] Demner-Fushman, D., Kohli, M. D., Rosenman, M. B., Shooshan, S. E., Rodriguez, L., Antani, S., & Thoma, G. R. (2016). Preparing a collection of radiology examinations for distribution and retrieval. *Journal of the American Medical Informatics Association*, 23, 304–310.

[57] Shih, G., Wu, C. C., Halabi, S. S., & Kohli, M. D. (2019). Evaluating the utility of a public pneumonia chest radiography dataset with deep learning algorithms. *PloS One*, 14, e0225846.

[58] Heath, M., Bowyer, K., Kopans, D., Moore, R., & Kegelmeyer Jr, W. (2001). The digital database for screening mammography. In: *Digital Mammography*. Springer, Berlin, pp. 212–218.

[59] Acharya, U. R., Ng, E. Y., Fujita, H., Bhat, S., Sree, S. V., & Suri, J. S. (2015). Automated diagnosis of glaucoma using digital fundus images. *Journal of Medical Systems*, 39(9), 92.

[60] Isensee, F., Jaeger, P. F., Kohl, S. A. A., Petersen, J., & Maier-Hein, K. H. (2021). nnU-Net: A self-configuring method for deep learning-based biomedical image segmentation. *Nature Methods*, 18(2), 203–211.

[61] Decencière, E., Zhang, X., Cazuguel, G., Laÿ, B., Cochener, B., Trone, C., & Atif, J. (2014). Feedback on a publicly distributed image database: The Messidor database. *Image Analysis & Stereology*, 33(3), 231–234.

[62] López, M. A., Celaya-Padilla, J. M., Toledo-Moreo, F. J., & González-Ortega, D. (2020). Multimodal medical image datasets for artificial intelligence applications: A systematic review. *Journal of Medical Systems*, 44(7), 123.

[63] Berget, I., Johansen, D., & Rønning, E. J. (2019). Kvasir: A multi-class image dataset for computer aided gastrointestinal disease detection. *Data in Brief*, 26, 104340.

[64] Kumar, S., Kim, S. H., & Pelletier, J. P. (2017). An automated computer-aided diagnostic system for bone scintigraphy images: A multi-stage approach. *Computer Methods and Programs in Biomedicine*, 140, 131–146.

[65] Washko, G. R., Dransfield, M. T., Estépar, R. S., Diaz, A., Matsuoka, S., Yamashiro, T., & Hatabu, H. (2014). Quantitative CT metrics of airway disease and emphysema in severe COPD. *Chest*, 146(2), 396–405.

[66] Kinahan, P. E., & Fletcher, J. W. (2010). Positron emission tomography-computed tomography standardized uptake values in clinical practice and assessing response to therapy. *Seminars in ultrasound, CT and MRI*, 31(6), 496–505.

[67] Riederer, S. J., Farzaneh, F., & Pennington, M. A. (1986). The NEMA MRI test phantom. *Medical Physics*, 13(2), 137–145.

[68] Prior, F. W., Clark, K., Commean, P., Freymann, J., Jaffe, C., Kirby, J., & Marquez, G. (2013). TCIA: An information resource to enable open science. *Proceedings of SPIE*, 8674, 86740.

[69] Depeursinge, A., Vargas, A., Platon, A., Geissbuhler, A., Poletti, P.A., & Müller, H. (2012). Building a reference multimedia database for interstitial lung diseases. *Computerized Medical Imaging and Graphics*, 36(3), 227-–238.

[70] Clark, K. et al. (2013). Brain Tumor: The cancer imaging archive (TCIA). *Journal of Digital Imaging,* 26(6), 1045–1057.

[71] Armato, S. G. et al. (2011). LIDC-IDRI: The lung image database consortium (LIDC) and image database resource initiative (IDRI): A completed reference database of lung nodules on ct scans. *Medical Physics,* 38(2), 915–931.

[72] Karssemeijer, N., Brake, G. T. (1998). Digital database for screening mammography. In: Karssemeijer, N.,Thijssen, M., & Hendriks, J., Erning, L. (eds) *Proceedings of the 4th International Workshop on Digital Mammography*, Part of the book series: Computational Imaging and Vision, vol. 13, pp. 95–102.

15 Deploying Machine Learning Methods for Human Emotion Recognition

Ansu Elsa Regi and A. Hepzibah Christinal
Karunya Institute of Technology and Sciences

15.1 INTRODUCTION

Facial expressions are among the simplest nonverbal cues that humans use to convey their emotions and intentions. In recent years, automatic facial expression recognition (FER), which attempts to investigate and interpret human facial activity in the domains of computer vision, artificial intelligence, and pattern recognition, among others, has become an increasingly popular research topic. FER can be used in a variety of fields, including healthcare, social robotics, human–computer interaction, and the perception of human emotions. In addition, gaze direction and biosignals, such as the electrocardiogram (ECG) and electroencephalogram (EEG), can be used to identify emotions. Instructors display a wide range of emotions that are brought on by various factors [2]. For instance, an instructor may feel happy when a goal of education is achieved or when pupils follow instructions. Disappointment arises when school students demonstrate a lack of enthusiasm and a refusal to understand an idea. Similarly, students who lack discipline exhibit anger. Teachers claim that regulating these facial expressions typically aids them in accomplishing their objectives because they frequently result from disciplinary classroom interactions. Human–robot interaction (HRI) that is intelligent uses emotional expressions. In order to optimize the educational environment, emotion analysis may also be used to monitor students' feelings. As a result, this strategy will help the learners study more effectively. In interpersonal interactions, 38% of affective information is communicated through speech tone, 55% by facial expressions, and 7% by words [1]. So, in HRI applications, facial emotion analysis is a dependable way for identifying human emotions. The proposed methodology consists of four stages: preprocessing, extraction and selection of features, feature decomposition, and classification. This work aims to give a real-time investigation on emotion recognition and application in robotic application domains. Four stages comprise the suggested methodology: data pre-processing, extraction and selection of features, feature deconstruction, and feature classification.

 DOI: 10.1201/9781003405368-15

15.2 LITERATURE SURVEY

Multiple researchers identified their frameworks for addressing HRI. The work reported in Reference [3] presents a conditional-generative-adversarial-network-based (CGAN-based) architecture for minimizing intraclass variations by individually controlling face expressions and concurrently learning generative and discriminative representations. This design has a generator G and three discriminators (Di, Da, and Dexp). Transformation of any face image query into a prototypical expression form while maintaining certain factors is done by the generator G. The accuracy achieved was 81.83%.

In Reference [4], a CNN-based model was proposed. It was designed to detect smiles, identify emotions, and classify gender. Hence, it is regarded as a model with multiple functions. It achieved a precision of 71.03%.

15.3 SEVERAL ATTEMPTS HAVE BEEN MADE TO DETECT EMOTIONS USING DEEP LEARNING

The work described in Reference [5] presented a deep CNN for FER. This system will automate the extraction of face expression characteristics for automated recognition. In addition, it includes modules for input, data pre-processing, detection, and outcome. In addition, it was employed to mimic and measure the identification capabilities on the Japanese Female Facial Expression (JAFFE) dataset and the Extended Cohn–Kanade (CK+) dataset as a function of network structure, rate when compared, and pre-processing. The authors utilized the K-nearest technique to make the findings more convincing. The precision rates for the JAFFE and CK+ datasets are 76.7442% and 80.03%, respectively.

In Reference [6], an alternative model was proposed. It was evaluated on a facial emotion HDR image dataset, taking into account a catalogue of faces under various lighting conditions. It is based on support vector machines (SVMs), local binary patterns (LBPs), and presence. To perform the emotion recognition task, it utilizes the Speeded-Up Robust Feature transform. This model demonstrated an accuracy of up to 80%.

In Reference [7], the authors proposed a model for submission to the fifth group-level emotion recognition sub-challenge of Emotion Recognition in the Wild (EmotiW 2017). They utilized a CNN to extract the feature from the face images detected. Another function of CNN is to be prepared for the task of face recognition, as opposed to traditional supervised training on emotion detection issues. An ensemble of random forest (RF) classifiers was learned in the final pipeline to indicate an emotion score using an available training set. This was accomplished by using the data from the previous pipeline. On the validation data, this model had an accuracy of 75.4%.

Another development in the field of emotion detection is the utilization of videos as a source of data. This approach aims to recognize and identify emotions by analyzing the facial expressions and movements of individuals captured on video.

A deep learning hybrid model was presented for emotion detection from videos by the authors of Reference [8]. When processing static facial images, a spatial CNN is utilized, whereas a temporal CNN is utilized when working with optical flow images. These two processing branches are utilized, independently, in order to recognize high-level spatial and temporal features on visual features. These two CNNs have been fine-tuned through the use of pretrained CNN models and target video face expression datasets. The features that have been gathered first from segment-level spatial and temporal branches are combined by a deep fusion network, which is built with a deep belief network (DBN) model as its basis. The face expression classification tasks are carried out by enrolling the obtained fused features into a linear SVM. The authors' work was accurate 75.39% of the time.

In addition, another algorithm for emotion detection from video-based was presented in Reference [9]. The authors looked into a variety of approaches to combine data that were both spatial and temporal. They came to the conclusion that combining information from different time periods and locations produced the best results when attempting to identify facial expressions using video. In contrast to the framework presented in Reference [8], the work presented here can be trained from beginning to end for whole-video recognition. With the help of CNNs and LSTMs, the goal of this framework is to build a framework for trainable deep neural networks for pattern categorization that can take into account both the spatial and temporal information contained within the video. The accuracy of this framework was determined to be 65.72%.

15.4 PROPOSED METHOD

The system needs to go through two primary stages before it can successfully recognize facial expressions and identify feelings. The first stage is known as the training phase. In this section, the input images that were obtained from the JAFFE database are subjected to digital processing with the assistance of various preprocessing tools and techniques. Additional processing is done to the preprocessed images in order to obtain the feature vectors. Once the neural network has processed the data, the resulting information is stored in the system's database. This allows for easy retrieval and use of the insights gained from the neural network's analysis. These data are used to train vectors, which can then be applied to the task of categorizing unidentified feelings.

The second portion of the research involves putting the collected facial expressions to the test after viewing the selected advertisement.

The unidentified input face pattern must be preprocessed before moving on to the next step, which is the extraction of the facial features using the cropped image of the face as a guide. After the data have been collected, they will be sorted so that the distinguishing characteristics of the seven fundamental feelings can be identified.

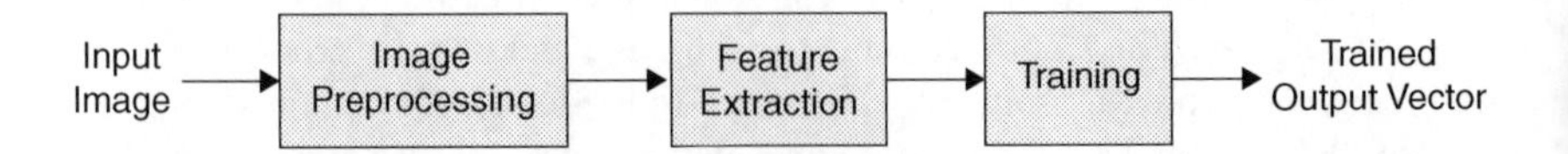

FIGURE 15.1 Task of categorizing unidentified feelings.

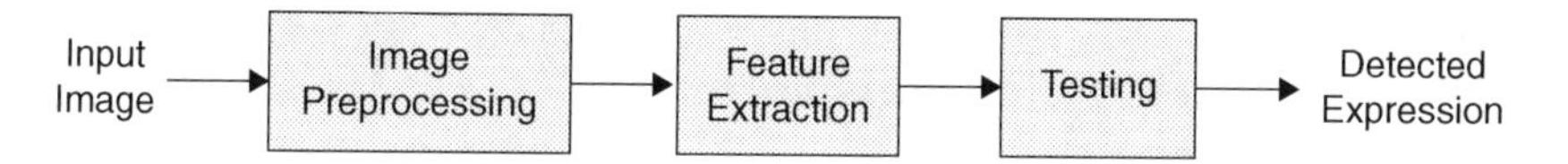

FIGURE 15.2 Unidentified input face pattern.

It is necessary to perform the preprocessing on the input images in order to lessen the amount of noise and variations in illumination. As part of the pre-processing, the colored image will have its dimensions reduced, and it will then be converted to gray scale in order to make the process run more quickly. The final stage of the FER system involves classifying the retrieved face features into one of the seven distinct emotions. These emotions are as follows: neutral, happiness, sadness, surprise, disgust, anger, and fear.

The classification of these feelings was accomplished through the application of machine learning. In addition, an artificial neural network was used in order to make comparisons between the images that were taken from the newly generated database.

15.5 PRINCIPAL COMPONENT ANALYSIS

Principal component analysis (PCA) makes extensive use of signal processing and statistical pattern recognition as one of its many methodologies. It involves the extraction of features and the reduction of data. Reducing the dimensionality of multidimensional data and then extracting the necessary number of major components is how the PCA method operates. It is essential to decrease the large inherent dimensionality of the data and transform it into a smaller inherent dimensionality. A dimension with a smaller range ought to be adequate and in a position to describe the data. There needs to be a robust correlation between the variables that have been observed and the data.

An application of information theory known as PCA involves the process of encoding and decoding images of people's faces in order to arrive at an intuitive understanding of the information that is contained in those faces. It brings attention to the essential regional and general characteristics of the face.

The primary objective that we have set for ourselves is to extract useful information from face images. A program encodes information effectively and provides test images derived from the database that is being created. The process of extracting information for a simple approach involves capturing the divergence of the images that have been gathered. A component of the program involves encoding and contrasting the characteristics of each individual face image. Eigenvectors are the set of collected features that distinguish the variation between different face images; in addition, it provides the location for a particular image.

15.6 CLASSIFICATION

For the purposes of robotic vision applications, we develop an automated facial expression identifier in the course of this work. Its purpose is to recognize human emotions. Classifiers are fed discriminant features that were extracted from a face in order to determine an individual's emotional state based on their appearance. The following classifiers are utilized during the classification process: Gaussian NB, DT, MLP with backpropagation, KNN, a multiclass SVM, QDA, RF, and LR. In order to determine the structure and hyperparameters of classifiers that work best, the trial and error method and search by grid [10] are both utilized. In addition, a ten-fold cross-validation is utilized in order to estimate the ideal combinations of hyperparameters in order to prevent overfitting. In this work, an investigation is conducted to determine which classifier hyperparameter settings are most effective.

The training part and the testing part of the dataset each contain a different subset of the dataset's images. The performance of the classifier is evaluated using the testing part, whereas the training phase serves to train and evaluate the classifier. The ratio of 80/20 has been decided upon for the division, as shown in Figure 12. The current model uses a method called ten-fold cross-validation, which involves further dividing up the training phase into ten separate parts (subsets). Then, nine folds are utilized in the training of the classifier, and the tenth fold is utilized to validate the training that was previously accomplished. This process will continue until each of the ten folds has been applied to the validation procedure precisely once. When it is time for testing, the optimal configurations that were determined during the training stage are put to test [11–15].

15.7 RESULTS AND DISCUSSION

The Facial Expression Recognition Algorithm was designed to identify seven fundamental feelings, including neutral, happy, disgusted, unhappy, afraid, surprised, and angry expressions on a person's face.

The training and testing purposes for each set of images were distinct from one another. The images were retrieved from the database in an iterative fashion by the algorithm, and the feature vector was derived through the application of PCA dimensionality reduction. By using a trained image dataset and a neutral network, one can classify the feelings associated with an image.

Currently, we are in the training phase, during which we are going to train the system using examples from the JAFFE database. As supplemental libraries, Math, Pandas, NumPy, Matplotlib, and OS were utilized. Five measures were employed to assess the proposed framework: accuracy, F1-score, precision, recall, and training duration. The KNN classifier beats other classifications in terms of precision, accuracy, recall, and F1-score, as shown by the results. On the CK+ and JAFFE datasets, it achieved the greatest accuracies of 97% and 95%, respectively.

Now that it has been generated, the system is able to identify the feelings conveyed in JAFFE database pictures. The system is now able to identify the seven fundamental feelings: sadness, anger, happiness, disgust, fear, and surprise in addition to neutral. It is necessary to perform a gray scale conversion on the image to be processed. The operation will proceed more quickly as a result of this because the image uses just one channel of color, which makes use of fewer bits than colored images do. The database will then receive the images that were chosen for inclusion in it. First, the neural network needs to be trained so that it can accurately recognize emotions. After that, it will use the database to assign a category to the image that you have selected. The work that is being proposed demonstrates how various classification algorithms and PCA can be applied to the task of recognizing facial expressions. The JAFFE dataset was utilized in the system's implementation in order to determine the seven fundamental expressions. After that, the system that recognizes facial expressions was used to correlate the results of the analysis of the various areas.

Our mission was to identify the necessary feeling that would lead to an outcome that is both successful and memorable. According to the findings, a commercial that emphasizes joy tends to perform better than those that focus on anger or sadness, which tend to receive lower ratings. This is in contrast to the fact that advertisements that focus on those emotions tend to receive lower ratings.

The implications of our findings are not without their caveats. Emotion is not the only component that contributes to the success of a commercial. It also does not mean that all commercials featuring happy people are successful.

Commercials that are both informative and serious have the potential to be at the top. In order to achieve a higher level of precision, the next phase of the research could involve consulting a number of facial expression databases, such as the Taiwanese Facial Expression Database, the Cohn-Kanade Facial Expression Database, and the Indian Facial Expression Database.

15.8 CONCLUSION

The topic of Human–Robot Interaction, abbreviated as HRI, has been covered in this chapter. The research paper offered a novel strategy for the recognition of facial expressions as a potential solution. This work consists of several stages, the first of which is the collection of data, followed by the preprocessing of that data, the calculation of a feature vector with the help of PCA dimensionality reduction, and the application of that feature vector to various machine learning algorithms. In addition, the generated features are classified by making use of a number of different classification algorithms, some of which include QDA, NB, SVM, KNN, RF, LR, DT, and MLP.

15.9 FUTURE WORK

The future work that may be derived from this research includes presenting a method for emotion identification from different modalities, such as films, spoken words, and written text. This work is expected to take place in the future. In addition, we are pursuing a research trend in hardware implementation of the suggested technique. In addition, further ML approaches, such as semi-supervised learning and dictionary learning, can be utilized to address this problem.

REFERENCES

1. R.R. Jurin, D. Roush, and J. Danter, Communicating Without Words. In: *Environmental Communication*. Second Edition. Springer, Dordrecht, 2010. https://doi.org/10.1007/978-90-481-3987-3_14.
2. F. Cubukcu, "The significance of teachers' academic emotions," *Procedia-Social and Behavioral Sciences*, vol. 70, pp. 649–653, 2013.

3. J. Deng, G. Pang, Z. Zhang, Z. Pang, H. Yang, and G. Yang, "cGAN based facial expression recognition for human-robot interaction," *IEEE Access*, vol. 7, pp. 9848–9859, 2019.
4. D. V. Sang, L. T. B. Cuong, and V. Van Thieu, "Multi-task learning for smile detection, emotion recognition and gender classification," In: *Association for Computing Machinery, New York, NY, United States-SoICT 2017: The Eighth International Symposium on Information and Communication Technology*, pp. 340–347, Nha Trang City Viet Nam, December 7–8, 2017,
5. K. Shan, J. Guo, W. You, D. Lu, and R. Bie, "Automatic facial expression recognition based on a deep convolutional-neural-network structure," In: *2017 IEEE 15th International Conference on Software Engineering Research, Management and Applications (SERA)*, pp. 123–128, London, UK, 2017. doi: 10.1109/SERA.2017.7965717
6. E. O. Ige, K. Debattista, and A. Chalmers, "Towards HDR based facial expression recognition under complex lighting," In: *Proceedings of the 33rd Computer Graphics International (CGI '16). Association for Computing Machinery*, pp. 49–52, New York, 2016. https://doi.org/10.1145/2949035.2949048
7. S. Berretti, B. Ben Amor, M. Daoudi, and A. Del Bimbo, "3D facial expression recognition using SIFT descriptors of automatically detected keypoints," *The Visual Computer*, vol. 27, no. 11, pp. 1021–1036, 2011.
8. S. Zhang, X. Pan, Y. Cui, X. Zhao, and L. Liu, "Learning affective video features for facial expression recognition via hybrid deep learning," *IEEE Access*, vol. 7, pp. 32297–32304, 2019.
9. X. Pan, G. Ying, G. Chen, H. Li, and W. Li, "A deep spatial and temporal aggregation framework for video-based facial expression recognition," *IEEE Access*, vol. 7, pp. 48807–48815, 2019.
10. P. M. Ferreira, F. Marques, J. S. Cardoso, and A. Rebelo, "Physiological inspired deep neural networks for emotion recognition," *IEEE Access*, vol. 6, pp. 53930–53943, 2018.
11. K. Crammer and Y. Singer, "On the algorithmic implementation of multiclass kernel-based vector machines," *Journal of Machine Learning Research*, vol. 2, pp. 265–292, 2001
12. L. Li, K. Jamieson, G. DeSalvo, A. Rostamizadeh, and A. Talwalkar, "Hyperband: a novel bandit-based approach to hyperparameter optimization," *Journal of Machine Learning Research*, vol. 18, no. 1, pp. 6765–6816, 2017.
13. S. Agarwal, B. Santra, and D. P. Mukherjee, "Anubhav: recognizing emotions through facial expression," *The Visual Computer*, vol. 34, no. 2, pp. 177–191, 2018.
14. Y. K. Bhatti, A. Jamil, N. Nida, M. H. Yousaf, S. Viriri, and S. A. Velastin, "Facial expression recognition of instructor using deep features and extreme learning machine," *Computational Intelligence and Neuroscience*, vol. 2021, 5570870, 2021.
15. S. Zhao, H. Cai, H. Liu, J. Zhang, and S. Chen, "Feature selection mechanism in CNNs for facial expression recognition," In: *Proceedings of the British Machine Vision Conference*, Newcastle, UK, 2018.

16 Maternal Health Risk Prediction Model Using Artificial Neural Network

Divya Kumari
Bennett University

16.1 INTRODUCTION

The health of a woman during the process of carrying a child in her womb till the child's birth and in the later phase of the pregnancy is referred to as maternal health. In the past year, around the globe, there have been around three lakh deaths of women during pregnancy, childbirth, and in the later phase. The major reasons for deaths are infection, loss of blood, high blood pressure, abortion, and other health-related issues. It is extremely important to have proper health facilities with skilled professionals who manage these issues in a timely manner. This could result in the prevention of such deaths by huge numbers in many parts of the world. Every woman across the globe needs to have respectful and good-quality maternal health care. During COVID-19, pregnant women were more prone to viral infection and special care was required to ensure the well-being of the mother as well as the child [1]. In developing and poor countries, pregnancy-related deaths are not preventable because of poor socioeconomic development. The maternal health care in African countries like Nigeria clearly reflects this fact and therefore there is a great need to improve the health facilities in such nations [2]. There is a need to eradicate poverty and improve the quality of life in these poor countries to ensure proper childbirth and later the well-being of mother and child. During the COVID-19 pandemic, there has been a significant expansion of telehealth services. In maternal health care, telehealth services have provided positive outcomes and various studies conducted between January 2015 and April 2022 have proved this [3]. In a rural country like India, the mortality rate related to maternal health care has always been very high because of poor infrastructure and improper health facilities. The country is divided in several ways like region, caste, living standard, etc. This makes a large population usually devoid of the health facilities. However, in recent decades, there has been a huge exposure by mass media, which has resulted in the reduction of inequalities prevalent in maternal health care [4]. The nations around the globe have been working hard to improvise maternal health care and thus decrease mortality rate. The African countries have used various strategies to achieve this goal. Gabon demographic and health surveys (GDHSs) was used in a study [5] to explore the relationship of financial status and maternal health care in Gabon using the logistic regression (LR) approach. It was found that the condition of women

DOI: 10.1201/9781003405368-16

plays a very significant role and the improvement in the financial status of a family is directly proportional to a better and satisfactory maternal health care service. The education of women has a great influence on the proper utilization of maternal health care services. A study [6] performed on the data collected from 2015 to 2016 National Family Health Survey showed that the more educated women utilize these services better. This study also indicates that the other factors are socioeconomic condition of the women and the place they hail from. COVID-19 affected pregnant women who required maternal health care. A study [7] performed on women who were admitted between April and August 2020, and who sought maternal healthcare, showed that there was a delay in these services. Modern technology has made the process of taking care of pregnant women easier. Internet of things (IoT)-enabled modern devices have done away with pregnancy-related complications and the related dangerous incidents [8]. World Health Organization has conducted surveys in previous years to determine the causes of maternal deaths. Recent studies have determined sepsis, hemorrhage, and hypertensive disorders as the main factors [9]. There are also several factors associated with risk of fatalities during pregnancy or during childbirth. A study [10] identified caesarean section delivery, improper education, hemorrhage, and hypertensive disorder as the major factors. Age is one of the most important factors and as the age of a woman advances, she is more prone to have complications during pregnancy. The comparison of the outcomes of pregnancy of women of different age groups like 30–34, 35–40, and greater than 40 confirms that women aged greater than 40 had the most complicated pregnancy and childbirth [11]. Diabetes has become common these days in pregnant women. It leads to harmful effects to both mother and her child in the womb. The increased health risks because of diabetes needs to be prevented, and it is therefore one of the major aspects of maternal care. The chances of death of the mother and the child increase because of various complications. Congenital abnormalities and stillbirths are very common in diabetic patients. It is required that the pregnant women with diabetes are given early antenatal care to avoid the complications caused in pregnancy because of diabetes [12]. Research has been carried out to determine the effective measures to take proper care of pregnant women during the pregnancy. IoT-enabled devices are used to monitor pregnant women in developing nations [13]. The study used machine learning algorithms to identify the health risks in pregnancy and the IoT-enabled devices based on these algorithms are used to send notifications to the pregnant women and their families about the health conditions of the pregnant women. The best accuracy, about 98%, is obtained using the logistic model tree (LMT).

16.2 MATERIAL AND METHODS

In this section, the details of the methodology adopted are given. It has three subsections. Section 16.2.1 describes the used dataset. Section 16.2.2 explains the type of pre-processing adopted in the work. Section 16.2.3 explains the proposed method with diagrammatical representation.

16.2.1 Dataset Details

The dataset [14] was collected from the UCI repository. It was collected from the villages of Bangladesh. The community clinics, maternal care centers, and various

TABLE 16.1
Description of the Input Attributes

Features	Description
Age	Woman's age in years
Systolic blood pressure (systolic BP)	Blood pressure's upper value expressed in mmHg
Diastolic blood pressure (diastolic BP)	Blood pressure's lower value expressed in mmHg
Blood sugar (BS)	Glucose levels of blood expressed in mmol/L (molar concentration)
Body temperature (body temp)	Temperature of body
Heart rate	Woman's heart rate in normal resting conditions expressed in beats per minute

hospitals of Bangladesh have contributed the data to create this dataset. The monitoring system based on IoT is used to create the dataset. The dataset has the information of 1,014 individuals. The total number of input features is six. The input features for any pregnant woman are her age, upper and lower values of blood pressure, blood glucose level, and heart rate. The output feature or the labels on the dataset are "high risk," "medium risk," and "low risk." This refers to the intensity of health risk during pregnancy for a woman. Thus, based on the age and different information related to the health of a woman, it can be predicted if she will have any kind of risk during pregnancy. According to the outcome, suitable preventive measures can be adopted, and a pleasant, healthy, and stress-free environment can be provided to a pregnant woman and her child during and after the pregnancy period.

The output feature is the "risk level." It has the following three classes: "low risk," "medium risk," and "high risk." To apply machine learning algorithms, these values are converted into numeric values. For "low risk," the value taken is 0; for "medium risk," it is 1; and for "high risk," it is 2.

16.2.2 Data Pre-processing

In the maternal health risk dataset, the output attribute has three class levels, which are "high risk," "low risk," and "medium risk." The total number of instances in the "high risk" class is 272, "low risk" class is 406, and "medium risk" class is 336. Table 16.2 depicts these values. To get an unbiased result on the application of classification-based machine learning algorithms, the balancing technique is applied. SMOTE with Tomek link is used to balance the data [15].

Figure 16.1 shows the bar plot of the various class labels before the application of the SMOTE–Tomek link technique. Before application of the balancing technique, the class labels have different number of instances.

SMOTE–Tomek link balances the number of instances in each class. The total number of instances in the dataset is increased. The application of classification-based machine learning algorithms becomes easier, and it gives an unbiased result. Figure 16.2 shows the bar plot of the output attribute "risk level" after applying the SMOTE–Tomek link.

TABLE 16.2
Description of the Output Attribute: Risk Level

Classes	No. of Instances
High risk	272
Low risk	406
Medium risk	336

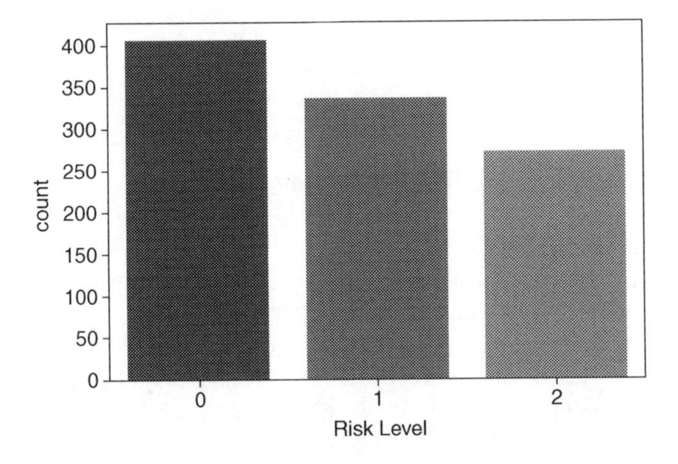

FIGURE 16.1 Bar-plot of risk level before applying the SMOTE–Tomek.

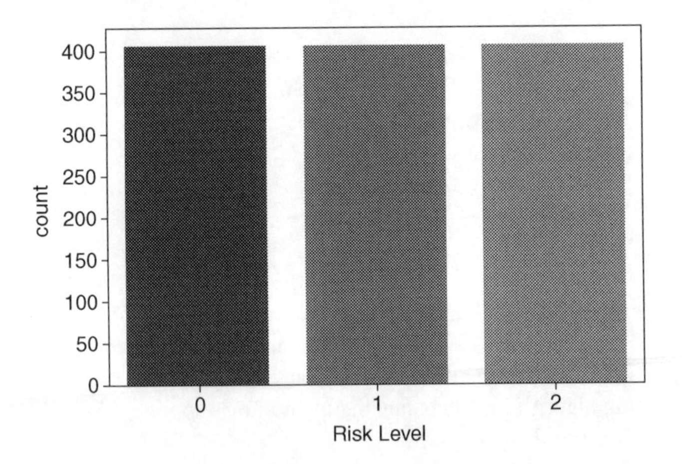

FIGURE 16.2 Bar-plot of risk level after applying the SMOTE–Tomek.

16.2.3 Proposed Method

Figure 16.3 shows the various steps of the proposed methodology. At first, the dataset [14] is collected from the UCI repository. It is quantified and pre-processed in the next step. The dataset has an uneven number of attributes in the various class

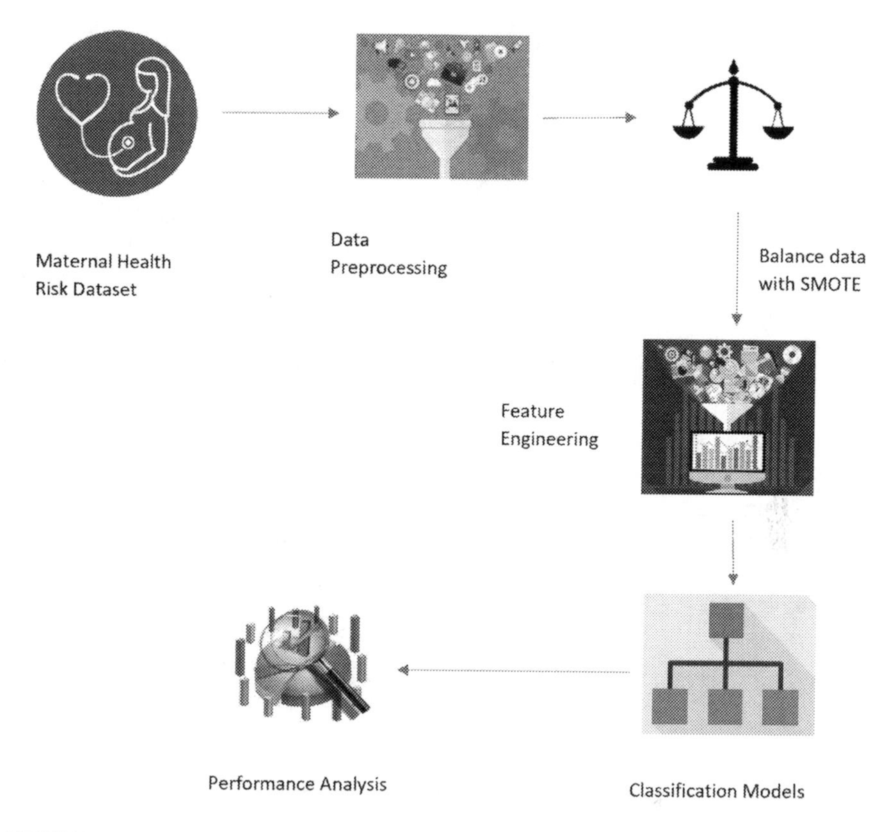

FIGURE 16.3 Various steps of the proposed methodology.

levels. To get unbiased result on the application of various classification methods, the dataset needs to balanced. SMOTE with the Tomek link is applied for the same purpose. The feature engineering is done in the later step where the association of every attribute with other attributes is found. Heatmap is created to find the correlation between various attributes. The feature importance score is also calculated to find the most important feature. In the next step, various classification models are created using algorithms, such as LR, k-nearest neighbors (k-NNs), random forest (RF), gradient boosting (GB), and artificial neural network (ANN). The performance analysis is done to find the algorithm that gives the best result. Sensitivity, specificity, F-score, and accuracy are the performance metrics used in the work. It is found that artificial neural network gives the best result.

16.3 METHODS USED

In this Section 16.3.1, the various classification algorithms used are discussed. The feature engineering technique used in the work is discussed in Section 16.3.2. The graphical representation is also presented. The different algorithms used to calculate the feature importance score is also given.

16.3.1 Supervised Machine Learning Techniques

Various supervised machine learning algorithms with suitable parameters are applied in this work. The logistic regression [16] technique is employed in the work where the parameters l2 norm of penalization is employed, 0.0001 is the stopping criteria's tolerance value, 1.0 is the inverse of the regularization used, and the random state employed is 123. k-NN [16] is also used in the work. For best results, the following parameters are employed in k-NNs: no. of neighbors taken are 5, 30 is the leaf size, Minkowski metric is employed, and uniform weight is taken. Gradient boosting (GB) algorithm [17] is used in the work where multinomial and binomial deviance is used as the loss function, 1,000 boosting stages are employed, and 4 is the maximum number of leaf nodes. In GB, the base learners are required to fit and for this purpose, all the samples are employed. Backpropagation neural network [18] is used where the number of hidden layers is three and 30–20–20 neurons are used. RF [19] is another algorithm used in the work. The number of trees used in the work are 100. While performing the algorithm, it is important to maintain the quality of split. The criterion employed for this purpose is Gini. While splitting the internal nodes, the minimum number of samples used is two, which can be referred to as the minimum split samples. The algorithm is employed such that while performing the split, the maximum depth can be reached. It is possible in following two cases: in the first case, the expansion of tree is done such that every leaf in the tree has samples less than minimum split samples. When the process of split is done, then to do it in the best way, the number of features that should be considered is calculated with the help of a parameter. This parameter finds the square root of the total number of features. All these algorithms are applied on the data with the optimal parameters. The output with the best accuracy is considered. Other performance metrices are also calculated for each used supervised machine learning algorithm.

16.3.2 Feature Importance

The total number of input features used in the work is six. For any study involving a classification problem, input features play a vital role. The most appropriate features help in determining most accurate results. There is also a correlation among each input feature in the study. The correlation between all input features is determined and presented in Figure 16.4 using heatmap. It shows the degree of association of all input features, which are age, level of blood sugar, temperature of body, both kinds of blood pressure: systolic and diastolic, and heart rate of the individuals. The values that are positive between two features in the heatmap show that with an increase in the value of one feature, the other feature value will also increase, and similarly, the decrease in one feature value will also lead to an increase in the second feature value. The negative correlation value between two features means an increase in the first feature value will result in the decrease of the second feature value and vice versa.

The feature importance score of each input attribute is obtained using methods, such as LR, GB, RF, and decision tree (Figure 16.5).

The feature importance score of each input attribute is calculated using the LR technique. In Python, there are inbuilt functions to calculate feature importance using

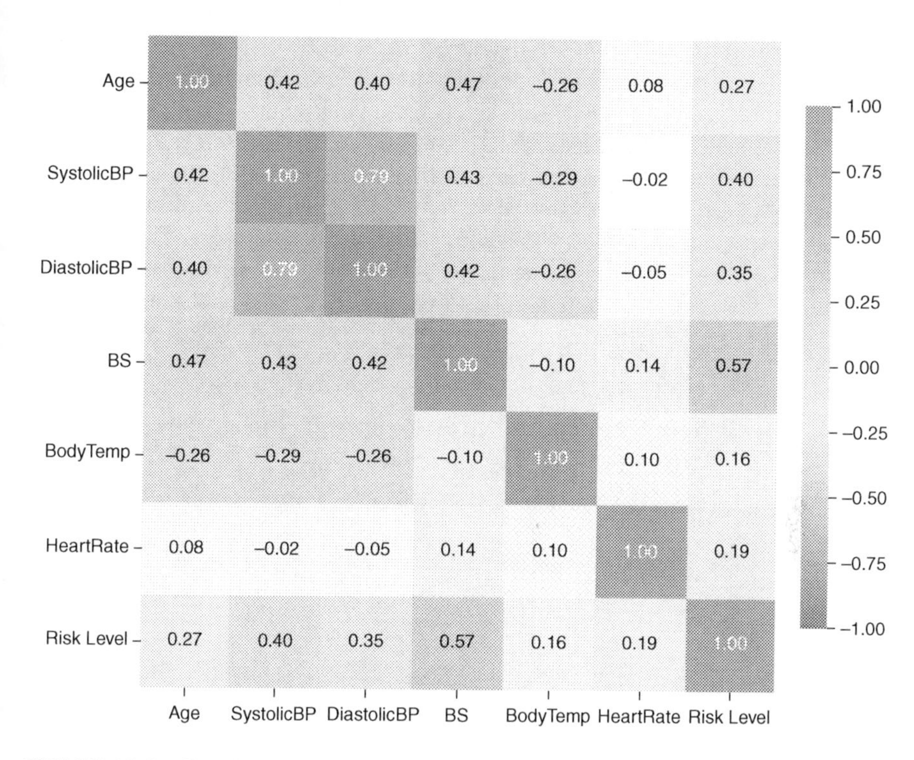

FIGURE 16.4 Correlation between all attributes in the maternal health risk dataset.

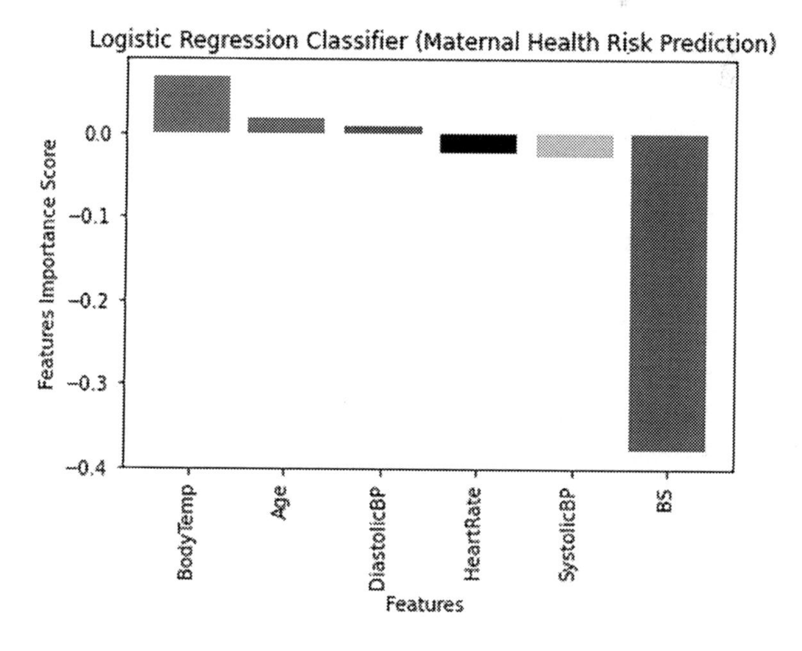

FIGURE 16.5 Logistic regression feature importance score.

few machine learning algorithms. Using the LR method, the best importance score is obtained for blood sugar level. Table 16.3 represents the values of these scores.

GB algorithm is also used to calculate the feature importance score [20–22]. The bar plot of the feature importance score of various machine learning algorithms is shown in Figure 16.6. Table 16.4 represents the exact score values in numerical form. This technique also determines that the level of blood sugar is the most important feature in the prediction of maternal health risk. More the sugar level of a woman, more she will have a high risk of maternal health. Other algorithms are also employed to find the most important feature in the risk prediction of maternal health.

TABLE 16.3
Values of Importance Score for Each Input Attribute Using Logistic Regression

Features	Importance Score
Age	0.01753
Systolic BP	−0.02699
Diastolic BP	0.00948
Blood sugar	−0.37790
Body temperature	0.06656
Heart rate	−0.02302

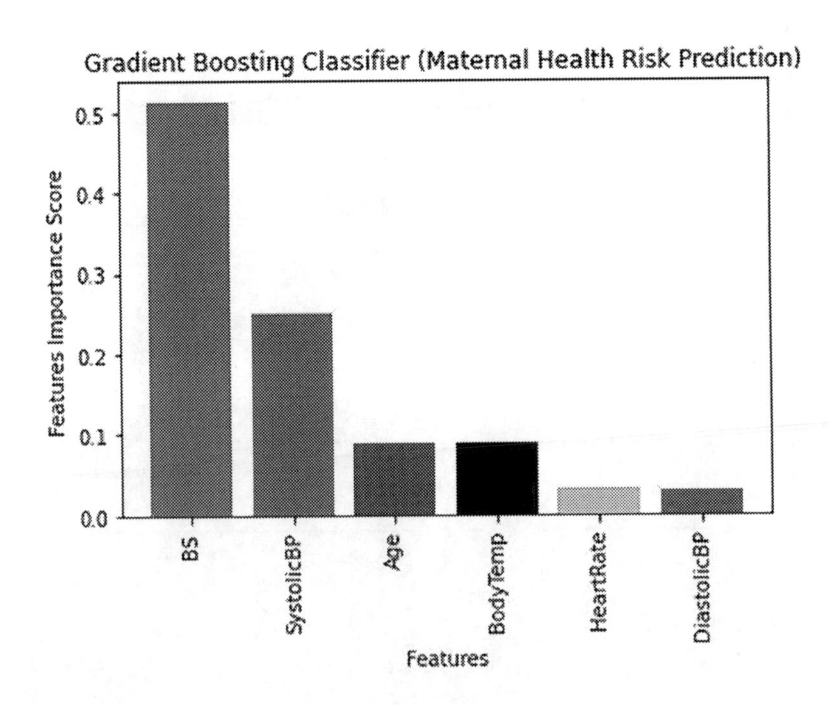

FIGURE 16.6 Gradient boosting feature importance score.

TABLE 16.4
Values of Importance Score for Each Input Attribute Using Gradient Boosting

Features	Importance Score
Age	0.08867
Systolic BP	0.24972
Diastolic BP	0.02981
Blood sugar	0.51269
Body temperature	0.08780
Heart rate	0.03131

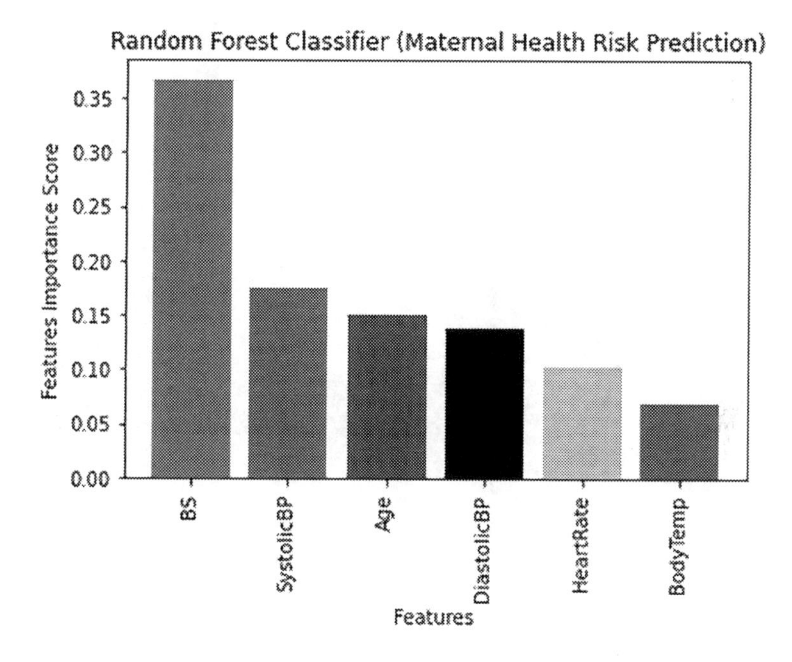

FIGURE 16.7 Random forest feature importance score.

The next algorithm used for the determination of the feature importance score is RF. It uses the bagging technique. First, a RF model [23–24] is built using the input attributes. Then, the feature importance score is calculated using the model. It gives results as blood sugar is the most important feature and determining body temperature as the least important feature. Figure 16.7 shows the graphical representation of the technique and bar plot of the input attributes and their feature importance score. The exact calculated values of the feature importance score are given in Table 16.5.

Decision tree algorithm is also applied to calculate the feature importance score of the input attributes in the dataset. Figure 16.8 shows the diagrammatical representation, and Table 16.6 represents the exact values of the importance scores of the features.

TABLE 16.5
Values of Importance Score for Each Input Attribute Using Random Forest

Features	Importance Score
Age	0.15995
Systolic BP	0.17913
Diastolic BP	0.12119
Blood sugar	0.37079
Body temperature	0.06780
Heart rate	0.10114

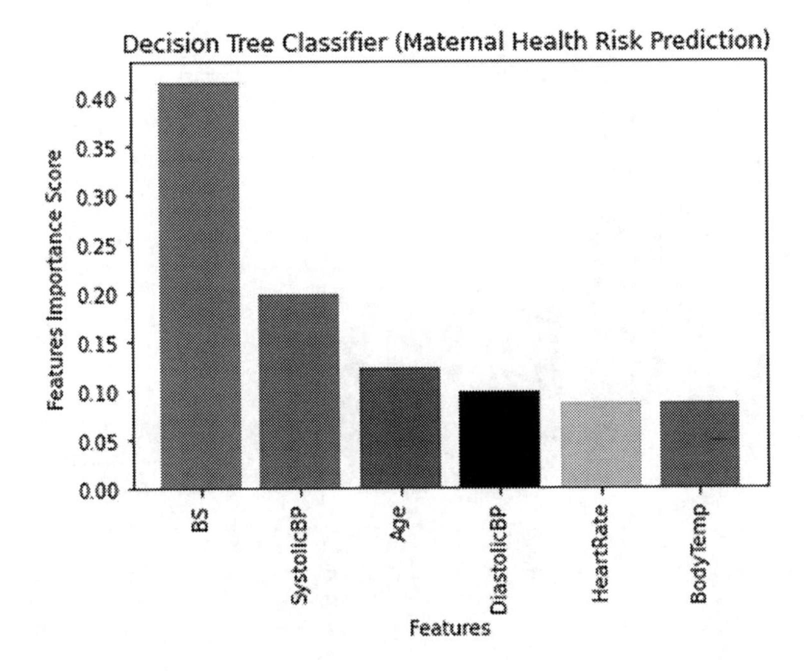

FIGURE 16.8 Decision tree feature importance score.

TABLE 16.6
Values of Importance Score for Each Input Attribute Using Decision Tree

Features	Importance Score
Age	0.11921
Systolic BP	0.20501
Diastolic BP	0.10348
Blood sugar	0.41165
Body temperature	0.07186
Heart rate	0.08878

16.3.2.1 Results

Various parameter metrices are calculated to analyze the performance of the used supervised machine learning algorithm to predict the risk of maternal health. Recall, specificity, precision, F-score, and accuracy are calculated. Table 16.7 summarizes the results of the prediction.

As the predicted class labels are multiple, multiclass prediction is done for each used algorithm. The employed supervised machine learning algorithms are LR, k-NNs, GB, RF, and ANN. For the purpose of validation, the test sample estimate method is used. The diagrammatical representation of the comparison of various performance metrices of all used supervised machine learning algorithms is given in Figure 16.9.

The best results are given by the artificial neural network. It gives an accuracy of 98.9% and sensitivity and specificity of 99.6% and 98.1%, respectively. The results show that the ensemble learning approaches used in the work (RF and GB) perform better than k-NN and LR methods.

TABLE 16.7
Maternal Health Risk Prediction

Methods used	Sensitivity/ recall (%)	Specificity (%)	Precision (%)	*F*-score (%)	Accuracy (%)
LR	61.40	60.0	59.32	60.34	60.68
k-NN	76.27	70.69	72.58	74.38	72.52
GB	86.67	81.67	82.54	84.55	85.71
RF	88.14	81.67	82.54	85.24	84.87
ANN	99.6	98.1	98.1	98.8	98.9

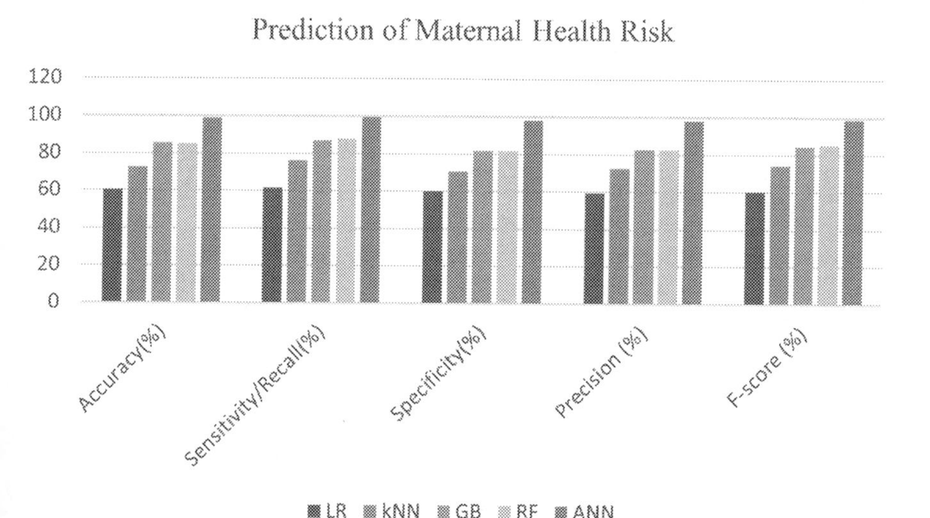

FIGURE 16.9 Performance analysis of various used supervised machine learning algorithms.

16.4 CONCLUSION

The World Health Organization considers maternal health as one of the most important concerns in the world, as a huge number of complications and deaths occur because of improper health facilities. In developing nations like India, health facilities need to improve, and no special care is given to pregnant women, especially in the rural areas. The prediction of risk of maternal health of women in advance will play a significant role in the improvement of quality of life of these women. Based on the health conditions of the women, the proper health facilities will be reachable to them and important preventive measures can be adopted. A healthy mother and her healthy child will be the outcome of the adaptation of these preventive measures.

16.5 FUTURE SCOPE

By using this model, maternal health risk prediction is possible in a very efficient manner. An android-based device can be created, and it will be helpful in the determination of the kind of risk a pregnant woman has during and after pregnancy. The employed supervised machine learning algorithms in this work can be used to develop this device. The input attributes based on the health conditions of a woman will be given as input to the device, and the employed classification algorithm will determine the outcome as to whether the maternal health of the woman will be at low risk, medium risk, or high risk. This will be possible in real time and can be adopted at a large scale in developing nations like India.

REFERENCES

1. Chen, Yu, Zhe Li, Yuan-Yuan Zhang, Wei-Hua Zhao, and Zhi-Ying Yu. "Maternal health care management during the outbreak of coronavirus disease 2019." *Journal of Medical Virology* 92, no. 7 (2020): 731–739.
2. Olonade, Olawale, Tomike I. Olawande, Oluwatobi Joseph Alabi, and David Imhonopi. "Maternal mortality and maternal health care in Nigeria: Implications for socio-economic development." *Open Access Macedonian Journal of Medical Sciences* 7, no. 5 (2019): 849.
3. Cantor, Amy G., Rebecca M. Jungbauer, Annette M. Totten, Ellen L. Tilden, Rebecca Holmes, Azrah Ahmed, Jesse Wagner, Amy C. Hermesch, and Marian S. McDonagh. "Telehealth strategies for the delivery of maternal health care: A rapid review." *Annals of Internal Medicine* 175, no. 9 (2022): 1285–1297.
4. Ali, Balhasan, and Shekhar Chauhan. "Inequalities in the utilisation of maternal health care in rural India: Evidences from National Family Health Survey III & IV." *BMC Public Health* 20, no. 1 (2020): 1–13.
5. Zhang, Tingkai, Xinran Qi, Qiwei He, Jiayi Hee, Rie Takesue, Yan Yan, and Kun Tang. "The effects of conflicts and self-reported insecurity on maternal healthcare utilisation and children health outcomes in the democratic republic of congo (Drc)." *Healthcare* 9, no. 7 (2021): 842.
6. Barman, Bikash, Jay Saha, and Pradip Chouhan. "Impact of education on the utilization of maternal health care services: An investigation from National Family Health Survey (2015-16) in India." *Children and Youth Services Review* 108 (2020): 104642.
7. Goyal, Manu, Pratibha Singh, Kuldeep Singh, Shashank Shekhar, Neha Agrawal, and Sanjeev Misra. "The effect of the COVID-19 pandemic on maternal health due to delay in seeking health care: Experience from a tertiary center." *International Journal of Gynecology & Obstetrics* 152, no. 2 (2021): 231–235.

8. Talpur, Mir Sajjad Hussain, Murtaza Hussain Shaikh, Riaz Ali Buriro, Hira Sajjad Talpur, Fozia Talpur, Hina Shafi, and Munazza Ahsan Shaikh. "Internet of things as intimating for pregnant women's healthcare: An impending privacy issues." *TELKOMNIKA Indonesian Journal of Electrical Engineering* 12, no. 6 (2014): 4337–4344.
9. Say, Lale, Doris Chou, Alison Gemmill, Özge Tunçalp, Ann-Beth Moller, Jane Daniels, A. Metin Gülmezoglu, Marleen Temmerman, and Leontine Alkema. "Global causes of maternal death: A WHO systematic analysis." *The Lancet Global Health* 2, no. 6 (2014): e323–e333.
10. Bauserman, Melissa, Adrien Lokangaka, Vanessa Thorsten, Antoinette Tshefu, Shivaprasad S. Goudar, Fabian Esamai, Ana Garces et al. "Risk factors for maternal death and trends in maternal mortality in low-and middle-income countries: A prospective longitudinal cohort analysis." *Reproductive Health* 12, no. 2 (2015): 1–9.
11. Kenny, Louise C., Tina Lavender, Roseanne McNamee, Sinead M. O'Neill, Tracey Mills, and Ali S. Khashan (2013). Advanced Maternal Age and Adverse Pregnancy Outcome: Evidence from a Large Contemporary Cohort PloS one, 8(2), e56583. https://doi.org/10.1371/journal.pone.0056583.
12. Ali, Sarah, and Anne Dornhorst. "Diabetes in pregnancy: Health risks and management." *Postgraduate Medical Journal* 87, no. 1028 (2011): 417–427.
13. Ahmed, Marzia, Mohammod Abul Kashem, Mostafijur Rahman, and Sabira Khatun. "Review and analysis of risk factor of maternal health in remote area using the internet of things (IoT)." In: *ECCE2019: Proceedings of the 5th International Conference on Electrical, Control & Computer Engineering*, Kuantan, Pahang, Malaysia, 29th July 2019, pp. 357–365. Springer, Singapore, 2020.
14. https://archive.ics.uci.edu/ml/datasets/Maternal+Health+Risk+Data+Set
15. Batista, Gustavo EAPA, Ana LC Bazzan, and Maria Carolina Monard. "Balancing training data for automated annotation of keywords: a case study." *Wob* 3 (2003): 10–18.
16. Hastie, Trevor, Robert Tibshirani, Jerome H. Friedman, and Jerome H. Friedman. *The Elements of Statistical Learning: Data Mining, Inference, and Prediction*, Vol. 2. Springer, New York, NY, 2009.
17. Mayr, Andreas, Harald Binder, Olaf Gefeller, and Matthias Schmid. "The evolution of boosting algorithms." *Methods of Information in Medicine* 53, no. 6 (2014): 419–427.
18. Hagan, M. T., H. B. Demuth, M. Beale, and O. De Jesus. *Neural Network Design*. PWS Pub. Co., Boston, 1996.
19. Biau, Gérard, and Erwan Scornet. "A random forest guided tour." *Test* 25 (2016): 197–227.
20. Gupta, Umesh, and Deepak Gupta. "Least squares structural twin bounded support vector machine on class scatter." *Applied Intelligence* 53, no. 12 (2023): 15321–15351.
21. Kumar, Sanjeev, Suneet Gupta, and Umesh Gupta. "Discrete cosine transform features matching-based forgery mask detection for copy-move forged images." In: *2022 2nd International Conference on Innovative Sustainable Computational Technologies (CISCT)*, pp. 1–4. IEEE, 2022.
22. Malviya, Lokesh, Sandip Mal, Radhikesh Kumar, Bishwajit Roy, Umesh Gupta, Deepika Pantola, and Madhuri Gupta. "Mental Stress Level Detection Using LSTM for WESAD Dataset." In *Proceedings of Data Analytics and Management: ICDAM 2022*, pp. 243–250. Singapore: Springer Nature Singapore, 2023.
23. Parihar, Ashish Singh, Umesh Gupta, Utkarsh Srivastava, Vishal Yadav, and Vaibhav Kumar Trivedi. "Automated Machine Learning Deployment Using Open-Source CI/CD Tool." In *Proceedings of Data Analytics and Management: ICDAM 2022*, pp. 209–222. Singapore: Springer Nature Singapore, 2023.
24. Kumar, Sanjeev, Suneet K. Gupta, Manjit Kaur, and Umesh Gupta. "VI-NET: A hybrid deep convolutional neural network using VGG and inception V3 model for copy-move forgery classification." *Journal of Visual Communication and Image Representation* 89 (2022): 103644.

Index

Printed in the United States
by Baker & Taylor Publisher Services